Kolmogorov's Heritage in Mathematics

Picture taken from the Kolmogorov Archive with kind permission of A.N. Shiryaev

Éric Charpentier · Annick Lesne
Nikolaï K. Nikolski (Eds.)

Kolmogorov's Heritage in Mathematics

With 22 Figures

 Springer

Éric Charpentier
Institut de Mathématiques
Université Bordeaux 1
351 cours de la Libération
33405 Talence Cedex
France
eric.charpentier@math.u-bordeaux1.fr

Nikolaï K. Nikolski
Institut de Mathématiques
Université Bordeaux 1
351 cours de la Libération
33405 Talence Cedex
France
Nikolai.Nikolski@math.u-bordeaux1.fr

Annick Lesne
Laboratoire de Physique Théorique
de la Matière Condensée
Université Pierre et Marie Curie
Case 121, Tour 24-25 - 4ème étage, pièce 407
4 Place Jussieu
75252 Paris
France

And

Institut des Hautes Études Scientifiques
Le Bois-Marie, 35 route de Chartres
91440 Bures-sur-Yvette
France
lesne@lptmc.jussieu.fr

Édition originale: *L'héritage de Kolmogorov en mathématiques* © Éditions Belin – 2004
Ouvrage publié avec le concours du Ministère français chargé de la culture – Centre national du livre
This edition was supported by the French Ministry of Culture – Centre national du livre

Cover picture taken from the Kolmogorov Archive with kind permission of A. N. Shiryaev

Library of Congress Control Number: 2007923591

ISBN 978-3-540-36349-1 Springer Berlin Heidelberg New York

Springer is a part of Springer Science+Business Media
springer.com
© Springer-Verlag Berlin Heidelberg 2007

Typesetting: by the authors and Integra, India using a Springer LATEX macro package
Cover design: WMX-Design, Heidelberg

Printed on acid-free paper   SPIN: 11775072   5 4 3 2 1 0

Contents

List of Contributors

Vasco Brattka
Laboratory of Foundational Aspects
of Computer Science
Department of Mathematics
and Applied Mathematics
University of Cape Town
South Africa
http://cca-net.de/vasco
BrattkaV@maths.uct.ac.za

Victor M. Buchstaber
Steklov Mathematical Institute of
Russian Academy of Sciences, and
Moscow State University, Russia
http://www.mi.ras.ru/~buchstab
buchstab@mi.ras.ru
and School of Mathematics, The
University of Manchester, UK,
Victor.Buchstaber@manchester.ac.uk

Loïc Chaumont
Laboratoire Angevin de Recherche
en Mathématiques (LAREMA)
University of Angers, France
http://math.univ-angers.fr/~
chaumont
loic.chaumont@univ- angers.fr

Thierry Coquand
Department of Computer Science
and Engineering

Chalmers University of Technology
and Göteborg University
Sweden
http://www.cs.chalmers.se/~coquand
coquand@cs.chalmers.se

Giuseppe Da Prato
Scuola Normale Superiore
Pisa, Italy
http://www.sns.it/scienze/menunews/
docentiscienze/daprato
daprato@sns.it

Bruno Durand
Laboratoire d'informatique fonda-
mentale de Marseille
University of Provence (Aix-
Marseille I), France
http://www.lif.univ-mrs.fr/~bdurand
Bruno.Durand@lif.univ-mrs.fr

Kevin Ford
Department of Mathematics
University of Illinois at Urbana-
Champaign, USA
http://www.math.uiuc.edu/~ford
ford@math.uiuc.edu

Étienne Ghys
Unité de mathématiques pures et
appliquées (UMPA)

CNRS and École normale supérieure
de Lyon, France
http://www.umpa.ens-lyon.fr/~ghys
etienne.ghys@umpa.ens-lyon.fr

John H. Hubbard
Department of Mathematics
Cornell University, Ithaca, NY, USA
http://www.math.cornell.edu/People/
            Faculty/hubbard.html
jhh8@cornell.edu

J.-P. Kahane
Département de Mathématiques
d'Orsay
University Paris-sud, Orsay, France
http://www.academie-sciences.fr/
            Membres/K/Kahane_JP.htm
Jean-Pierre.Kahane@math.u-psud.fr

Denis V. Kosygin
Courant Institute of Mathematical
Sciences
New York University, USA
kosygin@cims.nyu.edu

Laurent Mazliak
Laboratoire de Probabilités et
Modèles Aléatoires
Institut de mathématiques de Jussieu
University Paris VI, France
http://www.proba.jussieu.fr/users/
            lma/mazliak.html
mazliak@ccr.jussieu.fr

Mikhaïl Nikouline (M. Nikulin)
Statistique Mathématique et ses
Applications
University Victor Segalen (Bordeaux II), France
http://www.sm.u-bordeaux2.fr/stat
nikouline@sm.u-bordeaux2.fr

Karl Sigmund
Faculty for Mathematics
University of Vienna

and International Institute for
Applied Systems Analysis, Austria
http://homepage.univie.ac.at/
                    Karl.Sigmund
karl.sigmund@univie.ac.at

Yakov G. Sinai
Department of Mathematics
Princeton University, USA
sinai@Math.Princeton.EDU

Valentin Solev
Laboratory of Statistical Methods,
Steklov Institute of Mathematics,
RAS, Saint Petersburg, Russia
http://www.pdmi.ras.ru/staff/solev.html
solev@pdmi.ras.ru

Vladimir M. Tikhomirov
Moscow State University and
Independent University of Moscow,
Russia

Paul Vitanyi
Centrum voor Wiskunde en Infor-
matica (CWI), Amsterdam
The Netherlands
http://www.cwi.nl/~paulv
Paul.Vitanyi@cwi.nl

Marc Yor
Laboratoire de Probabilités et
Modèles Aléatoires
Institut de mathématiques de Jussieu
University Paris VI, France
http://www.academie-sciences.fr/
                    Membres/Y/Yor_Marc.htm

Alexander Zvonkin
Laboratoire Bordelais de Recherche
en Informatique (LaBRI)
University Bordeaux I, France
http://www.labri.fr/perso/zvonkin
zvonkin@labri.fr

Introduction

Éric Charpentier[1], Annick Lesne[2], and Nikolaï Nikolski[3]

[1] Institut de Mathématiques de Bordeaux, University Bordeaux I, France
   http://www.math.u-bordeaux1.fr/imb/
   eric.charpentier@math.u-bordeaux1.fr
[2] Laboratoire de physique théorique de la matière condensée, University Paris VI,
   and Institut des Hautes Études Scientifiques, Bures-sur-Yvette, France
   http://www.lptmc.jussieu.fr/users/lesne/
   lesne@lptmc.jussieu.fr
[3] Institut de Mathématiques de Bordeaux, University Bordeaux I, France
   http://www.math.u-bordeaux1.fr/imb/
   Nikolai.Nikolski@math.u-bordeaux1.fr

Translated from the French by Elizabeth Strouse

Andrei Nikolaevich Kolmogorov (Tambov 1903, Moscow 1987) was one of the most brilliant mathematicians that the world has ever known. Incredibly deep and creative, he was able to approach each subject with a completely new point of view: in a few magnificent pages, which are models of shrewdness and imagination, and which astounded his contemporaries, he changed drastically the landscape of the subject. Most mathematicians prove what they can, Kolmogorov was of those who prove what they want.

In this book we have asked several world experts to present (one part of[4]) the mathematical heritage left to us by Kolmogorov[5]. Each chapter treats one of Kolmogorov's research themes, or a subject that was invented as a consequence of his discoveries. We present here his contributions, his methods, the perspectives he opened to us, the way in which this research has evolved up to now, along with examples of recent applications and a presentation of the modern prospects.

We hope that this book can be read by anyone with a master's (or even a bachelor's) degree in mathematics, computer science or physics, or more

---

[4] In the book *Kolmogorov in perspective* (History of Mathematics, Vol. 20, American Mathematical Society, 2000) one can find a (more or less) complete bibliography of Kolmogorov's work, which consists of about 500 publications. A lot of these are at the heart of very active research going on today

[5] A book entitled *The Kolmogorov Legacy in Physics* has already appeared (Springer, 2004). It contains contributions to dynamical systems, complexity, turbulence and probability. We strongly recommend it to mathematicians

generally by anyone who likes mathematical ideas. Rather than presenting detailed proofs, we give the main ideas, and a bibliography for those who wish to understand the technical details. One can see that sometimes very simple reasoning (with the right interpretation and tools) can lead in a few lines to very substantial results.

Here is a quick summary of the themes that are treated here, with, for each, some significant examples of the "master's touch".

Fourier Series (Chap. 1). In 1922, at the age of 19, Kolmogorov managed to construct a Lebesgue integrable function on $[0, 2\pi]$ whose Fourier series diverges almost everywhere, that is, except on a set of Lebesgue measure 0. (Up to this point the Fourier series of all the functions that had been considered converged almost everywhere.) Three years later, he found an example of a function whose Fourier series diverged *everywhere*! In the same time, Kolmogorov obtained important results on lacunary Fourier series, harmonic conjugates...

Logic (Chap. 2). In 1925, Kolmogorov became interested in intuitionistic logic. For Brouwer (the father of Intuitionism), Intuitionism and Formalism were antinomic approaches. Moreover, intuitionistic logic was generally considered as a weakening of classical logic. But, baffling all expectations, *Kolmogorov managed to formalize intuitionistic logic* and to present it as an *extension* of classical logic. He then deduced that any "finitary" proposition with a classical proof could be proved using intuitionistic logic. All this in only two pages!

If intuitionistic logic is perceived as an extension of classical logic, obtained by adding connectors with no analogues in classical logic, one needs to find an interpretation of these connectors. This is what Kolmogorov did in 1932 when he interpreted intuitionistic logic as a "calculus of problems". This interpretation later turned out to be relevant to computer science.

Probabilities (Chaps. 3 and 4). It was also around 1925 that Kolmogorov began working on probability theory. He began with a clever generalization of the Bienaymé-Chebyshev inequality: this "Kolmogorov's inequality" quickly revealed its usefulness. He used it, with Khinchin, to obtain a famous convergence rule for series of random variables.

In 1930 he used the inequality to deduce a version of the strong law of large numbers ("almost sure" convergence – i.e. convergence with probability equal to one – of the empirical mean towards the mathematical expectation) which contained all prior versions of this law (Borel, Cantelli, Khinchin) like a set of Russian dolls: this is the "L^2 version" given in Chap. 3. A little later he obtained the "L^1 version" which is in a certain sense optimal since it holds whenever the variables (independant and with the same law) have a finite expectation. Kolmogorov announced this result in his book *Foundations of probability theory*, in 1933, without giving a proof; he explained the proof to Maurice Fréchet and let him have the honor of being the first to publish it in the first book of his *Recherches théoriques modernes sur le calcul des probabilités* (Gauthier-Villars, 1937).

In 1931, Kolmogorov began thinking about continuous time Markov processes. Instead of studying the trajectories of a process, he studied the transition probability densities and determined the differential equations they satisfied. These are, of course, the fundamental equations from which the modern theory of diffusion has been constructed. Because of the applications to modern physics, they have even been generalized to Hilbert spaces of infinite dimension: Chapter 4 describes this active area of research which has had several recent applications.

Lomnicki, Steinhaus, Cantelli had sketched out axiomatizations of probability theory, based on Borel's idea to formulate it in the langage of measure theory; but it is Kolmogorov, in 1933, who solved the problem: his axiomatization, both natural and powerful, is the one that is still used today.

Statistics (Chaps. 5 and 6). In 1933, Glivenko et Cantelli proved what is sometimes called the *fundamental theorem of statistics*: almost surely, the empirical distribution function of a real random variable converges uniformly towards the true distribution function when the size of the sample tends towards infinity. Soon afterwards, Kolmogorov gave the precise law for the convergence, in terms of a universal function (independant of the − sought after − law of the random variable) which leads to a very efficient goodness-of-fit test (a test for the unknown law).

The Chap. 5 retraces this discovery of Kolmogorov and presents certain improvements with a surprising application to number theory: it gives the asymptotic behavior (when n tends to infinity) of the probability that an integer chosen at random has at least one divisor between n and $2n$, and of the probability that it has exactly r such divisors.

Epsilon-entropy, a quasi-universal tool introduced by Kolmogorov in the 1950's (see Chap. 8) has become a fundamental tool in statistics for measuring the quality of estimators: the Chap. 6 gives a survey of the current state of knowledge, intended for master's degree students as well as for researchers.

Topology (Chap. 7). During the years 1934–1937, Kolmogorov was very excited about topology. From 1934 to 1935, at the same time as Alexander, he constructed the cohomology of topological spaces and discovered its ring structure. In 1935, Samuel Eilenberg asked if an open mapping (one that takes open sets to open sets) can increase the dimension of its domain; Kolmogorov answered this question with a three page article, which appeared in 1937, in which he gave a very clever construction of an open mapping which transforms a topological space of dimension 1 into a topological space of dimension 2. This result was later used to analyze groups' actions on topological spaces.

Geometry (Chap. 8). One of Kolmogorov's great gifts was his extraordinary geometrical intuition. This played an essential role in almost all his work, as shown in the fascinating Chap. 8. This chapter describes the contributions of Kolmogorov to geometry, approximation theory, the invention of ε-entropy, etc.

Mathematical Ecology (Chap. 9). In 1936, Kolmogorov published a short note on the predator-prey model of Lotka and Volterra: they had (independently) expressed the rates of growth of populations by explicit functions, depending on a small number of parameters, and they solved the equations explicitly. Clearly such a method can only apply to highly simplified models. Kolmogorov, on the other hand, did not choose specific functions, but instead concentrated on monotonicity properties (which give robust conditions); then he applied the qualitative methods of Poincaré to study the long-time behavior of populations (equilibria, limit cycles). Of course, this qualitative approach is the only one that can be used in realistic models.

Dynamical Systems. KAM Theorem (Chaps. 10, 11) and entropy (Chap. 12). Around 1953, Kolmogorov became interested in dynamical systems. He proved a fundamental theorem which explains why, when a system stays close enough to a system in which numerous laws of conservation hold, most movements of the former remain close to the regular movements of the latter, instead of capitalizing on the lack of laws to wander. Kolmogorov presented his proof in talks, but did not publish it. Arnold and Moser later published the first proofs of Kolmogorov's theorem (with different hypotheses) and this is where the name of the KAM (Kolmogorov, Arnold, Moser) theorem comes from. For more than 30 years, all known proofs of this theorem were extremely complicated, but in 1984 a wonderfully simple proof was published. This proof was improved upon in 2002. When one of the authors of this improved proof explained it in Moscow in 2002, members of the audience who had heard Kolmogorov in 1957 said that the new proof was actually the same as Kolmogorov's original one! (Chap. 11, p. 215.) Chapter 10 gives a historical survey of the problem of stability of movements in Celestial mechanics, explains the role of resonance and small divisors and presents an analogue of KAM theorem in a toy model where mathematical difficulties are weakened. Chapter 11 illustrates the KAM theorem in the cases of the solar system and of the forced pendulum; then, it gives a rigorous statement of the theorem and sketches the main ideas of the new proof.

In 1958, Kolmogorov introduced the idea of entropy of dynamical systems, a quantity which is invariant by metric isomorphisms. This turned out to be a very powerful tool which permits one to show that two dynamical systems are not metrically isomorphic. At first, Kolmogorov thought that deterministic systems determined by differential equations would necessarily have zero entropy (roughly: "no disorder") unlike probabilistic systems. But Kolmogorov and Sinai noticed that there exist deterministic systems with nonzero entropy. This discovery is one of the starting points of the modern theory of *deterministic chaos*.

The modern (and fundamental) notion of *hyperbolicity* of a dynamical system was developed as a result of the investigation of these deterministic systems with nonzero entropy. Chapter 12 describes this discovery and the latest results on the subject.

The superposition theorem (chaps. 8 and 13). The 13th Hilbert problem (Paris, 1900) was to determine if any continuous function of three variables could be constructed in a finite number of steps, each one an application of a continuous function of *one* or *two* variables. (This was an idealization of a method of graphical resolution of equations.) Hilbert expected a negative answer. In 1956 Kolmogorov showed that any continuous function of n variables on $[0, 1]^n$ is constructible if one permits the use of continuous functions of *three* variables as auxiliaries, and in 1957 his student Arnold proved that any continuous function of three variables is constructible using functions of *two* variables, thus resolving Hilbert's 13th problem (with a different result than that expected by Hilbert). Soon after, Kolmogorov showed that any continuous function of n variables on $[0, 1]^n$ is constructible using only continuous functions of one variable and additions: this is the *Kolmogorov's Superposition Theorem*.

In 1987, Hecht-Nielsen deduced that any continuous function could be implemented by a certain type of neural network with continuous activation functions and real weights. Constructive versions of the superposition theorem have recently made possible the transposition of this result to *computable* functions, activation functions and weights, and may give a way to construct networks corresponding to given functions. This application of the superposition theorem to neural networks would surely have pleased Kolmogorov: in fact, according to Arnold, one of Kolmogorov's last mathematical works was motivated by his curiosity about the structure of the brain (cf. chapitre 9, p. 177).

More recently other applications of the superposition theorem have appeared: Kolmogorov would surely have been happy to know that it applies to subjects such as the Radon transform and topological groups.

Complexity of description (Chaps. 14 and 15). In 1965, Kolmogorov defined the *complexity of description* of an object; this is, more or less, the length of the shortest algorithm which can describe this object (the exact definition is given in Chap. 14). This is a wonderful tool. One of its applications was a surprisingly simple and short proof by Gregory Chaitin of Gödel's incompleteness theorem (Chap. 14), which can be understood by everybody! Chaitin was very pleased to find this proof which, he explained, gives one the impression that the incompleteness phenomenon discovered by Gödel is natural — unlike the traditional proof, based on the "liar paradox" which makes the phenomenon seem rather pathological and uncommon.

Kolmogorov's complexity theory also makes sense of propositions such as "the sequence is random": for Kolmogorov this means more or less that the sequence has no regularity with which one can "summarize" it, it is "incompressible".

The existence of such incompressible objects gives a simple way to prove things in all branches of mathematics: this is the "incompressibility method". Two examples of this method are given in Chap. 15, one in number theory (how to show in a few lines, and almost without calculation, that the nth prime number has an order of magnitude less than $n\log^2 n$, for instance), and the other in graph theory (finding the maximum size of complete graphs contained in a random graph).

But the principal object of Chap. 15 is to propose a new approach to deterministic chaos: instead of studying the unpredictability in terms of the probabilities of ensembles of trajectories, one uses the complexity theory of Kolmogorov to express the fact that an *individual* trajectory is "unpredictable".

Acknowledgements

We of course thank all our authors for the wonderful texts they contributed.

Our gratitude also extends to the translators and to the colleagues whose comments and suggestions made this book possible: Michel Balazard, Pierre Castéran, Robert Deville, Étienne Matheron, Cyril Mauvillain, Michel Mendès France, Vesselin Petkov, Elizabeth Strouse, Philippe Thieullen, John Tromp and Alexander Zvonkin.

1

The youth of Andrei Nikolaevich and Fourier series

Jean-Pierre Kahane

Département de Mathématiques d'Orsay, University Paris-sud, Orsay, France
http://www.academie-sciences.fr/Membres/K/Kahane_JP.htm
Jean-Pierre.Kahane@math.u-psud.fr

Translated from the French by Elizabeth Strouse

Andrei Nikolaevich Kolmogorov was 14 years old in October 1917. In Moscow this was neither a time for peace nor for reflection. According to his biography in [Kol91], it seems that he was a railroad worker during the difficult years of 1919 and 1920. In 1920–21 he began his studies at the University of Moscow where he would become a student of N.N. Lusin. At the same time he was conducting a study, using land registers, of the evolution of land ownership during the 15th and 16th centuries in the region of Novgorod. During the spring of 1922, at the age of 19, he constructed a Lebesgue integrable function whose Fourier series diverged almost everywhere. In spite of his continuing love for history, this Fourier series would become an irresistible force bringing him into mathematics.

1.1 Convergence and divergence of Fourier series

In truth, the battle between convergence and divergence of Fourier series is also a part of history. For Daniel Bernoulli, in the 18th century, it was clear that a sound was a superposition of harmonics, and thus that a periodic function could be represented as the sum of a trigonometric series:

$$f(x) = \frac{1}{2}a_0 + \sum_{n=1}^{\infty}\left(a_n \cos \frac{2\pi n x}{T} + b_n \sin \frac{2\pi n x}{T}\right). \qquad (1.1)$$

Fourier at the beginning of the nineteenth century [DR98], believed that, using his formulas

$$a_n = \frac{2}{T}\int_0^T f(x)\cos\frac{2\pi n x}{T}dx, \quad b_n = \frac{2}{T}\int_0^T f(x)\sin\frac{2\pi n x}{T}dx, \qquad (1.2)$$

he had totally justified Bernoulli's belief. He had verified the convergence of the series in certain cases – in particular for the characteristic function of an interval – and declared that this convergence would hold for an arbitrary function as long as the Fourier coefficients were calculated using his formulas. It was Dirichlet, in 1829, who produced the first general convergence theorem, concerning functions which were piecewise continuous and piecewise monotone[1]; thinking that he could come back and eliminate the assumption of monotonicity. It turns out that this assumption can be weakened, as was done by Jordan ([Jor81], 1881) when he introduced functions of bounded variation, but continuity alone does not guarantee convergence as shown by the amazing (at the time) example furnished by Paul du Bois-Reymond of a continuous function whose Fourier series diverges at a point ([DuB73], 1873). For the next quarter century Fourier series had a bad reputation and were thought to be useful only for producing monsters like the continuous functions without derivatives which made Hermite turn away in "dread and horror".

The scene changed in 1900 when the very young Fejér Lipót[2] [Fej00] showed that the arithmetic means of the partial sums converged to the (bounded) function's value at any point where the function was continuous. The extensions and applications of this result revived interest in Fourier series. Then Lebesgue gave a new framework with his integral. After the publication of his *Leçons sur les séries trigonométriques* in 1906, when one spoke of a "Fourier series" one meant a Fourier-Lebesgue series, that is, a Fourier series whose coefficients are obtained using the Lebesgue integral. This framework

---

[1] Let us mention an exception. Camille Deflers, in 1819, in a *Note sur quelques intégrales définies, et application à la transformation des fonctions en séries de quantités périodiques* (Bulletin de la Société Philomatique de Paris, november 1819, pp. 161–166), proved what we call today the "Riemann-Lebesgue Lemma" using integration by parts (his proof works for C^1 functions) and introduced the "Dirichlet kernel" ten years before Dirichlet; then he deduced that a sufficiently regular function was the sum of its Fourier series (his argument works in particular for C^2 functions, but Deflers was not precise about the conditions under which his results were true). By using a version of "Riemann-Lebesgue Lemma" for continuous functions (which follows from the C^1 case by uniform approximation: a tool which is now elementary, but was discovered after Deflers), the argument of Deflers proves that a periodic *differentiable* function is the sum of its Fourier series: it is this result which is taught today and wrongly attributed to Dirichlet. In fact the Dirichlet theorem works for functions f which are *piecewise monotone* (and continuous, if one wants the sum to be equal to the function). There is certainly a tie between the two theorems: a periodic C^1 function is continuous and of bounded variation, and so, on any interval, it can be expressed as the difference of two continuous increasing functions (Jordan, 1881), to which the Dirichlet theorem can be applied: the Deflers theorem, in the case where f' is continuous, can thus be deduced from that of Dirichlet; but Deflers' proof (essentially the one which is taught today) is much more simple. (Editor's note.)

[2] Or Léopold Fejér: it is, in fact, with this gallicized form of his name that he signed his note in *Comptes rendus*. (Editor's note.)

would soon obtain a name, it would be the space L^1. The Riesz-Fischer theorem in 1907 established the isomorphism of the spaces L^2 and l^2 by using Fourier's formulas. Once again it seemed worthwhile to study Fourier series.

Divergence phenomena interested and inspired Fejér and Lebesgue in very different ways. They both worked with continuous functions and with Fourier series which diverged at a point. Kolmogoroff (as he signed his articles at the time) published an example in 1923 which made quite a sensation. It was complemented by a note in *Comptes rendus* in 1926, with an even more elaborate example: that of a (Lebesgue-)integrable function whose Fourier series diverged everywhere.

The example of Kolmogorov is even more interesting today. Indeed, at the time one could wonder if a stronger result could not be obtained, that is, if there existed a continuous function whose Fourier series diverged everywhere. But we know since Carleson ([Car66], 1966) and Hunt ([Hun67], 1967), that this is not possible: all functions belonging to a space L^p with $p > 1$ are limits almost everywhere of the partial sums of their Fourier series. The best result in this direction is that of Antonov ([Ant96], 1996): if $|f| \log^+ |f| \log^+ \log^+ \log^+ |f|$ is integrable on the circle (where $\log^+$ is equal to the positive part of log: $\log^+ x = \max\{\log x, 0\}$), the Fourier series of f converges to f almost everywhere.

If one assumes only that f is integrable on the circle the best that one can hope to obtain for the partial sums is thus an estimation of type $S_n(f) = o(l(n))$ almost everywhere. Hardy, in 1913, established such a formula with $l(n) = \log n$, and conjectured that it was the best possible result of this type. His conjecture is still open.

What is left to do today is thus, either to improve Hardy's formula (and to contradict Hardy's conjecture), or to improve the construction of Kolmogorov by giving a result with Ω (the opposite of o) for the S_n almost everywhere, if not everywhere. We have known for a long time that a result with Ω almost everywhere holds for any sequence $l(n)$ growing more slowly than $\log \log n$, and the result has also been established everywhere (Chen 1962). A big step was completed in 1999 by Konyagin when he constructed an integrable function whose partial Fourier sums are everywhere $\Omega(l(n))$ where the squares of the $l(n)$ are $o(\log n / \log \log n)$. One can try to improve Konyagin's estimation, as did Bochkarev in the case where the circle was replaced by the Cantor group (2003). But it would be difficult to attain $l(n) = o((\log n)^p)$ with $p > 1/2$.

1.2 Harmonic conjugates and Fourier series

In February of 1923 (when he had not yet turned twenty), Andrei Nikolaevich made his second major contribution to the theory of Fourier series, the paper *"Sur les fonctions harmoniques conjuguées et les séries de Fourier"* published in 1925 in the review *Fundamenta Mathematica*. The operation which takes a harmonic function to its conjugate can be interpreted in many ways: changing the real part of an analytic function into its imaginary part, or applying the Hilbert transform; or, in terms of Fourier series, changing the sign of the coefficients which correspond to negative frequencies and setting the constant term to zero. Today Kolmogorov's theorem can be stated in the following fashion: the operation of harmonic conjugation takes the space L^1 into weak L^1. (The functions discussed here are always real-valued functions on the circle.)

What is weak L^1 ? It is the set of functions f verifying the following condition: the inverse image of the ray $]y, +\infty[$ has a measure $m(y)$ bounded by C/y where C is a constant which depends only on the function f. In the case of L^1, $m(y)$ is integrable, and, since it is also decreasing, it verifies the given condition.

What was known about harmonic conjugation in 1923? First of all, by the Riesz-Fisher theorem (1907) it was known that it maps L^2 into L^2. Then, as noted in Lusin's work[3] which appeared in 1915, that it does not map L^1 into L^1 (this is rather clear to us today when we think of the conjugate of the Dirac measure). It was later, in 1927, that Marcel Riesz, the young brother of Frederic Riesz, showed that, for all $p > 1$, conjugation maps L^p into L^p.

In 1925, when it was published, the theorem of Kolmogorov seemed to be only an interesting curiosity. The weak L^1 space was not yet defined (in fact, it was first defined because of this theorem). Kolmogorov gives an application to the partial sums of Fourier series as a consequence of his result: they converge to the function in L^p for $p < 1$, and so they converge in measure. It was, according to Zygmund, the first appearance of the idea that the partial sums could be expressed in terms of harmonic conjugates ([Zyg59], Vol. 1, p. 381, §6). This curiosity, and especially its consequences for the partial sums, immediately attracted attention, as witnessed by Littlewood's article in 1926 "On a theorem of Kolmogoroff " [Lit26], which brings complex variable methods to bear on the problem.

Today this theorem is an important one because the Hilbert transform is the prototype of a whole class of singular integrals which have the same property. And this property of mapping L^1 into weak L^1 together with that of mapping L^2 into L^2, leads to the proof of Marcel Riesz's theorem by using a powerful tool, the interpolation theorem of Marcinkiewicz ([Mar39], 1939). In the second edition of Zygmund's book *Trigonometric series* [Zyg59] (1959) Kolmogorov's theorem is treated as a major result and two proofs of it are given, one using real analysis and the other using complex analytic functions.

1.3 Fourier series, integration and probability

Kolmogorov's first publications, which I will discuss a bit later, are about Fourier series. From 1925 on, he worked on a host of other subjects, in particular integration and probability. These two subjects are related to Fourier series.

---

[3] "The integral and the trigonometric functions" (in Russian), Moscow, 1915

1.3.1 The integral of the harmonic conjugate of an integrable function

Historically, every definition of the integral generates a new class of Fourier series; those whose Fourier coefficients are obtained with the Fourier formula when the integral is evaluated using the given definition. In fact, Fourier series have often served as a testing ground for different definitions of the integral. We thus can define Fourier-Riemann series, Fourier-Lebesgue series, Fourier-Stieltjes series, Fourier-Denjoy series, Fourier-Schwartz series, and on and on. Kolmogorov himself was essentially interested in Fourier-Lebesgue series. At the same time, he asked the question: while the harmonic conjugate of a Lebesgue-integrable function is not necessarily Lebesgue-integrable, could it be integrable in a more general sense? This question is answered by an article which appeared in 1928. This article is in French and entitled *"Sur un procédé d'intégration de M. Denjoy"*. This title is translated in the *Selected Works* by *"On the Denjoy integration process"*, which is a mistake (it should be *"On a Denjoy integration process"*). The English title makes one think that the article concerns Denjoy's totalization, either the first [Den12], which showed how to integrate any differentiable function, or the second [Den33, Den41], which gave a method for calculating the coefficients of an everywhere convergent trigonometric series given its sum. But Kolmogorov's article had nothing to do with all this. In 1919, Denjoy had published in *Comptes rendus* a note, *"Sur l'intégration riemannienne"* [Den19], in which he gave three generalizations of the Riemann integral, all three of them containing the Lebesgue integral (which he called in this note *"besguienne"*; this would not have pleased Lebesgue, who, they say, had threated Denjoy that he would call the totalization *"l'intégrale joyeuse"*, that is "joyous integral"; in any case the name *"besguien"* did not survive). These three generalization were called A), B) and C). Kolmogorov is quite clear: he is talking about the integral B). It has no relationship, as far as I know, with either the first or the second totalization. As far as I know, this is the first and the only interesting usage of the integral B). There is an explanation of all this and Kolmogorov's result on pp. 262–263 of the first volume of [Zyg59].

1.3.2 Series of independent random variables and lacunary Fourier series

The knowledge and the ideas of Kolmogorov concerning measure and integration were formed at this time, and later had a major impact on his interpretation of probabilities. But we must say that his first work in probability theory came essentially from Rademacher series, considered as series of independent random variables. In 1922, Rademacher [Rad22] published his theorem about the Rademacher orthogonal systems: L^2 convergence implies convergence almost everywhere. The article [KK25] of Khinchin (or Khintchine) and Kolmogorov in 1925 refines and extends this result: L^2 convergence and almost everywhere convergence are equivalent for a series of independent random variables, as are L^2 divergence and almost everywhere divergence. This article is mentioned in the *List of Works* of the *Selected Works* where it says that it appears in Vol. 1 along with all of the articles about integration and orthogonal series; in fact, it appears in Vol. 2, along with the other articles which are dedicated to probability. This article motivated many later studies of lacunary Fourier series, while, on the other hand, the study of lacunary Fourier series and of random Fourier series furnished new tools to be used in probability. The classical reference for this subject, as for the others, is Zygmund's book.

It is wonderful that Kolmogorov himself had established the theorem about almost everywhere convergence of lacunary Fourier series in 1922, before any mention had been made of independent random variables and at the same time as Rademacher's article appeared. This work appeared in 1924, before the article with Khinchin. He deals with more than the L^2 case; he shows in a few lines that almost everywhere convergence holds for any Fourier series which is lacunary in the sense of Hadamard, that is, in the sense that the ratio of a frequency to the preceding frequency is always greater than some number $q > 1$.

1.4 The descendants of the articles of young Kolmogorov

In 1923 and 1924 young Andrei Nikolaevich had only published articles about Fourier series. We have already discussed two aspects of these articles: divergence and lacunary Fourier series. It seems appropriate to complete this visit before finishing our promenade.

All of these articles are written in French and announce both the date when they were written and the date when they were published. The first, written on June 2nd, 1922, and published in 1923 in *Fundamenta Mathematica*, is the one concerning almost everywhere divergence which I said above made a great sensation at the time. Although the article was short, five pages, the technique seemed so elaborate that Zygmund decided not to put it into his treatise. *A fortiori* the proof of the existence of an everywhere divergent Fourier-Lebesgue series, of which a short indication had been given in a note to *Comptes rendus* in 1926, seemed to him to be unpublishable in his book. Today these proofs have been simplified, and, in particular, the reduction of the second result to the first has become part of a rather general framework. In Katznelson's book, a new edition of which has appeared recently, he gives the scope of this result with all the details and the description of the divergence sets for other classes of functions, such as the continuous functions, in a few pages at the end of his chapter about the convergence of Fourier series.

The second article, in the order in which they were written, is that of October 7, 1922, published in *Fundamenta Mathematica* in 1924. This is a very short article, two pages, which contains two theorems. The second is the one about lacunary Fourier series which I have already discussed. It has generated a remarkable number of articles, because the subject of lacunary Fourier series is related to all of functional analysis and, in particular, with the geometry of Banach spaces simultaneously with probability. One gets an idea of the work in this area (in particular that of Banach, Sidon, Zygmund, Kaczmarz and Steinhaus, Rudin, Marcus and Pisier, Bourgain) by looking at the chapter *"Lacunes et randon"* of my book with Pierre-Gilles Lemarié-Rieusset.

Today, that is, after the work of Carleson in 1966, the first theorem is totally outdated. But it seems in any case worth discussing as the beginning of a beautiful story. Kolmogorov shows that, if f is an L^2 function and if $m(n)$ is a lacunary sequence in the Hadamard sense, then the partial sums of order $m(n)$ converge almost everywhere to f. This date by which this result should be applied has passed, but we describe the simple and interesting method of its proof anyway; first to decompose the series into blocs $B(n)$ corresponding to frequencies situated between $m(n)$ and $m(n+1)$, then to regroup the $B(n)$ into two series according to whether n is even or odd. For each of these series the convergence of the Fejér sums implies the convergence of the partial sums with large gaps at the end, thus, the convergence of either the partial sums of order $m(2n)$ or those of order $m(2n+1)$ (depending on the series). Thus, because of the gaps, this gives for each of the two series convergence almost everywhere of the partial sums of order $m(n)$. Until 1966 this was the best known result on convergence for the L^2 class, and many specialists thought that it could not be improved (as far as I remember Zygmund was one of them). It could then have been tempting to extend it to L^p, $p > 1$, by using an analogous decomposition of the Fourier series. This was the monumental work of Littlewood and Paley in the 1930's; this decomposition is now called the Littlewood-Paley decomposition, and we just saw that it is exactly that of Kolmogorov. But it was necessary to develop an arsenal of weapons to do it, which is justifiably called the Littlewood-Paley theory. This theory was not yet finished when the first edition of Zygmund's book appeared in 1935, but there is an excellent explanation of it in the second edition which appeared in 1959.

The justification at the time was the extension to L^p of Kolmogorov's L^2 theorem. In 1966 and 1967, the theorems of Carleson and Hunt changed things completely: a much better result was obtained without using the decomposition and the theory of Littlewood-Paley. If Zygmund had wished to keep everything up to date, he should have included these new results in a new edition and renounced the old methods. Instead, he renounced the preparation of a 3rd edition and settled for several *"reprints"* of the second edition, with only slight additions. A "third edition" was produced recently by *Cambridge University Press*. Luckily it is nothing more than the second edition with the addition of an introduction by Robert Fefferman. The *Bulletin of the American Mathematical Society* asked me to review this book for them, a review I was quite pleased to write, as the book is a masterpiece, and because the Littlewood-Paley theory, although it no longer serves its initial purpose, has become an essential tool in other areas of analysis particularly in the study of partial differential equations. The exposition given by Zygmund is one of the best that exist. It is sometimes good, in mathematics, to think twice before throwing old things away.

The third article, concerning the order of magnitude of Fourier coefficients, was written on December 3, 1922, and published in 1923 in the *Bulletin of the Polish Academy of Sciences*. It establishes that, if the coefficients of a cosine series form a convex sequence converging to zero, then the series is a Fourier-Lebesgue series and, therefore, there exists Lebesgue-integrable functions whose Fourier coefficients converge to zero as slowly as anyone could want them to. This fact was discovered over and over, independently, by different mathematicians. The priority for the discovery should surely be given to W. H. Young, in his article *"On the Fourier series of bounded functions"* from 1912. For Fourier integrals, the corresponding result, that a continuous even function which converges to zero at infinity and is convex to the right of zero is equal to the Fourier transform of a positive integrable function, is attributed to Pólya 1949 (the functions in question are called *"Pólya functions"*, and it would be easy to justify calling the corresponding sequences *"Young sequences"*). But Kolmogorov adds a necessary and sufficient condition for the cosine series that he treats to converge towards the function in the L^1 metric, that is, the condition that the coefficients be $o(1/\log n)$: this is totally new.

The fourth article, written in common with Seliverstov (or Seliverstoff), is a note to *Comptes rendus* from January 14th, 1924; it was a time when the *Comptes rendus* attracted some very good works because they were published so quickly. It concerns almost everywhere convergence and an improvement of the conditions given by Hardy for trigonometric series and by Mensov (or Menchoff) for orthogonal series. It is the first of a series of articles of Kolmogorov and Seliverstov (1926), Plessner (1926), Hardy and Littlewood (1944), whose content is found in Zygmund's book from 1959 and which are only interesting from a historical point of view because of Carleson's theorem. The methods are in any case very clever and might be useful some day for other purposes.

I now stop the promenade in 1924, when Andrei Nikolaevich was not yet twenty one years old. He had been a precocious genius, in tune with the latest developments of a theory which had been copiously "plowed", and he left a very deep mark on it. It is true that in the beginning of the twentieth century some very young men made huge contributions to the rise of mathematics. The beauty of mathematics can attract young people, even in difficult and troubled times, and it seems to me that other Kolmogorovs are in the process of appearing, men or women, individuals or groups of individuals, on one continent or on another. Perhaps in Europe, perhaps in France, and there is no better way to leave Andrei Nikolaevich than to wish them welcome.

Appendix:
Two other aspects of Kolmogorov's results concerning harmonic conjugates

by Nikolaï Nikolski

I The A-integral of Kolmogorov

The A-integral was introduced by Kolmogorov in his classic book [Kol33] (p. 65 of the American edition) under the name of generalized expected value of a random variable (the name "A-integral" was proposed later by Kolmogorov himself). A function f is A-integrable if

1. f belongs to weak L^1 (that is, if $\lim_{t \to \infty} t|\{|f| > t\}| = 0$), and:

2. the limit $\lim_{t \longrightarrow \infty} \int_{\{|f| < t\}} f(x)dx$ exists (we call this limit $(A) \int f dx$).

Here dx is an arbitrary finite measure. One can view this notion as a lebesguian analogue of the "principal value" of Cauchy $\lim_{t \longrightarrow \infty} \int_{-t}^{t} f(x)dx$ on the real axis. In the case of the interval $[0, 2\pi]$ the definition had been proposed before Kolmogorov, by E. Titchmarch [Tit29], at first without the condition (1). But, as was proved in [Tit29], in this case the "integral" is not additive. On the other hand, the combination of conditions (1)-(2) has become useful in probability, in harmonic analysis and in complex analysis. Dozens of articles have been published on the subject. In particular:

1. E. Titchmarch [Tit29] showed (1929) that if $f \in L^1([0, 2\pi])$, then its harmonic conjugate $\tilde{f}$ is A-integrable and the conjugate Fourier series is the Fourier series of $\tilde{f}$ in the sense of the A-integral.

2. P. Ulyanov [Uly57] showed that a Cauchy transform $f(z) = \frac{1}{2\pi i} \int_{\gamma} \frac{h(t)dt}{t-z}$ of a function $h \in L^1(\gamma)$ on a smooth contour γ is a Cauchy A-integral of its boundary values (this result is not true if, e.g. γ has angular points).

3. A. B. Aleksandrov [Ale81] showed that Ulyanov's result extends to functions f of the Smirnov class Nev$^+$ whose boundary values satisfy condition (1) of the definition. He also proved analogues of this result for harmonic functions on $\mathbb{R}^n$, and he used the A-integral to implement the duality $\mathrm{CI}(\mathbb{D})^* = H^\infty(\mathbb{D})$ between the space of Cauchy integrals $\mathrm{CI}(\mathbb{D}) = L^1/H^1_-$ and the algebra $H^\infty(\mathbb{D})$.

4. Other important publications on this subject are [ABGHV95], [Luk82], [Sal88], [Mad84]; in particular, [ABGHV95] contains an application of the A-integral to the description of those subsets of the unit circle for which one can arbitrarily choose the Cauchy data for the Laplace equation.

II On the weak (1,1) type of the harmonic conjugate

S. A. Vinogradov [Vin81] strengthened Kolmogorov's theorem in the following way: the map $\mu \mapsto \tilde{\mu}$ is continuous as a map $U^* \to$ weak L^1 where U^* is the dual of the space of uniformly convergent Fourier series (replacing $C(\mathbb{T})^* \to$ weak L^1 in Kolmogorov). This theorem has many applications, see [Kis87] for the general panorama.

References

[ABGHV95] Aleksandrov, A.B., Bourgain, J., Giesecke, M., Havin V., Vymenets, Yu.: Uniqueness and free interpolation for logarithmic potentials, and the Cauchy problem for the Laplace equation in $\mathbb{R}^2$. GAFA, **5**:3, 530–571 (1995)

[Ale81] Aleksandrov, A.B.: A-integrability of boundary values of harmonic functions (in Russian). Mat. Zametki, **30**, 59–72 (1981)

[Ant96] Antonov, N.Yu.: Convergence of Fourier series. East J. Approximation, **2**, 187–196 (1996)

[Boc03] Bochkarev, S.V.: Everywhere divergent Fourier-Walsh series. Dokl. Russian Akad Nauk, **390**:1 (2003), 11–14. Doklady Math., **67**:3, 307–310 (2003)

[Car66] Carleson, L.: Convergence and growth of partial sums of Fourier series. Acta Mathematica, **116**, 135–157 (1966)

[Che62] Chen, Y.M.: A remarkable divergent Fourier series. Proc. Japan Acad., **38**, 239–244 (1962)

[Den12] Denjoy, A.: Une extension de l'intégrale de M. Lebesgue. Comptes Rendus Acad. Sciences Paris, **154**, 859–862 (1912)

[Den19] Denjoy, A.: Sur l'intégration riemannienne. Comptes Rendus Acad. Sciences Paris, **169**, 219–221 (1919).

[Den33] Denjoy, A.: Sur le calcul des coefficients des séries trigonométriques. Comptes Rendus Acad. Sciences Paris, **196**, 237 (1933)

[Den41] Denjoy, A.: Leçons sur le calcul des coefficients d'une série trigonométrique. Gauthier-Villars, Paris (1941)

[DR98] Dhombres, J., Robert, J.-B.: Fourier, créateur de la physique mathématique. Coll. "Un savant, une époque", Belin, Paris (1998)

[DuB73] Du Bois-Reymond, P.: Über die Fourierschen Reihen. Nachr. Akad. Wiss. Göttingen, 571–584 (1873)

[Fej00] Fejér, L.: Sur les fonctions bornées et intégrables. Comptes Rendus Acad. Sciences Paris **131**, 984–987 (1900)

[Fis07] Fischer, E.: Sur la convergence en moyenne. Comptes Rendus Acad. Sciences Paris, **144**, 1022–1024 (1907)

[Har69] Hardy, G.H.: Collected Papers, Vol. 3. Oxford (1969)

[Hun67] Hunt, R.A.: On the convergence of Fourier series. Proc. Conference Edwardsville, Illinois, 235–255 (1967)

[Jor81] Jordan, C.: Sur la série de Fourier. Comptes Rendus Acad. Sciences Paris, **92**, 228–230 (1881)

[Kat04] Katznelson, Y.: An introduction to harmonic analysis. Wiley (1968) Cambridge University Press (2004)

[Kis87]   Kislyakov, S.V.: Classical themes of Fourier analysis. In Khavin, V.,
           Nikolski, N. (ed) "Itogi Nauki i Techniki, Contemp. Problems Math."
           (Russian Math. Encyclopaedia), Vol. 15, 135–196 (1987), VINITI,
           Moscow (in Russian). Engl. transl.: *Encyclopaedia of Math. Sci.*, Vol.
           15, 113–166, Springer, Berlin Heidelberg New York (1991)
[KK25]   Khinchin, A.Y., Kolmogorov, A.N.: Über Konvergenz von Reihen,
           deren Glieder durch den Zufall bestimmt werden. Mat. Sb., **32**,
           668–677 (1925)
[KL95]   Kahane, J.-P., Lemarié-Rieusset, P.G.: Fourier series and wavelets.
           Gordon and Breach, London (1995) French transl.: Séries de Fourier
           et ondelettes. Cassini, Paris (1998)
[Kol23]   Kolmogorov, A.N.: Une série de Fourier-Lebesgue divergente presque
           partout. Fund. Math., **4**, 324–328 (1923)
[Kol25]   Kolmogorov, A.N.: Sur les fonctions harmoniques conjuguées et les
           séries de Fourier. Fund. Math., **7**, 23–28 (1925)
[Kol26]   Kolmogorov, A.N.: Une série de Fourier-Lebesgue divergente partout.
           Comptes Rendus Acad. Sciences Paris, **183**, 1327–1329 (1926)
[Kol33]   Kolmogorov, A.N.: Grundbegriffe der Wahrscheinlichkeitrechnung.
           Ergebnisse der Mathematik (1933) Russian transl.: ONTI, Moscow
           (1936) Engl. transl.: Chelsea, NY (1950)
[Kol91]   Kolmogorov, A.N., Selected Works, Volume 1. Kluwer (1991)
[Kon00]   Konyagin, S.V.: On divergence of trigonometric Fourier series every-
           where. Comptes Rendus Acad. Sciences Paris, **239**, série 1, 693–697
           (1999). Sb. Math., **191**, 97–120 (2000)
[Lit26]   Littlewood, J.E.: On a theorem of Kolmogoroff. J. London Math. Soc.,
           1, 229–231 (1926)
[Luk82]   Lukashenko, T.P.: On the A-integral representation of the Hilbert
           transform and conjugate function. Anal. Math., **8**:4, 263–275 (1982)
[Mad84]   Madan, S.: On the A-integrability of singular integral transforms. Ann.
           Inst. Fourier, **34**:2, 53–62 (1984)
[Mar39]   Marcinkiewicz, J.: Sur l'interpolation des opérations. Comptes Rendus
           Acad. Sciences Paris, **208**, 1272–1273 (1939)
[Rad22]   Rademacher, H.: Einige Sätze über Reihen von allgemeinen Orthogo-
           nalfunktionen. Math. Ann., **87**, 112–138 (1922)
[Rie60]   Riesz, F.: Œuvres, Vol. 1. Gauthier-Villars, Paris, Akadémiai kiadó
           Budapest (1960)
[Rie07a]   Riesz, F.: Sur les systèmes orthogonaux de fonctions. Comptes Rendus
           Acad. Sciences Paris, **144**, 615–619 (1907)
[Rie07b]   Riesz, F.: Sur une espèce de géométrie analytique des systèmes de
           fonctions sommables. Comptes Rendus Acad. Sciences Paris, **144**,
           1409–1411 (1907)
[Rie24]   Riesz, M.: Sur les fonctions conjuguées et les séries de Fourier.
           Comptes Rendus Acad. Sciences Paris **178**, 1464–1467 (1924)
[Rie28]   Riesz, M.: Sur les fonctions conjuguées. Math. Zeit., **27**, 218–244
           (1928)
[Sal88]   Salimov, T.S.: The A-integral and boundary values of analytic func-
           tions (in Russian), Mat. Sb. (N.S.), **136 (178)**:1, 24–40 (1988). Engl.
           transl.: USSR-Sbornik, **64**:1, 23–39 (1989)
[Tit29]   Titchmarch, E.: On conjugate functions. Proc. London Math. Soc.,
           29, 49–80 (1929)

[Uly57]   Ulyanov, P.L.: On A-integrals for contours. Doklady Acad. Sci. USSR, **112**:3, 383–385 (1957)
[Vin81]   Vinogradov, S.A.: A refinement of Kolmogorov's theorem on the conjugate function and interpolation properties of uniformly convergent power series. Trudy Mat. Inst. Steklova, **155**, 7–40 (1981). Engl. transl.: Proc. Steklov Inst. Math., **155**, 3–37 (1983)
[You00]   Young, W.H.: Selected Papers. Lausanne (2000)
[Zyg59]   Zygmund, A.: Trigonometric series, Volumes 1 et 2. Cambridge Univ. Press (1959)

2

Kolmogorov's contribution to intuitionistic logic

Thierry Coquand

Department of Computer Science and Engineering, Chalmers University
of Technology and Göteborg University, Sweden
http://www.cs.chalmers.se/~coquand
coquand@cs.chalmers.se

Translated from the French by Emmanuel Kowalski

Kolmogorov published only two papers related to mathematical logic. They
are concerned with aspects of intuitionism and contain simple and fundamental contributions. Although they were written at the beginning of his
mathematical career, it seems that Kolmogorov's interest for mathematical
logic was long-lasting, as shown for instance by his work on the closely related topic of complexity[1]. We will start by explaining the content of these
works, presenting their historical context, and discussing their current rele-
vance. We conclude with a presentation of a recent development which involves
the results from both papers. To write this survey, I have been helped by the
references [Hei67] and [KolI], to which I direct the reader for further information[2]. I wish to thank Hugo Herbelin and Per Martin-Löf for their many
comments on a preliminary version of this text.

2.1 The first paper (1925). Formalization
of intuitionistic logic

2.1.1 Historical context

In this section, we will present Kolmogorov's first paper [Kol25] concerning
mathematical logic, which was published in Russian in 1925 and only translated in English in 1967 [Hei67]. The context in which it appears needs to
be explained in order to fully understand the subject of this paper. At that

---

[1] See the Chaps. 14 (by B. Durand and A. Zvonkin) and 15 (by P. Vitanyi) in this
volume. (Editor's note.)

[2] The references [Hes03] and [Man98], that I discovered after writing this survey,
are also directly relevant

time, the debate between Brouwer and Hilbert concerning the foundations of mathematics was raging. Weyl, who was Hilbert's most brilliant disciple, had just written a paper [Wey21] where, in dramatic language[3], he allied himself to Brouwer's cause, and strongly criticized Hilbert's Axiomatic approach. The paper of Hilbert [Hil23] quoted by Kolmogorov, although less virulent than the previous one [Hil22], was partly an answer to this article of Weyl and to Brouwer's criticisms. Many eminent mathematicians, such as Borel and Lebesgue, insisted on the doubtful status of transfinite reasonings [Bor50]. It may be guessed that such questions were subjects of animated discussions among Lusin's students[4], among whom were Khintchine, P. S. Novikov and Kolmogorov. Echoes of such discussions can be found in the examples quoted by Kolmogorov at the end of his paper[5].

Hilbert's article [Hil23] raises the problem of justifying the rules of quantification (both existential and universal) over an infinite domain, in particular the following principle

$$(\neg\forall x.A) \rightarrow \exists x.\neg A,$$

which follows from the Principle of Excluded Middle, and may be used to deduce the existence of an element $\exists x.\neg A$ from a proof of the impossibility of its non-existence $\neg\forall x.A$[6]. This is a typical instance of what Hilbert calls a *transfinite argument*, a terminology which is also used in Kolmogorov's paper (these terms may be somewhat surprising, since the adjective "transfinite" is associated nowadays with the use of the class of countable ordinals, or more generally of uncountable classes). Hilbert remarks that this rule is justified in the case of quantification applied to a finite domain. But, accepting the critique of Brouwer and Weyl, he admits that the intuitive meaning of $\forall x.A$ and $\exists x.A$ is far from clear in general. As, in analysis, it is not legitimate to "extrapolate to infinite sums and products the theorems which are valid for finite sums and products", it is not possible here to treat quantifiers applied to an infinite domain with the same semantics used in the finite case. Hence, the general idea will be to treat *formally* such quantifications as $\forall x.A$ and $\exists x.A$, stating precise rules of inference that apply to them, and show that those

---

[3] As shown by this sample quote: *"In this coming dissolution of the empire of analysis, even if few are forewarned, I was looking for a firm ground...* Because this state is not tenable, *as I convinced myself, and Brouwer, here is the revolution!"*. This polemic culminated in 1928 when Hilbert fired Brouwer, and the editors opposing this measure, from the editorial board of the prestigious journal *Mathematische Annalen* [vanD90]

[4] Lusin, following Borel and Lebesgue, was also rather critical of purely formal uses of transfinite reasonings [Lus30]

[5] Indeed, Khintchine [Khi26] and Kolmogorov [Kol29] both published general articles presenting the debate on intuitionism

[6] In this text, we use the following notation: $\neg A$ for the negation of A, $A \rightarrow B$ for A implies B, and we denote $A_1 \rightarrow (A_2 \rightarrow B)$ by $A_1 \rightarrow A_2 \rightarrow B$. The notation $A(x)$ simply means that x may be a free variable in A and we denote by $A(t/x)$, or simply $A(t)$, the result of substituting t for x in A

rules can only lead to correct results. (This is quite similar to the treatment of possibly divergent series in analysis.) As Hilbert stated: "By staying within finitary territory, the goal is to succeed in handling freely the transfinite and in dominating it entirely!". The method that Hilbert suggests is extremely original. It consists first in defining the quantifiers from a symbol τ subject to a unique axiom, namely

$$A(\tau_x A) \rightarrow A(x),$$

which expresses a strong form of the Axiom of Choice; then, one must show that this symbol may be eliminated from any proof of a finitary result[7].

But Hilbert's paper is only a programme, and contains no definitive result[8]. In this context, the main result of Kolmogorov's first paper is very remarkable, since it purports to prove conclusively that transfinite methods can only yield correct finitary results. Moreover, this proof is different from the one suggested by Hilbert, and is very natural. It confronts directly the problem of interpretation of quantifiers, and avoids considering the τ symbol (in fact, we will see below that Kolmogorov's method does not apply directly to an interpretation of the strong form of the Axiom of Choice that is derived from this symbol).

2.1.2 Kolmogorov's formalization of intuitionistic logic

Kolmogorov's first contribution in this paper is a complete formalization of *minimal* propositional calculus (a strict subset of intuitionistic logic which is usually attributed to Johansson [Joh36]) and minimal predicate calculus. As indicated by Wang, Kolmogorov's formalization is no less remarkable than Heyting's [Hey30]. The very possibility of such a formalization is already quite surprising, if we reflect that the motivations behind intuitionism were opposed to the process of formalization[9]. Kolmogorov's work is final concerning propositional calculus, but less precise with respect to predicate calculus.

---

[7] Using Hilbert's example, if A is the predicate "to be corruptible", then $\tau_x A$ would be a man "of such absolute integrity that if he turned out to be corruptible, then all men would be corruptible" [Hil23]. It is then possible to define $\forall x.A$ as equivalent to $A(\tau_x A)$. Afterwards, the symbol τ will be replaced by a dual symbol ϵ, designating a choice function. Bourbaki uses the symbol τ in its formulation of set theory, but with the dual meaning of the choice symbol ϵ. Hilbert's method is explained very suggestively in the reference [Wey44]

[8] A general result of elimination of the symbol τ in analysis was published by Ackermann [Ack24]; von Neumann [vonN27] found a problem in this proof and showed that its range of applicability is much more restricted than stated, in fact similar to Kolmogorov's result. It seems that the problem of elimination of the τ symbol in analysis remains open [Kre65]

[9] According to Wang [Wan87], Brouwer considered this result to be more remarkable and surprising than Gödel's celebrated incompleteness theorem [Göd31]

The formalization is directly inspired from Hilbert [Hil23], who had suggested the following axioms for implication and negation:

1. $A \to B \to A$
2. $(A \to A \to B) \to A \to B$
3. $(A \to B \to C) \to B \to A \to C$
4. $(B \to C) \to (A \to B) \to A \to C$
5. $A \to \neg A \to B$
6. $(A \to B) \to (\neg A \to B) \to B$

The only inference rule is the *modus ponens*: one may deduce B from $A \to B$ and A. For example, taking $B = A$ in the first two axioms (i.e. *by instantiation of B by A*), we obtain the following two special cases (two *instances*):

$$A \to A \to A$$

$$(A \to A \to A) \to A \to A$$

and by means of *modus ponens*, it follows that we also have $A \to A$.

The notions of semantic were not yet widely developed and it is interesting to notice how Kolmogorov states the property of *completeness* of this axiomatic system: it is not possible to add a new axiomatic schema without contradiction. More precisely, a new axiom, such as

$$(\neg\neg(A \to B)) \to \neg\neg A \to \neg\neg B,$$

is either provable from the axioms above, or leads to a contradiction, i.e. one may then deduce an arbitrary proposition[10].

The first four axioms deal only with implication, and Kolmogorov raises the pertinent question of their completeness for implication. As explained by Wang (p. 416 of [Hei67]), this system is not complete, but it becomes so if *Peirce's law*

$$((A \to B) \to A) \to A$$

is added to it.

The last two axioms concerning negation are both rejected by Kolmogorov, and replaced by a unique axiom called *principle of contradiction*

$$(A \to B) \to (A \to \neg B) \to \neg A$$

The last of Hilbert's axioms must be rejected because it may be seen as a formulation of the principle of excluded middle $A \vee \neg A$, as explained by Kolmogorov in his Note 9[11]. As noticed by Kolmogorov in his Note 3, the principle

$$(\neg \forall x.A) \to \exists x.\neg A$$

---

[10] In this example, one may deduce this new axiom from the ones given previously
[11] The numbering of notes refers to that used in the English translation in [Hei67]

would follow from this axiom, combined with others which are valid from the intuitionistic point of view. However, this is definitely problematic when quantification over an infinite domain is concerned, and this is acknowledged by Hilbert himself [Hil23].

Things are not so clear concerning Hilbert's Axiom 5. It will be considered as admissible by Heyting in his formalization of intuitionistic logic [Hey30] (and accepted by Kolmogorov in his paper [Kol32]). Today, it is common to introduce a special proposition $\perp$ representing an absurdity and, following Brouwer, to *define* $\neg A$ as being $A \to \perp$. (With this definition, Kolmogorov's principle of contradiction becomes *provable* from the axioms for implication.) An elegant formulation of Hilbert's debatable Axiom 5 is then

$$\perp \to A$$

which expresses the principle that *ex falso quodlibet*. This principle has been the subject of much discussion in intuitionistic mathematics. If it is rejected, there is no special condition on the proposition $\perp$, and it may be considered that there are no negative propositions. Griss, among others, has argued for the rejection of Axiom 5 [Gri55], and hence for considering only affirmative propositions. In answer to these developments, Brouwer constructed an example [Bro48] showing that it is necessary in intuitionistic mathematics to consider negative propositions in an essential way, and hence that Hilbert's Axiom 5 must be admitted. As we will see later on, the fact that Kolmogorov makes no use of Axiom 5 in his interpretation of classical logic, so that negation $\neg A$ may be defined as $A \to X$ where X is an arbitrary proposition, may be exploited in a non-trivial way.

2.1.3 Negative interpretation

The main result is that it is not possible to prove new finitary results with non-constructive methods. In particular, one can not introduce contradictions.

The idea is very simple and is well summarized by the following imaginary dialogues (taken from E. Nelson [Nel92, Nel04]) between an intuitionist mathematician I and a classical mathematician C, who starts the dialogue:

C: I have just proved $\exists x.A$.
I: Congratulations!. What is it?
C: I don't know. I assumed $\forall x.\neg A$ and derived a contradiction.
I: Oh. You proved $\neg \forall x.\neg A$.
C: That's what I said.

Here is a similar dialogue:

C: I have proved $A \lor B$.
I: Good. Which did you prove?
C: What?

I: You said you proved A or B; which did you prove?

C: Neither; I assumed $\neg A \wedge \neg B$ and derived a contradiction.

I: Oh; you proved $\neg[\neg A \wedge \neg B]$.

C: That's right. It's another way of saying the same thing.

So, the idea is that the intuitionistic mathematician may very well follow a classical reasoning, at the price of reformulating some statements. It is the principle which is put to work in Kolmogorov's paper, for propositional calculus based on $\rightarrow$ and $\neg$, and then for predicate calculus. This paper contains another idea, which is the key to a semantic understanding of the negative interpretation. Kolmogorov notices that in order to have a calculus which is equivalent to Hilbert's, it is enough to add the axiom of double negation

$$\neg\neg A \rightarrow A$$

To show this, he proves that Hilbert's Axioms 5 and 6 can both be derived from this axiom. Then Kolmogorov considers all propositions A which satisfy the implication $\neg\neg A \rightarrow A$. Let us call such propositions *regular*[12]. A first fundamental remark is that any finitary proposition is regular. Brouwer, in the paper [Bro25] quoted by Kolmogorov, had shown that any negative proposition $\neg A$ is regular. The second fundamental remark is that Hilbert's axioms are valid for regular propositions. In semantic terms, regular propositions form a model of classical propositional calculus (a boolean algebra). Since finitary propositions are regular, it is now clear, in a semantic way, that any finitary proposition which is classically provable will also be intuitionistically provable. Kolmogorov proves this explicitly by describing an interpretation A^* of each proposiion A in such a way that if A is classically provable, then A^* is intuitionistically provable (and A^* is equivalent with A for a finitary proposition).

The interpretation $A \longmapsto A^*$ is defined by induction on the formula A. If A is atomic (i.e. does not involve any logical connector), then A^* is $\neg\neg A$. If A is a composite formula $F(A_1, \ldots, A_n)$ then A^* is $\neg\neg F(A_1^*, \ldots, A_n^*)$. Kolmogorov only considers the connectors $\rightarrow$ and $\neg$ in this paper, but his definition can be applied directly to the connectors $\vee$ and $\wedge$, and those authors which refer to "Kolmogorov's interpretation" (e.g. the reference [Mur90]) use freely this extension

$$(A \vee B)^* = \neg\neg(A^* \vee B^*) \qquad (A \wedge B)^* = \neg\neg(A^* \wedge B^*)$$

It may be remarked that this interpretation is far from being "minimal": one might take, as Gentzen does for instance, $(A \wedge B)^* = A^* \wedge B^*$, $(A \rightarrow B)^* = A^* \rightarrow B^*$. Concerning negation, the definition gives $(\neg A)^* = \neg\neg\neg A^*$ whereas $(\neg A)^* = \neg A^*$ is sufficient. But Kolmogorov's interpretation is very systematic.

---

[12] A natural interpretation of intuitionistic logic is to use the *open sets* of a topological space X as truth values. Regular propositions are then those which are interpreted by *regular* open sets, i.e. by those equal to the interior of their closure

2.1.4 Predicate calculus

Kolmogorov's treatment of predicate calculus is less detailed. He never defines explicitly the negative interpretation of the formulas $\forall x.A$ and $\exists x.A$. It seems, however, that he takes them to be special cases of his general definition, so that the definition used is

$$(\forall x.A)^* = \neg\neg\forall x.A^* \qquad\qquad (\exists x.A)^* = \neg\neg\exists x.A^*$$

He remarks that the translation of *Aristotle's Axiom*

$$(\forall x.A) \to A(x)$$

is valid, using implicitly such an interpretation. Although this axiom is mentioned, he forgets to insert it in his list of axioms for quantification (numbered from I to IV), as noticed by Wang [Hei67] (which is surprising because the dual Axiom IV is present).

As we have already seen, the paper of Hilbert quoted by Kolmogorov uses the Axiom $A(\tau_x\ A) \to A(x)$ and it turns out that, for the τ symbol, adding this Axiom to propositional calculus is sufficient, and there is no need to add any further inference rule to the rule of *modus ponens*. But this is not the case for universal quantification. Kolmogorov is aware of this point, and states that what he calls *Principle P*: if $A(x)$ is deductible, then $\forall x.A(x)$ is also. It may be noted that with Axiom II and *modus ponens*, one recovers also the rule stating that one may infer $A \to \forall x.B(x)$ from $A \to B(x)$ if x is not a free variable in A.

One would expect to see a justification of the problematic rule

$$(\neg\forall x.A) \to \exists x.\neg A$$

using the fact that its translation is valid from the intuitionistic point of view. Instead, Kolmogorov shows that this rule is an intuitionistic consequence of the rule of double negation.

All this may be compared to the treatments by Gödel [Göd33] and Gentzen [Gen69]. As remarked by Wang [Hei67], the only element in Kolmogorov's paper which is missing to make it a complete system of the minimal predicate calculus is an explicit statement of Aristotle's Axiom.

Although Kolmogorov's paper, published in Russian only, was unknown to Gödel and Gentzen, one may speculate about an indirect influence [vonP94], since Gödel refers to a paper of Glivenko [Gli29], who was certainly aware of Kolmogorov's results[13].

---

[13] It is probably because Kolmogorov's paper was in Russian that Glivenko does not refer to it in [Gli28, Gli29]. Similarly, Kolmogorov himself does not cite his own paper in his article on the interpretation of intuitionistic logic, although he does refer to Heyting's paper [Hey30]. In his correspondence with Heyting [Kol88] concerning Gödel's paper [Göd33], Kolmogorov mentions to Heyting that his paper [Kol25] contains similar results, and that he believes that this method may be applied to a large part of classical mathematics

2.1.5 References to the article of 1925

Kolmogorov never came back to the contents of his paper [Kol25], and in particular he did not refer to it in 1932 in the paper [Kol32]. As remarked by J. von Plato [vonP94], there are nevertheless some echoes of this paper in Kolmogorov's book on probability theory [Kol33]. Events which are infinite disjunctions or conjunctions are considered as "ideal events", which correspond to no empirical event. And he notes that if the probability of an factual event is found using ideal events, then such a computation must also be valid empirically.

2.2 Classical and intuitionistic mathematics

2.2.1 The problem

Kolmogorov's result had been foreseen in a large measure by Brouwer. Already in 1908 [Bro08], he clearly notices that the use of the excluded middle can not lead to a contradiction, essentially because we have $\neg\neg(A \vee \neg A)$, since it is contradictory to have both $\neg A$ and $\neg\neg A$: *"So that for the moment we can not trust the excluded middle principle for infinite systems. Neither will we ever have to confront a contradiction, and discover in this manner the unfounded nature of our arguments, by an unjustified use of this principle. Indeed for this both the affirmation and the contradictory character of a statement would be contradictory, and this is forbidden by the principle of contradiction."* In 1923 [Bro23], referring to this remark, he considers very plausible the proof of intuitionistic non-contradiction of excluded middle, but he adds: *"nothing of mathematical value will thus be gained: an incorrect theory, even if it cannot be inhibited by any contradiction that would refute it, is none the less incorrect, just as a criminal policy is none the less criminal even if it cannot be inhibited by any court that would curb it"!*

This indicates the limitations of this type of results in the intuitionistic framework. One may however wonder whether Kolmogorov's goal is achieved, independently of this particular criticism. Is it possible to prove, in an intuitionistic way, that classical mathematics are not contradictory? This is a very interesting question, which is not easy to answer in a simple manner, since it depends crucially on the range of intuitionistic mathematics. According to Myhill [Myh68], and following Gödel, it is necessary to make a distinction between two forms of intuitionism, whether impredicative definitions are permitted or not[14].

---

[14] A definition is called impredicative if it defines an object by means of a quantification over a class containing this object. This notion arose from a debate between Russell and Poincaré [Hei86]. Poincaré [Poi09] pointed out the surprising aspect of such definitions. These intuitions have been confirmed by demonstration theory, which shows that definitions of this type have great logical force [Kre68, Gir72]. It

2.2.2 Predicative intuitionism

Gödel's paper [Göd33] implicitly states that intuitionistic mathematics rejects impredicative definitions. According to Gödel, it is not possible to conclude that, in general, intuitionistic mathematics is an extension of classical mathematics from the fact that the double-negation method proves that classical *arithmetic* can be interpreted in intuitionistic *arithmetic*[15]. This may be shown using arguments of demonstration theory: the logical strength of an intuitionistic system such as the one in the book of Kleene and Vesley [KV65], is comparable with that of the theory ID_1 of a general inductive definition, and is therefore much weaker than classical analysis. Hence, it is absolutely impossible to prove in this system that classical analysis is non-contradictory, by Gödel's Second Incompleteness Theorem [Göd31].

Current intuitionistic mathematics considers rather strong forms of the reflection principle (see, e.g. [Pal98]), but those theories remain logically weaker than the comprehension principle Π_2^1 and therefore cannot suffice to justify classical analysis, exactly as predicted by Gödel [Göd33]. It remains an open problem to give a constructively valid semantic of classical analysis.

2.2.3 Impredicative intuitionism

If the free use of impredicative definitions is allowed, purely formally, one obtains what may be called "impredicative intuitionism", which may be defined simply by the refusal of the double negation rule. The objection concerning the lack of logical strength does not apply anymore, and the problem of justifying classical methods is raised anew in this setting.

Around the end of the 1960's, this impredicative form of intuitionism was analyzed in detail, in works culminating with the proof of normalization by J.-Y. Girard [Gir71] (a result which is a very strong form of relative

---

is interesting to know that such definitions can be represented using the τ symbol [Spe62], and one of Hilbert's (and later Gentzen's) goal was precisely to show that such definitions were consistent [Hil29]. On the other hand, they are considered as being unjustified from the constructive point of view, see e.g. Gödel [GödIV] and P. Martin-Löf [Mar84]

[15] There is also the problem of understanding how the result of Kolmogorov-Gödel does not contradict the theorem that states that the consistency of arithmetic can not be proved in finitary manner. Indeed, as explained very clearly in [Nel92], and as foreseen in Kolmogorov's paper, this interpretation is in particular an intuitionistic proof of non-contradiction of classical arithmetic. Gödel's paper [Göd33] is the source of a distinction between *finitary* and *intuitionistic*. Thus Kolmogorov's result proves the consistency of classical methods with respect to intuitionism, but not with respect to finitism. All this is still subject to debate, since the range of the finitary method remains very vague [Zac98]. As indicated by Gödel himself [Göd31], it is possible to argue that his incompleteness theorem does not in fact contradict Hilbert's programme

consistency), then the discovery, also due to J.-Y. Girard, that the most general form of impredicative intuitionism leads to a contradiction [Gir72, Miq01]. It seems that one of the motivations for these researches was precisely the remark that the negative interpretation of Kolmogorov-Gödel may be extended, almost trivially, to second-order arithmetic[16] [Kre68]. Indeed, the comprehension principle for analysis

$$\exists X \forall y. \ (y \in X \leftrightarrow A(y))$$

is an example of a mathematical axiom which implies its own negative interpretation. So it is a paradigmatic example of Note 15 of Kolmogorov's paper, according to which the axioms of mathematics satisfy the implication $A \to A^*$, although there may be problems in their intuitionistic justification[17].

This interpretation was then extended to the theory of types [Myh71] and to set theory [Pow75]. The idea behind those extensions is easily understood in semantic terms [Law71]: one may see the set of intuitionistic truth values as a complete Heyting algebra. The sub-algebra of regular truth values is a complete boolean algebra, and it is then possible to follow the general method of construction of boolean models of type or set theory (see, e.g. [Bel85]).

2.2.4 The problem of the Axiom of choice

In Note 15 to his paper, Kolmogorov remarks that the axioms of mathematics, including the axiom of choice, have the property that they imply their negative interpretation. This assertion depends on the formulation of Zermelo's Axiom of Choice which is used (note that a precise formulation in the predicate calculus of set theory had only recently been given by Skolem [Sko22]). It seems that taking a formulation close to Zermelo's original one, according to which one may always find a a set of representatives for an arbitrary equivalence relation on a given set, yields indeed a proposition which implies its own negative interpretation[18]. On the other hand, if the functional form of the Axiom of Choice is considered, namely

$$(\forall x \exists y. A) \to \exists f \forall x. A(x, f(x))$$

---

[16] Recall that *first order* arithmetic allows only quantifications over integers ($\forall n$, $\exists m$), and that *second order* arithmetic allows also quantifiers over properties, or what amounts to the same thing, over subsets of $\mathbb{N}$ ($\forall A$, $\exists B$)

[17] In the reference [GödIV], it appears in the course of the correspondence between Bernays and Gödel that Gentzen, and Gödel, had realized by 1939 that it was possible to extend the negative interpretation to the theory of types. However, Gödel and Bernays both considered that such a reduction is not constructively satisfactory

[18] I owe this remark to P. Martin-Löf. As noticed by Kolmogorov, this axiom in not intuitionistically justified, and in fact one can even show that it implies the principle of excluded middle [Dia75]

which requires quantification over function symbols, then this form *does not imply* its negative interpretation. What would be required is the implication

$$(\forall x. \neg\neg A(x)) \rightarrow \neg\neg \forall x. A(x)$$

which is not justified intuitionistically. This problem is the origin of works such as [Spe62, BBC98, BO05, Kri03, Miq03]. It is in fact while studying this question that it was realized, in the 1960's, that intuitionistic logical systems such as the one described in [KV65] are much weaker logically than classical analysis [Kre68].

2.3 Refinements of Kolmogorov's result

As we have seen, Kolmogorov like Weyl [Wey21] and Hilbert [Hil23], does not consider existential propositions as finitary. As stated by Weyl [Wey21] an existential proposition is not a "genuine judgement that affirms a state of things", but only a "judgement abstract". This is formally analogous to the fact that an existential proposition $E = \exists x. A(x)$ is not usually a regular proposition, even if $A(x)$ is one, because the implication $E^* \rightarrow E$ requires the use of the principle

$$\neg(\forall x. \neg A(x)) \rightarrow \exists x. \neg\neg A(x).$$

Kolmogorov's result does not extend to non-finitary propositions in general[19]. It is therefore rather surprising that it does extend to purely existential propositions, i.e. those of the form $A = \exists x. B(x)$ where $B(x)$ is a finitary proposition: if such a proposition A can be proved by classical means, then A may also be proved intuitionistically.

This result, in a weaker form, follows from the finitary interpretation of classical proofs (due to Gentzen [Gen35]). The general case may be proved using the *Dialectica* interpretation of Gödel [Göd58]. H. Friedman [Fri78] has given a particularly simple proof, which uses the negative interpretation. It can profitably be presented using the fact that Kolmogorov's interpretation is an interpretation of classical logic in *minimal* logic.

Indeed, let us replace the proposition $\bot$ indicating absurdity by an arbitrary proposition X, and let us write denote by $\neg_X C$ the implication $C \rightarrow X$. It follows from Kolmogorov's paper that is $A = \exists x. B(x)$ is classically provable, then A^* is intuitionistically provable. If $B(x)$ is a finitary proposition, we have $B(x)^* \equiv (B(x) \rightarrow X) \rightarrow X$ and therefore

$$A^* \equiv (\forall x. B(x)^* \rightarrow X) \rightarrow X \equiv (\forall x. B(x) \rightarrow X) \rightarrow X \equiv \neg_X \neg_X A$$

---

[19] The proposition $\exists n. \forall m. f(n) \leq f(m)$ concerning the non-negative integers, for instance, has a direct classical proof, but is not intuitionistically valid

Hence $\neg_X\neg_X A$ is intuitionistically provable. Since, in this argument, the proposition X is arbitrary, it can be replaced by the proposition A itself, and we obtain

$$(A \to A) \to A$$

and hence A is intuitionistically provable.

In algebraic terms, this result appears as Funayama's Theorem [Fun59], which states that an arbitrary complete Heyting algebra may be embedded in a boolean algebra, the embedding preserving upper bounds. This result has since been refined in the form of Barr's Theorem [Joh77] which affirms that there exists a surjective morphism from a boolean topos to an arbitrary elementary topos (the proof of which, in [Joh77], is indeed reminiscent of the argument just described).

Excepting the formulation of the axiom of choice previously mentioned, one can give natural examples of axioms A which do not satisfy the principle of Note 15 of Kolmogorov, according to which A implies its own negative interpretation A^*. If the general induction axiom does satisfy this principle, it is clear for instance that the Σ_1^0 axiom of induction does not, where the induction

$$B(0) \to (\forall x.B \to B(x+1)) \to \forall x.B$$

is restricted to purely existential propositions B. Indeed, the negative interpretation becomes

$$B^*(0) \to (\forall x.B^* \to B^*(x+1)) \to \forall x.B^*$$

and B^* is not purely existential. An analogue problem appears in the induction principle of the system ID_1. There exists a variant of the negative interpretation avoiding this problem, which is presented in the references [Avi00] and [CH99].

2.4 A calculus of problems

As indicated by E. Nelson [Nel92], Kolmogorov's result [Kol25] may be interpreted as showing that intuitionistic mathematics is not a *restriction* of classical mathemaics, by rather an *extension* (at least in the propositional case and for arithmetic). Moreover the operations of classical mathematics can be interpreted in the fragment of intuitionistic logic that uses only the connectors $\wedge$, $\to$ and $\neg$, and universal quantification, for instance by defining

$$(\exists x.A)^* = \neg(\forall x.\neg A^*) \qquad (A \vee B)^* = \neg(\neg A^* \wedge \neg B^*)$$

It is therefore appropriate to say that intuitionistic mathematics introduces a *new* connector $\vee$, with a constructive interpretation, and a *new* existential quantifier (also constructive), both of which have no classical counterpart. A natural question is then to describe clearly the semantics of those new operations. This is the objective of Kolmogorov's second article on intuitionistic logic, which presents a general calculus of problems, and shows that this calculus provides the desired semantics. This paper [Kol32] is short and informal. Kolmogorov doesn't define the notion of *problems* precisely, but contents himself with some examples. A problem can thus be

Find four positive integers x, y, z, n such that $x^n + y^n = z^n$ and $n > 2$

but one may also have "conditional" problems such as

If a solution of $ax^2 + bx + c = 0$ is given, find the other root

or

Assuming that π is rational, $\pi = m/n$,
find a similar expression for e.

The last example is one of a conditional problem with a false premise. In this case, Kolmogorov considers that a proof that the premise is false is a solution of the problem.

Assuming at least an informal understanding of the notion of problem, Kolmogorov introduces the following notation, where A, B denote arbitrary problems

$A \wedge B$ is the problem of solving problem A *and* problem B.
$A \vee B$ is the problem of solving *at least one* of problems A *or* B
$A \rightarrow B$ is the problem of reducing a solution of problem B to a solution of problem A (conditional problem)
$\neg A$ is the problem: assuming a solution of problem A exists, deduce a contradiction.

In his Note 4, Kolmogorov gives an example to clarify the last case: it should not be mistaken with the task of proving that A can not be solved. For instance, let A be the problem of the Continuum Hypothesis: it is not a solution of problem $\neg A$ to show that it can not be solved (as will indeed be shown in the 1960's).

It is very remarkable that the calculus of problems thus defined coincides with intuitionistic logic as formalized by Heyting [Hey30].

Consider for example the Axiom

$$(A \wedge (A \rightarrow B)) \rightarrow B$$

Its justification in the calculus of problems is as follows. The solution of problem B must be reduced to the solution of the problem $A \wedge (A \rightarrow B)$. So, suppose a solution of this problem is given. This means that one has both a solution of problem A, and a solution of problem $A \rightarrow B$, i.e. one may reduce the solution of problem B to that of problem A. Hence, as desired, we can solve problem B.

On the other hand, the principle of excluded middle in the form

$$A \vee \neg A$$

can not be justified. Indeed, a solution would be a general method which, given a problem A, would either give a solution or show that a contradiction follows from the assumption that problem A is solvable. As Kolmogorov states, unless one is bold enough to consider oneself omniscient, it must be recognized that there is no solution to this problem.

It is now also very clear why the problem

$$\neg\neg(A \vee \neg A),$$

is solvable. Indeed, a solution of $\neg(A \vee B)$ is a solution of problem $\neg A$, *and* a solution of problem $\neg B$. If $B = \neg A$, a contradiction is reached. This is exactly Brouwer's reasoning [Bro08] recalled above, but formulated in the calculus of problems.

So we can not hope to solve the problem $\neg\neg B \to B$, since by instantiation of B by $A \vee \neg A$, we would obtain a solution of

$$\neg\neg(A \vee \neg A) \;\to\; A \vee \neg A$$

and, since a solution of the premise is available, this would give a solution of the problem corresponding to the principle of excluded middle.

Thus, this paper proposes a very suggestive informal interpretation of intuitionistic calculus, which is clearer than the one developed by Heyting [Hey31].

2.5 Some recent developments

The calculus of problems has been refined in type theory [Mar82, Mar84, NPS90], which essentially adds to Kolmogorov's calculus an explicit notation for the solutions of problems. One introduces the notation $a \in A$ which may be read in one of the following manners:

a is a *solution* of *problem A*
a is a *proof* of *proposition A*
a is a *program* which is correct with respect to *specification A*
a is an *element* of the *set A*

Thus $f \in A \to B$ may be interpreted as the statement that f is a function of the set A to the set B, or as the statement that f is a method that reduces the solution of problem B to that of problem A, or as the statement that f is a proof of the implication $A \to B$. In this form, the interpretation obtained is known as the *Brouwer-Heyting-Kolmogorov* semantics of intuitionistic logic. This interpretation is closely related to the notion of *realizability* in proof theory [Nel92, Tro73], and to the *Curry-Howard* correspondence [How80, Mar82, Kri98]. It extends naturally to predicate calculus, to arithmetic, and even to set theory [Fri73]. It would be impossible to summarize even partially the researches related to this topic, and we will content ourselves with a few general remarks.

First of all, why is this interpretation relevant to computer science? As clearly expressed by Dijkstra [Dij76], an efficient programming methodology is to develop a computer program in parallel with the proof of its correctness, the correctness proof leading the development. This methodology can be expressed elegantly in Kolmogorov's calculus of problems: thus, a solution to a problem of the type

$$\forall x.P(x) \rightarrow \exists y.R(x,y)$$

may be seen as a computer program which, when executed with an input a satisfying the precondition $P(a)$, outputs a result b satisfing $R(a,b)$. An intuitionistic proof of a proposition A may therefore be seen as a program which is correct with respect to the specification A by construction. This is explained in references [Mar82, NPS90, Nel92].

Next, there is a very useful notation for solutions of problems, namely λ-calculus [Bar97], which provides a convenient formalism for the representation of programs [Bar97, Lan65]. It may be described by the following grammar[20]:

$$t ::= x \mid t\, t \mid \lambda x.t$$

If t is the solution of a problem B constructed from a hypothetical solution x of problem A, we can represent by an *abstraction* $\lambda x.t$ the solution of problem $A \rightarrow B$. Finally, if f is a solution of the problem $A \rightarrow B$ and a is a solution of problem A, then the *application* $f\ a$ represents the solution of problem B. The λ-calculus is a very practical notation to express functions. For instance, the identity function $x \longmapsto x$ is represented by the term $\lambda x.x$, whereas the function

$$A \rightarrow (A \rightarrow X) \rightarrow X$$

$$x \longmapsto (f \longmapsto f(x))$$

which takes x as argument and outputs the functional which evaluations an input function f with the argument x is represented by the term $\lambda x \lambda f.f\ x$.

A very simple model of functional programming is obtained using the λ-calculus with the single reduction rule

$$(\lambda x.u)\ t = u(t/x)$$

---

[20] Which means that terms are formed using the following rules: a term is either a variable x, or what is obtained by substitution of a term in another ($t\ t$, for $t_1(t_2)$), or the function which maps x to t (denoted $\lambda x.t$), where x is a free variable in the term t

which expresses precisely that the result of applying $(\lambda x.u)$ t is obtained by replacing x by t in u. For instance, if we apply the function $\lambda x \lambda f.f\ x$ to the arguments z and the identity function $\lambda y.y$, we get

$$((\lambda x \lambda f.f\ x)\ z)\ (\lambda y.y) = (\lambda f.f\ z)\ (\lambda y.y) = (\lambda y.y)\ z = z$$

as expected.

This notation is extremely suggestive when it comes to representing solutions in Kolmogorov's calculus of problems. For instance, Hilbert's four axioms for implication admit the following solutions:

1. $\lambda x \lambda y.y \in A \to B \to A$
2. $\lambda f \lambda x.f\ x\ x \in (A \to A \to B) \to A \to B$
3. $\lambda f \lambda y \lambda x.f\ y\ x \in (A \to B \to C) \to B \to A \to C$
4. $\lambda g \lambda f \lambda x.g\ (f\ x) \in (B \to C) \to (A \to B) \to A \to C$

Notice that the solution for Axiom 2 is the diagonalization operator, and that the solution for Axiom 4 is the composition operator.

It may also be noted that the term $\lambda x \lambda f.f\ x$ described a solution of the problem

$$A \to (A \to \bot) \to \bot$$

whereas it is not possible to construct a solution for the problem

$$((A \to \bot) \to \bot) \to A$$

Impredicative intuitionism also takes a very simple form in this framework [Gir72]. As remarked by J.-L. Krivine [Kri03], the comprehension scheme is essentially interpreted by the identity $\lambda x.x$! It is also possible to extend such an interpretation to an intuitionistic theory with the same logical strength as the Zermelo-Fraenkel system [Kri01, Miq03] (and one can even analyze some inconsistent systems [CH94]).

2.6 A calculus of problems for classical logic?

We have seen that one can not hope to solve the problem corresponding to the double-negation principle

$$\neg\neg A \to A$$

A relatively recent discovery (1990) [Gri90, Kri96] tempers this remark, giving hope for what can be described as a calculus of problems for classical logic. Moreover, this discovery has a direct link with Kolmogorov's negative interpretation [Mur90].

In what follows, we consider a special problem denoted $\bot$ and we define $\neg A$ as being the problem $A \to \bot$. A suggestive interpretation is to think of $\bot$ as the type of *executable* programs [Kri96]: typically, the type of the program *print* which prints an integer, of type N, will be $\neg N = N \to \bot$, since this program takes an integer as input, and produces no output but a side-effect (printing the integer). The idea will be to extend the λ-calculus with a new unary operation $C(t)$ such that $C(t)$ is a solution of problem A if t is a solution of problem $\neg\neg A$. The term $\lambda x.C(x)$ solves $\neg\neg A \to A$ and is therefore a formal solution to the problem corresponding to the principle of double negation. This new constant is defined by a unique rewriting rule

$$E[C(t)] = t \ (\lambda x.E[x]) \tag{1}$$

which is not "local" anymore, since it may only be applied to a term of type $\bot$. For instance, if t is of type $\neg\neg(N \to N)$ and u of type N, we will get

$$print \ (C(t) \ u) = t \ (\lambda x.print \ (x \ u))$$

This reduction preserves types: $\lambda x.print \ (x \ u)$ is of type $\neg(N \to N)$ and hence $t \ (\lambda x.print \ (x \ u))$ is indeed of type $\bot$. But to ensure this preservation of types, it is essential that the context of evaluation $E[x]$ be of type $\bot$.

It may then be held that the addition of $C(t)$ with this reduction rule provides a solution to the problem of the exluded middle. What is most important, however, is that such mechanisms occur naturally in computer science [SW74, Gri90, FF86, Kri96], for instance in the implementation of schedulers for operating systems. The effect of those instructions is to keep in memory a reference state of a program being executed, in order to be able to come back to it, in case of problem for example. This is well represented by the rule (1) where the term $\lambda x.E[x]$ corresponds to the reference state which is remembered. It is truly surprising that such a rule appears completely independently of the problem of the semantics of the principle of excluded middle.

Following the fundamental works of Strachey and Wadsworth [SW74, Fis72, Plo75], one may also explain such operations by a translation into pure λ-calculus. Such a translation is [Gri90]

$$(t \ u)^* = \lambda k.t^* \ (\lambda m.m \ u^* \ k) \qquad (\lambda x.t)^* = \lambda k.k \ (\lambda x.t^*)$$

$$x^* = x \qquad (C(t))^* = \lambda k.t^* \ (\lambda m.m \ (\lambda z.z \ (\lambda f \lambda d.f \ k)) \ (\lambda x.x))$$

One may then show that if t is a solution of problem A, which may involve the classical operation C, then t^* is an intuitionistic solution of problem A^*, where A^* is precisely Kolmogorov's negative interpretation [Kol25][21]. What is used here is the fact that Kolmogorov's translation is very systematic, since this allows a simple translation of terms. (If a negative interpretation like

---

[21] The proof uses the fact that one may write $A^* = \neg\neg A^+$ where A^+ is defined inductively using the clauses $X^+ = X$ and $(A \to B)^+ = \neg\neg A^+ \to \neg\neg B^+$. This is a substitutive transformation, i.e. it satisfies $(A(B/X))^+ = A^+(B^+/X)$ and is used for the definition of the negative interpretation in higher-order logics [Gir72, CH94]. It it then straightforward to show by induction on t that if t is of type A, then t^* if of type $\neg\neg A^+$

Gödel's or Gentzen's were used, the translation of terms would have to take their types into account[22].) It is also possible [Mur90] in this framework to give a computer science meaning of Friedman's refinement [Fri78] as described above. The negative interpretation described in Kolmogorov's paper [Kol25] is therefore formally identical to the interpretation used in computer science to derive the semantics of escape operations [SW74].

However, it is not clear how one could think of such terms involving the constant C as solutions of problems, as is the case in the intuitionist situation, and more generally, whether this interpretation of classical logic is fundamentally different from the negative interpretation. Such a link, reinforcing the the correspondence described above between proofs, programs and solutions of problems, seems to be fundamental nevertheless, and it is remarkable that it be so closely related to the pionnering works of Kolmogorov.

References

[Ack24]   Ackermann, W.: Begründung des "tertium non datur" mittels der Hilbertschen Theorie der Widerspruchsfreiheit. Math. Ann., **93**, 1–36 (1924)

[Avi00]   Avigad, J: Interpreting classical theories in constructive ones. J. Symbolic Logic, **65**:4, 1785–1812 (2000)

[Bar97]   Barendregt, H.: The impact of the lambda calculus in logic and computer science. Bull. Symbolic Logic, **3**:2, 181–215 (1997)

[BBC98]   Berardi, S., Bezem, M., Coquand, T.: On the computational content of the axiom of choice. J. Symbolic Logic, **63**:2, 600–622 (1998)

[Bel85]   Bell, J.L.: Boolean-valued models and independence proofs in set theory (second edition). With a preface by Dana Scott. Oxford Logic Guides, 12 (1985)

[BO05]   Berger, U., Oliva, P.: Modified bar recursion and classical dependent choice. Lecture Notes in Logic, **20**, 89-107 (2005)

[Bor50]   Borel, E.: Leçons sur la théorie des fonctions (fourth edition), Paris, Gauthier-Villars (1950)

[Bro08]   Brouwer, L.E.J.: De Onbetrouwbaarheid der logische Principes. Tijdschrift voor Wijsbegeerte, 152–158 (1908)

[Bro23]   Brouwer, L.E.J.: Über die Bedeutung des Satzes vom ausgeschlossenen Dritten in der Matematik, insbesondere in der Funktiontheorie. Journal für die reine und angewandte Mathematik, **154**, 1–7 (1923). English translation in [Hei67].

[Bro25]   Brouwer, L.E.J.: Intuitionistische Zerlegung mathematischer Grundbegriffe. Jahresber. Deutschen Math., Veringigung, **33**, 251–256 (1925)

[Bro48]   Brouwer, L.E.J.: Essentially negative properties. Indagationes Math., **10**, 322–323 (1948)

[CH94]   Coquand, T., Herbelin, H.: A-translation and looping combinators in pure type systems. J. Funct. Programming, **4**:1, 77–88 (1994)

---

[22] A direct translation of terms is also possible using a slight modification of Kuroda's interpretation [Kur51], see [Mur90]

[CH99]   Coquand, T., Hofmann, M.: A new method for establishing conservativity of classical systems over their intuitionistic version. Mathematical Structures in Computer Science, **9**, 323–333 (1999)

[Dia75]   Diaconescu, R.: Axiom of choice and complementation. Proc. Amer. Math. Soc., **51**, 176–178 (1975)

[Dij76]   Dijkstra, E.W.: A Discipline of Programming. Prentice-Hall (1976)

[FF86]   Felleisen, M., Friedman, D.: Control operator, the SECD-machine and the λ-calculus. In: Formal Description of Programming Concepts III, 131–141. North-Holland (1986)

[Fis72]   Fischer, M.J.: Lambda-calculus schemata. In: Proceedings ACM Conference on Proving Assertions about Programs, 104109, Los Cruces (1972). SIGPLAN Notices **7**:1 (January 1972)

[Fri73]   Friedman, H.: Some applications of Kleene's methods for intuitionistic systems. In: Mathias, A.R.D., Rogers, H. (ed): Cambridge Summer School in Mathematical Logic. Lecture Notes in Math., vol. 337, 113–170. Springer, Berlin (1973)

[Fri78]   Friedman, H.: Classically and intuitionistically provably recursive functions. In: Müller, G.H., Scott, D.S. (ed): Higher set theory. Lecture Notes in Math. vol. 669, 21–27. Springer, Berlin (1978)

[Fun59]   Funayama, N.: Embedding infinitely distributive lattices completely isomorphically into Boolean algebras. Nagoya Math. J., **15**, 71–81 (1959)

[Gen69]   Szabo, M.E. (ed): The collected papers of Gerhard Gentzen. North-Holland (1969)

[Gen35]   Gentzen, G.: Die Widerspruchsfreiheit der reinen Zahlentheorie. Math. Annal. **112**, 493–565 (1935) English translation in [Gen69].

[Gir71]   Girard, J.-Y.: Une extension de l'interprétation de Gödel à l'analyse, et son application à l'élimination des coupures dans l'analyse et la théorie des types. In: Proceedings of the Second Scandinavian Logic Symposium, 63–92. North-Holland, Amsterdam (1971)

[Gir72]   Girard, J.-Y.: Interprétation fonctionnelle et élimination des coupures de l'arithmétique d'ordre supérieur. PhD Thesis, Université Paris VII (1972)

[Gli28]   Glivenko, V.: Sur la logique de M. Brouwer. Bulletin Acad. Bruxelles, 14:5, 225–228 (1928)

[Gli29]   Glivenko, V.: Sur quelques points de la logique de M. Brouwer. Bulletin Acad. Bruxelles, 15:5, 183–188 (1929)

[Göd31]   K. Gödel, Über formal unentscheidbare Sätze der Principa mathematica und verwabdter Systeme I. Monatshefte für Mathematik und Physik, **38**, 173–198 (1931). English translation in [GödI].

[Göd33]   Gödel, K.: Zur intuitionistichen Aritmetik und Zahlentheorie. Ergebnisseneis mathematischen Kolloquiums, **4**, 34–38 (1933). English translation in [GödI].

[Göd58]   Gödel, K.: Über eine noch nicht benützte Erweiterung des finiten Standpunktes. Dialectica, **12**, 280–287 (1958). English translation in [GödII].

[GödI]   Gödel, K.: Collected works, Vol. I. Publications 1929–1936. Edited and with a preface by Solomon Feferman. The Clarendon Press, Oxford University Press, New York (1986)

[GödII]   Gödel, K.: Collected works, Vol. II. Publications 1938–1974. Edited and with a preface by Solomon Feferman. The Clarendon Press, Oxford University Press, New York (1990)

[GödIV]   Gödel, K.: Collected works, Vol. IV. Correspondance A-G. The Clarendon Press, Oxford University Press, New York (1986)

[Gri90]   Griffin, T.G.: A formulae-as-types notion of control. Conf. Record 17th Annual ACM Symp. on Principles of Programming Languages (1990)

[Gri55]   Griss, G.F.C.: La mathématique intuitionniste sans négation. Nieuw Arch. Wisk., **3**:3, 134–142 (1955)

[Hei67]   van Heijenoort, J.: From Frege to Gödel. A source book in mathematical logic, 1879–1931. Third printing of the 1967 original, edited by Jean van Heijenoort. Harvard University Press, Cambridge, MA (2002)

[Hei86]   Heinzmann, G.: Poincaré, Russell, Zermelo et Peano. Textes de la discussion (1906–1912) sur les fondements des mathématiques: des antinomies la prédicativité. Librairie Scientifique et Technique Albert Blanchard, Paris (1986)

[Hes03]   Hesseling, D.E.: Gnomes in the fog. The reception of Brouwer's intuitionism in the 1920s. Science Networks. Historical Studies, 28. Birkhäuser Verlag, Basel (2003)

[Hey30]   Heyting, A.: Die formalen Regeln der intuitionistischen Logik, I, II, III. Sitzungsberichte Akad. Berlin, 42–56, 57–71, 158–169 (1930)

[Hey31]   Heyting, A.: Die intuitionistiche Grundlegung der Mathematik. Erkenntnis, **2**, 106–115 (1931)

[Hil22]   Hilbert, D.: Neubegründung der Mathematik. Erste Mitteilung. Abh. aus dem Math. Sem. d. Hamb. Univ., **1**, 157–177 (1922)

[Hil23]   Hilbert, D.: Die logischen Grundlagen der Mathematik. Mathematische Annalen, **88**, 151–165 (1923)

[Hil29]   Hilbert, D.: Probleme der Grundlegung der Mathematik. Mathematische Annalen, **102**, 1–9 (1929)

[How80]   Howard, W.: The formulae-as-types notion of construction. In: To H.B. Curry: essays on combinatory logic, lambda calculus and formalism, 480–490, Academic Press, London-New York (1980)

[Joh36]   Johansson, I.: Der Minimalkalkül, ein reduzierter intuitionistischer Formalismus. Compositio Mathematica 4, 119–136 (1936)

[Joh77]   Johnstone, P.: Topos theory. London Mathematical Society Monographs, Vol. 10. Academic Press (1977)

[Khi26]   Khintchine, A.: Ideas of intuitionism and the struggle for content in contemporary mathematics (Russian). Vestnik Kommunisticheskaya Akademiya, **16**, 184–192 (1926)

[Kol25]   Kolmogorov, A.N.: On the principle of excluded middle (Russian), Matematicheski Sbornik, **32**, 646–647 (1925) English translation in [Hei67].

[Kol29]   Kolmogorov, A.N.: Contemporary disputes on the nature of mathematics (Russian) Nauch. Slovo, **6**, 41–54 (1929)

[Kol32]   Kolmogorov, A.N.: Zur Deutung der intuitionistischen Logik. Mathematische Zeitschrift, **35**, 58–65 (1932)

[Kol33]   Kolmogorov, A.N.: Grundbegriffe der Wahrscheinlichkeitsrechnung. Springer, Berlin (1933)

[Kol88]   Kolmogorov, A.N.: Letters of A. N. Kolmogorov to A. Heyting (Russian), Translated from the German and commented by V.E. Plisko. Uspekhi Mat. Nauk, **43**:6, 75–77 (1988). Translated in Russian Math. Surveys, **43**:6, 89–93 (1988)

[Kol1]   Kolmogorov, A.N.: *Selected works of A. N. Kolmogorov. Vol. I. Mathematics and Mechanics*. Mathematics and its Applications (Soviet Series), 25. Kluwer Academic Publishers Group, Dordrecht (1991)

[Kre65]   Kreisel, G.: Mathematical logic. Lectures on Modern Mathematics, Vol. III, 95–195. Wiley, New York (1965)

[Kre68]   Kreisel, G.: Functions, ordinals, species. In: Logic, Methodology and Philos. Sci. III, 145–159. North-Holland (1968)

[Kri96]   Krivine, J.L.: About classical logic and imperative programming. Ann. of Math. and Artif. Intell., **16**, 405–414 (1996)

[Kri98]   Krivine, J.L.: Ensembles et preuves. Quadrature, **33**, 9–16 (1998)

[Kri01]   Krivine, J.L.: Typed lambda-calculus in classical Zermelo-Fraenkel set theory. Arch. Math. Logic, **40**:3, 189–205 (2001)

[Kri03]   Krivine, J.L.: Dependent choice, 'quote' and the clock. Theoret. Comput. Sci., **308**:1–3, 259–276 (2003)

[Kur51]   Kuroda, S.: Intuitionistische Untersuchungen der formalistishen Logik. Nagoya Math. J., **2**, 35–47 (1951)

[KV65]   Kleene, S.C., Vesley, R.E.: The foundations of intuitionistic mathematics, especially in relation to recursive functions. North-Holland Publishing Co., Amsterdam (1965)

[Lan65]   Landin, P.J.: A correspondence between ALGOL 60 and Church's lambda-notation. I. Comm. ACM, **8**, 89–101 (1965)

[Lar92]   Largeault, J.: Intuitionnisme et théorie de la démonstration. Librairie Philosophique J. Vrin, Paris (1992)

[Law71]   Lawvere, W.: Quantifiers and sheaves. In: Actes du Congrès International des Mathématiciens (Nice, 1970), Tome 1, 329–334. Gauthier-Villars, Paris (1971)

[Lus30]   Lusin, N.: Leçons sur les ensembles analytiques et leurs applications. With a note by W. Sierpiński, and a preface by Henri Lebesgue. Gauthier-Villars, Paris (1930)

[Man98]   Mancosu, P.: From Brouwer to Hilbert. The debate on the foundations of mathematics in the 1920s. With the collaboration of Walter P. van Stigt. Reproduced historical papers translated from the Dutch, French and German. Oxford University Press, New York (1998)

[Mar82]   Martin-Löf, P.: Constructive mathematics and computer programming. In: Logic, Methodology and Philosophy of Science, VI, 1979, 153–175. North-Holland (1982)

[Mar84]   Martin-Löf, P.: Intuitionistic Type Theory. Bibliopolis (1984)

[Miq01]   Miquel, A.: Le calcul des constructions implicite: syntaxe et sémantique. PhD Thesis, Université Paris VII (2001)

[Miq03]   Miquel, A.: A strongly normalising Curry-Howard correspondence for IZF set theory. In: Computer Science Logic, CSL'03. Lecture Notes in Computer Science, vol. 2803. Springer-Verlag (2003)

[Mur90]   Murthy, C.R.: Extracting constructive content from classical proofs. PhD thesis, Department of Computer Science, Cornell University (1990)

[Myh68]   Myhill, J.: Formal systems of intuitionistic analysis, I. In: Logic, Methodology and Philos. Sci. III, 161–178. North-Holland (1968)

[Myh71]   Myhill, J.: Embedding classical type theory in "intuitionistic" type theory. In: Axiomatic Set Theory, 267–270. Amer. Math. Soc., Providence, R.I. (1971)

[Myh74]   Myhill, J.: Embedding classical type theory in "intuitionistic" type theory: a correction. In: Axiomatic Set Theory, 185–188. Amer. Math. Soc., Providence, R.I. (1974)

[Nel92]   Nelson, E.: Mathematical mythologies. In: Le labyrinthe du continu, 155–167. Springer, Paris (1992)

[Nel04]   Nelson, E.: review of [Hes03], Bull. Amer. Math. Soc., 41, 545–549 (2004)

[NPS90]   Nordström, B., Petersson, K., Smith, J.M.: Programming in Martin-Löf Type Theory. Clarendon Press, Oxford (1990)

[Pal98]   Palmgren, E.: On universes in type theory. In: Twenty-five years of constructive type theory, 191–204. Oxford Logic Guides, vol. 36 (1998)

[Plo75]   Plotkin, G.D.: Call-by-name, call-by-value and the lambda-calculus. Theoretical Computer Science, 1, 125–159 (1975)

[Poi09]   Poincaré, H.: La logique de l'infini. Revue de Métaphysique et de Morale, 17, 461–482 (1909)

[Pow75]   Powell, W.C.: Extending Gödel's negative interpretation to ZF. J. Symbolic Logic, 40, 221–229 (1975)

[Sko22]   Skolem, T.: Einige Bemerkungen zur axiomatischen Begründung der Mengenlehre. In: Proceedings of the 5th Scand. Math. Congress. Helsinki, 217–232 (1922). English translation in [Hei67].

[Spe62]   Spector, C.: Provably recursive functionals of analysis: a consistency proof of analysis by an extension of principles formulated in current intuitionistic mathematics. In: Proc. Sympos. Pure Math., vol. V, 1–27. American Mathematical Society, Providence, R.I (1962)

[SW74]   Strachey, C., Wadsworth, C.P.: Continuations: A mathematical semantics for handling fill jumps. Monograph PRG-11, Oxford University, Programming Research Group (1974)

[Tro73]   Troelstra, A.S.: Metamathematical investigation of intuitionistic arithmetic and analysis. Edited by Troelstra, A.S. Lecture Notes in Mathematics, vol. 344. Springer-Verlag, Berlin-New York (1973)

[vanD90]   van Dalen, D.: The war of the frogs and the mice, or the crisis of the Mathematische Annalen. Math. Intelligencer, 12:4, 17–31 (1990)

[vonN27]   von Neumann, J.: Zur Hilbertschen Beweistheorie. Math. Z., 26, 1–46 (1927)

[vonP94]   von Plato, J.: Creating modern probability. Its mathematics, physics and philosophy in historical perspective. Cambridge University Press, Cambridge (1994)

[Wan87]   Wang, H.: Reflections on Kurt Gödel. A Bradford Book. MIT Press, Cambridge, MA (1987)

[Wey21]   Weyl, H.: Über die neue Grundlagenkrise der Mathematik. Math. Z., 10, 39–79 (1921). English translation in [Man98].

[Wey44]   Weyl, H.: David Hilbert and his mathematical work. Bull. Amer. Math. Soc., 50, 612–654 (1944)

[Zac98]   Zach, R.: Numbers and functions in Hilbert's finitism. Taiwanese J. Philos. Hist. Sci., 10, 33–60 (1998)

3

Some aspects of the probabilistic work

Loïc Chaumont[1], Laurent Mazliak[2], and Marc Yor[3]

[1] Laboratoire Angevin de Recherche en Mathématiques (LAREMA), University
of Angers, France
http://math.univ-angers.fr/~chaumont
loic.chaumont@univ-angers.fr
[2] Laboratoire de Probabilités et Modèles Aléatoires, Institut de mathématiques
de Jussieu, University Paris VI, France
http://www.proba.jussieu.fr/users/lma/mazliak.html
mazliak@ccr.jussieu.fr
[3] Laboratoire de Probabilités et Modèles Aléatoires, Institut de mathématiques
de Jussieu, University Paris VI, France
http://www.academie-sciences.fr/Membres/Y/Yor_Marc.htm

Translated from the French by Kathleen Qechar

3.1 Introduction

Anyone reading about the mathematical works of Kolmogorov must naturally expect to find broad considerations about the axiomatization of probability which Kolmogorov developed at the beginning of the nineteen thirties and which forms the contents of his famous publication *Grundbegriffe der Wahrscheinlichkeitsrechnung* (*Foundations of the theory of probabilities*) [Kol33], published by Springer in 1933. It is certain that among all the works of the Soviet mathematician, this small opuscule of about sixty pages is the most famous part, and is often the only one to which his name is attached for a quite large public but also for some mathematicians. Without wanting in any way to diminish the importance of this work, it is nevertheless quite astonishing that the attention was thus focused on what does not constitute the most original creation of Kolmogorov in the field of probability. The aim of this part, devoted to certain aspects of the probabilistic works of the scientist, is precisely to highlight some of his most remarkable works in that domain. In this imposing monument, a drastic choice was necessary and we chose to focus on the two purely probabilistic directions that Kolmogorov worked on, namely on the one hand the study of the various types of convergence for sums of independent random variables, which enabled him to continue the studies of his Russian predecessors Markov and Lyapounov, and on the other hand literally revolutionary considerations about processes in continuous time, whose branches extend ahead in time until some discoveries which go back to hardly thirty years. Nevertheless, as we found it difficult, and almost impossible, that a chapter devoted to the probabilistic works of the Soviet mathematician does not refer to the axiomatization of probability, we will begin with a short glance of his main contributions to this topic, inviting the reader to refer to the numerous articles dealing with the question in a more detailed way (see e.g. [vonP94], [SV06]). We will also refer to the essential text of Shiryaev [Shi89] for a more complete chart of Kolmogorov's works. One will find in the article [Maz03] some indications on the life of the mathematician and the status of the discipline in the stalinist USSR.

3.2 The axiomatization of probability calculus

3.2.1 An abstract framework

As mentioned above, Kolmogorov's publication *Grundbegriffe der Wahrscheinlichkeitsrechnung* [Kol33] is a modest monograph of 60 pages published in 1933 along with several articles devoted to the modern probability theory. The Russian translation of the text is dated 1936, and it was mostly achieved due to some political reasons at the time when an important pressure was put on Soviet scientists so that they publish their works in Russian and in USSR rather than abroad. As for the first English translation, it is dated 1950. This relatively important delay shows that the axiomatization suggested by the Russian scientist wasn't as generally accepted as we usually think it was. Several probabilists, and among the most eminent ones, such as Paul Lévy, will never use the axiomatization of Kolmogorov, which will not prevent them in any way having extraordinary ideas. In fact, outside the USSR, before the 50's, more or less only Cramér's treatise [Cra37] refers to this field. Besides, this author doesn't give any detailed explanation; he only uses Kolmogorov's axiomatization because it is the most practical one among those available at that time (in particular the *theory of collectives* suggested by von Mises). However, from the 50's, it will definitely be adopted by the younger generation. What is attractive in the formal framework proposed by this axiomatization, is the fact that it provides e.g. a global explanation of the multiple paradoxes which had in the past plagued this discipline (like those of Joseph Bertrand, Emile Borel etc.) : each time, the precise definition of the probability space as a description of the considered random experiment allows to suppress the ambiguity. (On this subject, see *infra*, as well as [SV06] and [Szé86]. One can also refer to Itô's comments in the foreword of [Itô86].)

The great force of Kolmogorov's treatise is to voluntarily consider a completely abstract framework, without seeking to establish bridges with the applied aspects of the theory of probability, beyond the case of finite probabilities. In general, the search for such bonds inevitably brings to face delicate philosophical questions and is thus likely to darken mathematical modelling.

While speaking about the questions of application only in the part devoted to the finite probabilities[4], Kolmogorov is released from this constraint and can avoid the pitfalls that von Mises had not always circumvented. Indeed, the *theory of the collectives* also claimed to establish a discrimination between the experiments for which the application of the probabilities was legitimate, and others. But Kolmogorov, who presents a purely mathematical theory, does not have such an ambition, and thus not such a limitation. Within the abstract framework which he defines, any mathematical work is legitimate, and its validation for applications is a matter for other fields of knowledge. In particular, he allows himself to consider sets not having any topological structure, whereas for finer studies (like the phenomena of convergence), he will be free to work on better spaces through the use of images of probability laws. Besides this fact will place the axiomatization of Kolmogorov in opposition with the Bourbaki topological set-up concerning measure theory. The very general character of his theory will make it possible to the Russian mathematician to use in all its force the measure theory of Borel and Lebesgue, which is still relatively new at that time since its abstract version was mainly developed by Fréchet (quoted in the *Grundbegriffe* as the one who liberated measure theory from geometry) then by the Polish school (Banach, Sierpiński, Kuratowski...) in the 20's.

From the beginning of the twentieth century, Borel had been a promoter of the use of the measure theory and integral of Lebesgue for the treatment of questions of probability. In 1909, he publishes a revolutionary paper where such a method enables him to obtain a first strong version of the law of large numbers and interpretations on the distribution of real numbers. Undoubtedly his moderate consideration for probabilistic mathematics and serious doubts as for the legitimacy of their applications prevented him from fully reaping the crops from the seeds which he had sown.

Kolmogorov introduces the by now classical concept of a probabilisty space in the form of a triplet $(\Omega, \mathcal{F}, P)$ composed of a set Ω provided with σ-algebra (which he calls a *set field*) $\mathcal{F}$ and a normalized measure (probability) P. The random variables are simply functions X with real values defined on Ω such that for all $a \in \mathbb{R}$, $\{\omega \in \Omega, X(\omega) < a\} \in \mathcal{F}$ and their laws are the image measures of the probability P defined by $P^{(X)}(A) = P(X^{-1}(A))$, for all $A \in \mathcal{B}(\mathbb{R})$, the Borelian σ-algebra of $\mathbb{R}$.

The major contributions of Kolmogorov's work in the clarification of probabilistic concepts are incontestably the construction of a probability measure on an infinite product of spaces, which plays an important part in the theory of stochastic processes, and the formalization of the conditional law via the use of the Lebesgue-Radon-Nikodym theorem (the abstract version of this theorem had been published by Nikodym in 1930). Let us note incidentally

---

[4] Which is not without reminding one about the way in which, in his article of 1931 on Markov processes, he had proposed in a long introduction to defer to other works some reflections about the applicability of his theories

that it wasn't the first time that a probability on a product space was built: the most famous example is given by Wiener [Wie23] who since 1923, by applying techniques that Daniell had developed a few years before to extend Lebesgue's integral to spaces of infinite dimensions, built the probability measure − now called Wiener's measure − associated with Brownian motion (see [RY91]).

3.2.2 Construction of the conditional law

Let us present in a few words the construction of the conditional law, while following the text of Kolmogorov but with modernized notations for the sake of clarity of our exposition.

First of all let us recall the definition of the elementary conditional probability of an event C (i.e. of an element of the σ-algebra $\mathcal{F}$) by an event D such as $P(D) > 0$, by $P(C \mid D) = P(C \cap D)/P(D)$. Now, let U denote a real-valued random variable and B an event. We try to build a random variable $\omega \mapsto \pi(U(\omega); B)$, a Borelian function of U, so called conditional probability of B knowing U, and such that, for all $A \in \mathcal{B}(\mathbb{R})$ with $P(U \in A) > 0$, we have:

$$P(B \mid U \in A) = \int_{\Omega} \pi(U; B) dP(. \mid U \in A).$$

For any $A \in \mathcal{B}(\mathbb{R})$, we write $Q_B(A) = P(B \cap U^{-1}(A))$. Let us note that if $P^{(U)}$ is the law of U defined by $P^{(U)}(A) = P(U^{-1}(A))$, then $P^{(U)}(A) = 0$ implies $Q_B(A) = 0$ and thus, according to the Lebesgue-Radon-Nikodym theorem, we can find a Borelian function f_B such that $\forall A \in \mathcal{B}(\mathbb{R})$,
$Q_B(A) = \int_{\mathbb{R}} \mathbb{1}_A f_B dP^{(U)}$, i.e.

$$P(B \cap U^{-1}(A)) = \int_{\mathbb{R}} \mathbb{1}_A f_B dP^{(U)} = \int_{\Omega} \mathbb{1}_{U \in A} f_B \circ U dP$$

and thus

$$P(B \mid U \in A) = \int_{\Omega} f_B \circ U dP(. \mid U \in A)$$

and we set $\pi(U; B) = f_B \circ U$. From this point, Kolmogorov will recover all classical properties of conditional probabilities. He nicely illustrates the strength of his formalism by explaining the paradox of Borel related to the random drawing of a point on a sphere: the interested reader can refer to e.g. [Bil95], p. 462 and [SV06].

3.2.3 The 0-1 law (or *the all or nothing law*)

As mentioned previously, in 1933, measure theory isn't yet very commonly accepted, in any case not in its abstract form, and when Fréchet will discover the monograph of the Russian mathematician, he will be disconcerted by the very abstract form taken by certain arguments, like that of the law known as the 0-1 law which Kolmogorov placed in the appendix of his work. This law was stated independently, in particular by Lévy in 1934 (when he didn't yet know the *Grundbegriffe*), and it is interesting to compare the two approaches of this result, which we will do by way of illustration of the strongly synthesizing character of the axiomatic proposed by Kolmogorov. For an easy reading, we will use today's vocabulary and notations, keeping only the spirit of the two proofs.

Theorem 1. *Let $(X_n)_{n \geq 1}$ denote a sequence of independent real-valued random variables.*

We introduce $\mathcal{G}_n = \sigma(X_n, X_{n+1}, \dots)$ (the σ-algebra generated[5] by X_n, $X_{n+1}, \dots$), and $\mathcal{G} = \bigcap_{n \geq 1} \mathcal{G}_n$ (the "tail σ-algebra").

Then, any element of $\mathcal{G}$ is of probability 0 or 1.

Kolmogorov's proof: It is the most common proof taught today. Let A an element of $\mathcal{G}$. Let us suppose that $P(A) > 0$ and note P_A the conditional probability knowing A. According to the independence hypothesis made on the X_k's, for all $B \in \mathcal{F}_n = \sigma(X_1, X_2, \dots, X_n)$, B is independent from the elements of $\mathcal{G}_{n+1}$ and thus from A, and we have:

$$P_A(B) = \frac{P(A \cap B)}{P(A)} = P(B).$$

Therefore, the probabilities P_A and P coincide on all $\mathcal{F}_n$'s, and hence on Boole's algebra $\bigcup_{n \geq 1} \mathcal{F}_n$ and thus, according to the monotone class theorem, on the σ-algebra thus generated, i.e. $\mathcal{F} = \sigma(X_1, X_2, \dots, X_n, \dots)$. Since in particular $A \in \mathcal{F}$, we have $P_A(A) = P(A)$, i.e. $P(A) = 1$. □

Lévy's proof: In fact, Lévy contents himself to prove the result when variables X_n follow a uniform law on $[0,1]$. In this case, the obtention of a realization of the sequence $(X_n)_{n \geq 1}$ can be conceived as the one of a point in a cube of size 1 with an infinity of dimensions, the law of probability being given by Lebesgue's measure. Lévy's argument is then based on an observation which he affirms being obvious and which is equivalent in fact to a monotone class result: he points out (we employ the modern formalism) that for any event A of the σ-algebra $\mathcal{F} = \sigma(X_1, X_2, \dots, X_n, \dots)$ (which may therefore be written as follows $[(X_1, X_2, \dots, X_n, \dots) \in B]$, where B is a measurable set of $\mathbb{R}^{\mathbb{N}}$), and for all $\varepsilon > 0$, we can find $n > 0$ and $D_n \in \mathcal{F}_n = \sigma(X_1, X_2, \dots, X_n)$ such that[6] $P(D_n \Delta A) < \varepsilon$. In fact, his explanation is to say that measurable sets inside the infinitely dimensional cube are obtained by "*M. Lebesgue's constructions*" from the "intervals" of the cube, which are the sets of the

---

[5] i.e. the smallest σ-algebra which allows all of them to be measurable

[6] Δ indicates the symmetric difference: $D_n \Delta A = (D_n \cup A) \setminus (D_n \cap A)$

type $]a_0, b_0[\times]a_1, b_1[\times \cdots \times]a_n, b_n[\times[0,1] \times [0,1]\ldots$, in the same way as the Borel sets of $\mathbb{R}$ are built from the real-valued open intervals; and we know that for any element $A \in \mathcal{B}([0,1])$, there exists a finite collection of intervals $]\alpha_0, \beta_0[, \ldots,]\alpha_m, \beta_m[$ such that $\lambda(A \Delta \cup_{k=0}^m]\alpha_k, \beta_k[) < \varepsilon$ (where λ is Lebesgue's measure on $\mathbb{R}$).

Let now E an element of $\mathcal{G}$. Let us note that the independence of the $(X_n)_{n\geq 1}$'s allows us to write that $\forall n, P(E \mid \mathcal{F}_n) = P(E)$. Let N and $D_N \in \mathcal{F}_N$ such that $P(E \Delta D_N) < \varepsilon$ (and so, in particular, $P(D_N) > P(E) - \varepsilon$). Then, we have:

$$\varepsilon > P(D_N \cap E^c) = P(D_N)P(E^c \mid D_N).$$

However, $P(E^c \mid \mathcal{F}_N) = 1 - P(E \mid \mathcal{F}_N) = 1 - P(E) = P(E^c)$, thus $P(E^c \mid D_N) = P(E^c)$, hence

$$\varepsilon > P(D_N)P(E^c) > (P(E) - \varepsilon)(1 - P(E)) > P(E)(1 - P(E)) - \varepsilon$$

and so $P(E)(1 - P(E)) < 2\varepsilon$. This is true for all $\varepsilon > 0$, thus $P(E)(1 - P(E)) = 0$. □

3.3 Limit theorems and series of independent random variables

The direction in which Kolmogorov will develop his first works in Probability, undoubtedly guided by his elder Khinchin[7], may be found to be in continuity with the former studies which had specified the conditions of validity of the limit theorems (in particular the law of large numbers) for sums of random variables throughout the nineteenth century. The first paper, which goes back to 1925 [KK25] and which is the only article jointly written by Kolmogorov and Khinchin, is remarkable in the way that it introduces a number of techniques which will be at the base of some later developments of the theory of probability, in particular in the study of the results of convergence for martingales. This first work relates to the convergence of series of independent random variables. The main result is stated as follows (in modern terms):

Theorem 2. *Let $(X_n)_{n\geq 1}$ a sequence of centered (i.e. of 0 expectation) independent real-valued random variables. Let us suppose that $\sum_{n\geq 1} E(X_n^2) < +\infty$.*

Then $\sum_n X_n$ converges almost surely (a.s.), i.e. with probability 1.

The proof suggested by Kolmogorov is based on a famous inequality which is named after him today:

---

[7] One also spells: Khintchine

Lemma 1 (Kolmogorov's inequality). *Let* $S_n = X_1 + \cdots + X_n$. *Then*

$$P(\max_{1 \leq k \leq n} | S_k | \geq \varepsilon) \leq \frac{E(S_n^2)}{\varepsilon^2}. \tag{3.1}$$

Proof. We write

$$\{ \max_{1 \leq k \leq n} | S_k | \geq \varepsilon \} = \bigcup_{p=1}^{n} A_p$$

where

$$A_p = \{| S_1 | < \varepsilon, | S_2 | < \varepsilon, \ldots, | S_{p-1} | < \varepsilon, | S_p | \geq \varepsilon \}.$$

Note that the A_p's form a partition of the entire probability space and that for all $1 \leq p \leq n$, $S_n - S_p$ is independent from $S_p \mathbb{1}_{A_p}$ and has 0 expectation. Therefore, for $1 \leq p \leq n$,

$$E(S_n^2 \mathbb{1}_{A_p}) = E((S_n - S_p)^2 \mathbb{1}_{A_p}) + E(S_p^2 \mathbb{1}_{A_p}) \geq \varepsilon^2 P(A_p)$$

and summing with respect to p, $E(S_n^2) \geq \varepsilon^2 P(\max_{1 \leq k \leq n} | S_k | \geq \varepsilon)$. □

It is only a few years later, in a note for the *Comptes Rendus de l'Académie des Sciences* (CRAS) of Paris in 1930 [Kol30], that Kolmogorov will obtain from the previous result his most famous consequence, that is to say this nowadays classical version of the strong law of large numbers:

Corollary 1. *Let us suppose that the variables X_n are independent and centered. We write $E(X_n^2) = b_n$ and we suppose that $\sum_{n=1}^{\infty} \dfrac{b_n}{n^2} < +\infty$. Then*

$$\sigma_n = \frac{S_n}{n} \to 0, \ a.s.$$

Proof. First of all, let us note that for all fixed $N > 0$,

$$\overline{\lim}_n | \sigma_n | = \overline{\lim}_n \frac{| S_n - S_{2^N} |}{n},$$

where $\overline{\lim}_n$ indicates the superior limit for $n \to +\infty$. Moreover, it is obvious that:

$$\{\overline{\lim}_n | \frac{S_n - S_{2^N}}{n} | > \varepsilon\} \subset \cup_{n \geq 2^N} \{| \frac{S_n - S_{2^N}}{n} | > \varepsilon\}$$

$$\subset \cup_{m \geq N} \{\max_{2^m \leq k \leq 2^{m+1}} | \frac{S_k - S_{2^N}}{k} | > \varepsilon\}$$

$$\subset \cup_{m \geq N} \{\max_{2^m \leq k \leq 2^{m+1}} | \frac{S_k - S_{2^N}}{2^m} | > \varepsilon\}.$$

Therefore, for all $\varepsilon > 0$ and all $N > 0$:

$$P\left(\overline{\lim}_n \mid \sigma_n \mid > \varepsilon\right) = P\left(\overline{\lim}_n \frac{\mid S_n - S_{2^N}\mid}{n} > \varepsilon\right)$$

$$\leq \sum_{m=N}^{\infty} P\left(\max_{2^m \leq k \leq 2^{m+1}} \mid S_k - S_{2^N}\mid > \varepsilon 2^m\right) \leq \frac{1}{\varepsilon^2}\sum_{m=N}^{\infty}\frac{1}{2^{2m}}\sum_{k=2^N+1}^{2^{m+1}} b_k$$

$$= \frac{1}{\varepsilon^2}\sum_{i=N}^{\infty}\left(\sum_{m\geq i}\frac{1}{2^{2m}}\right)\sum_{k=2^i+1}^{2^{i+1}} b_k \leq \frac{16}{3}\frac{1}{\varepsilon^2}\sum_{n\geq 2^N}\frac{b_n}{n^2}.$$

By hypothesis, the last term can be made as small as we like, and so $\overline{\lim}_n \sigma_n = 0$ a.s. $\square$

Remark 1. The independence of the X_n variables occurs twice in this proof: first when we apply Kolmogorov's inequality, and secondly when we write

$$E\left((S_{2^{m+1}} - S_{2^N})^2\right) = \sum_{k=2^N+1}^{2^{m+1}} b_k$$

(using the additivity of the variance for non-correlated variables).

As mentioned above, the result of Lemma 1 and its proof can directly be applied to the case of discrete martingales[8]:

Corollary 2. *Let $(M_n)_{n\geq 1}$ denote a square integrable martingale such that $E(M_n) = 0$. Then*

$$P(\max_{1\leq k\leq n} \mid M_k \mid \geq \varepsilon) \leq \frac{E(M_n^2)}{\varepsilon^2}. \tag{3.2}$$

It is easily proven that (3.2) may be strengthened a little, in the form of the important *Doob inequality*: if $(M_n)_{n\geq 1}$ is a square integrable martingale such that $E(M_n) = 0$, we have

$$E[\max_{1\leq k\leq n}(M_k)^2] \leq 4E[M_n^2].$$

As we know, the theory of martingales has, after Doob's works, invaded the scene of contemporary probability theory. In order to illustrate the strength of the inequality (3.2) and the notion of martingale, let us prove the following result and its corollary.

---

[8] Let us recall that a discrete martingale is a sequence $(M_n)_{n\geq 1}$ of integrable random variables such as, for all n, the expectation of M_{n+1}, knowing all the previous values, is equal to the last one: $E\left(M_{n+1}\big|M_1,\ldots,M_n\right) = M_n$. (The martingale is called square integrable, or etc. if each M_n satisfies this property.)

Proposition 1. *Let (M_n) be a L^2 martingale (i.e. a square integrable martingale) such that $\sup_n E(M_n^2) < +\infty$. Then M_n converges, at the same time in L^2 and a.s., towards a random variable M.*

Proof. It is easy to show the convergence in L^2 by proving that $(M_n)_{n\geq 1}$ is a Cauchy sequence in L^2. The interested reader can refer to any elementary course on martingales. Let us now deduce the a.s. convergence.

Let $\varepsilon > 0$. For each $p > 0$, let us write $V_p = \sup_{n\geq p} \mid M_n - M_p \mid$. As $(M_n - M_p)_{n\geq p}$ is a square integrable martingale, we have according to (3.2) for all N

$$P(\max_{p\leq k\leq N} \mid M_k - M_p \mid \geq \varepsilon) \leq \frac{E((M_N - M_p)^2)}{\varepsilon^2}. \qquad (3.3)$$

As mentioned above, $(M_k)_{k\geq 0}$ is a Cauchy sequence in L^2 and so, for each given $m > 0$, we can choose p_m such that $\forall N \geq p_m$, $E((M_N - M_{p_m})^2) \leq \varepsilon^2/2^m$. Therefore, passing to the limit in (3.3) when $N \to +\infty$: $P(V_{p_m} \geq \varepsilon) \leq 1/2^m$. According to the Borel-Cantelli Lemma, a.s. there exists $m > 0$ such that $V_{p_m} < \varepsilon$ i.e. $\forall n \geq p_m$, $\mid M_n - M_{p_m} \mid < \varepsilon$, which is equivalent to saying that a.s. (M_n) is a Cauchy sequence in $\mathbb{R}$. $\square$

Corollary 3. *Let $(Z_n)_{n\geq 1}$ a sequence of independent random variables with the same Bernoulli law $P(Z = 1) = P(Z = -1) = \frac{1}{2}$. We consider the random walk on $\mathbb{Z}$: $S_0 = 0, S_n = Z_1 + \cdots + Z_n$. Let $a > 0$ an integer and $\tau = \inf\{n \geq 0, S_n = a\}$ the first passage time of a. Then the Laplace transform of the law of τ is given for all $\theta \geq 0$ by*

$$E[(\cosh\theta)^{-\tau}] = e^{-\theta a}.$$

Proof. We only indicate the broad outline of the proof, leaving the details to the interested reader (see e.g. [BMP01]). Let $X_n^\theta = \dfrac{e^{\theta S_n}}{(\cosh\theta)^n}$. We verify that it is a martingale and that[9] $(X_{n\wedge\tau}^\theta)_{n\geq 1}$ is a L^2 martingale, which converges a.s. and in L^2 towards the variable $W^\theta = \dfrac{e^{\theta a}}{(\cosh\theta)^\tau} \mathbb{1}_{\tau<+\infty}$. Passing to the limit as $\theta \to 0$, which is made possible by dominated convergence, we obtain that $P(\tau < +\infty) = 1$, and thus the desired result. $\square$

In 1924, Khinchin [Khi24] had proven a result which brought a radical precision to the law of large numbers, the *law of the iterated logarithm*. The generalization of Khinchin's result by Kolmogorov in 1929 ([Kol29]) was one of his greatest achievements.

Theorem 3. *Let $(X_n)_{n\geq 1}$ a sequence of independent real-valued random variables. Let us suppose that for all n, $E(X_n) = 0$ and $b_n = E(X_n^2) < +\infty$. We*

---

[9] $n \wedge \tau$ means $\min\{n, \tau\}$, the smallest number of n and τ

let $B_n = \sum_{k=1}^{n} b_k$ (that is $B_n = E(S_n^2)$). If $B_n \to +\infty$ and $\mid X_n \mid \le M_n = o(\sqrt{\frac{B_n}{\ln \ln B_n}})$, we have a.s.

$$\overline{\lim} \frac{S_n}{\sqrt{2B_n \ln \ln B_n}} = 1.$$

Today, Kolmogorov's proof still remains very much up to date, as it introduces techniques, in particular of large deviations, which became fundamental in the study of many limiting phenomena in probability theory. We will only show the less technical part of the result, leaving the reader consult one of the innumerable texts which present the complete proof (e.g.[Bil95]).

Let us note, for $\varepsilon > 0$, $\phi^{\varepsilon}(n) = (1 + \varepsilon)\sqrt{2B_n \ln \ln B_n}$; we shall prove that $P(S_n \ge \phi^{\varepsilon}(n)$, infinitely often$) = 0$. According to Borel-Cantelli's Lemma, it suffices to prove that for a well-chosen subsequence $n_k \uparrow +\infty$, we have

$$\sum_{k=1}^{\infty} P(\max_{n \le n_k} S_n \ge \phi^{\varepsilon}(n_{k-1})) < \infty. \tag{3.4}$$

As mentioned above, we will obtain the result thanks to the following lemma which gives some large deviations estimates for the sequence (S_n).

Lemma 2. *Let $x \ge 0$.*

(i) If $x \le B_n/M_n$ then $P(S_n > x) < e^{-(\frac{x^2}{2B_n})(1 - \frac{xM_n}{2B_n})}$
(ii) If $x \ge B_n/M_n$, then $P(S_n > x) < e^{-\frac{x}{4M_n}}$
(iii) $P(\max_{1 \le k \le n} S_k \ge x) \le 2P(S_n > x - \sqrt{2B_n})$.

Proof.

Let us fix n and in order to simplify the writing, this index n will be omitted in the next lines.

Let $a > 0$ such that $aM \le 1$. Then,

$$E\left(e^{aX_k}\right) = 1 + \sum_{r \ge 2} E\left(\frac{a^r X_k^r}{r!}\right) \le 1 + \frac{a^2 b_k}{2} \sum_{r \ge 2} 2 \frac{a^{r-2} M^{r-2}}{r!}$$

$$\le 1 + \frac{a^2 b_k}{2}\left(1 + \frac{aM}{2}\right) < \exp\left[\frac{a^2 b_k}{2}\left(1 + \frac{aM}{2}\right)\right],$$

and thus:

$$E(e^{aS}) = \prod_{k=1}^{n} E(e^{aX_k}) < \exp\left[\frac{a^2 B}{2}\left(1 + \frac{aM}{2}\right)\right].$$

As $P(S > x) < E(\frac{e^{aS}}{e^{ax}})$ (for all $a > 0$), we obtain the inequality $P(S > x) < \exp[-ax + \frac{a^2 B}{2}(1 + \frac{aM}{2})]$ from which we easily deduce the points (i) and (ii) by taking successively $a = x/B$ and $a = 1/M$.

As for the point (iii), let us note $U = \max_{1 \leq k \leq n} S_k$, and that $(U \geq x)$ is the union of the events $E_k = (S_1 < U, \ldots, S_{k-1} < U, S_k = U \geq x)$ for $1 \leq k \leq n$.

Thus, we have

$$P(S > x - \sqrt{2B}) \geq \sum_{k=1}^{n} P(E_k \cap (S > x - \sqrt{2B})) \geq \sum_{k=1}^{n} P(E_k \cap (S > U - \sqrt{2B}))$$

$$= \sum_{k=1}^{n} P(E_k) P(S > U - \sqrt{2B} \mid E_k) = \sum_{k=1}^{n} P(E_k) P(S - S_k > -\sqrt{2B} \mid E_k).$$

But $S - S_k$ is independent of E_k, and therefore this last expression is also

$$= \sum_{k=1}^{n} P(E_k) P\left(\sum_{i=k+1}^{n} X_i > -\sqrt{2B} \right) \geq \sum_{k=1}^{n} P(E_k) P\left(\left(\sum_{i=k+1}^{n} X_i \right)^2 < 2B \right).$$

Now,

$$1 - P\left(\left(\sum_{i=k+1}^{n} X_i \right)^2 < 2B \right) = P\left(\left(\sum_{i=k+1}^{n} X_i \right)^2 \geq 2B \right) \leq \frac{1}{2B} \sum_{i=k+1}^{n} b_i \leq \frac{1}{2},$$

and hence

$$P(S > x - \sqrt{2B}) \geq \frac{1}{2} \sum_{k=1}^{n} P(E_k) = \frac{1}{2} P(U \geq x). \quad \square$$

From Lemma 2, we deduce (3.4) for some well-chosen subsequence (n_k). Indeed, let us choose these integers such as for all k, $B_{n_{k-1}} \leq (1+\tau)^k \leq B_{n_k}$. From Lemma 2 (i)-(ii), we obtain, by using the hypotheses, for all $\mu > 0$, and k large enough (such that $M_{n_k} < \frac{\mu}{1+\varepsilon} \sqrt{2B_{n_k}/\ln\ln B_{n_k}}$):

$$P(S_{n_k} > \phi^\varepsilon(n_k)) < [k\ln(1+\tau)]^{-(1+\varepsilon)^2(1-\mu)}.$$

Then, choosing μ such that $(1 + \varepsilon)^2(1 - \mu) > 1$, we have $\sum_{k=1}^{\infty} P(S_{n_k} > \phi^\varepsilon(n_k)) < \infty$. Thus, we conclude by applying Lemma 2 (iii) and the fact that $\frac{\sqrt{2B_{n_k}}}{\phi^\varepsilon(n_k)} \to 0$. $\quad \square$

3.4 Processes in continuous time

In the beginning of the years 1930, a great number of probabilistic works of the Soviet school are related to the study of the stochastic processes in continuous time, meeting thus in particular the needs in physics or aiming at describing some "social phenomenons". The axiomatization due to Kolmogorov, which we commented on above, brought an essential element to the establishment of this theory. The theorem of construction of probability measures on a space of infinite dimension shows that the law of a stochastic process in continuous time is given in a unique way starting from the family of the finite-dimensional marginal laws of the process in question.

3.4.1 Chapman-Kolmogorov's equation

The first family of processes to which Kolmogorov, as a good heir to the Russian school of probability, turns naturally to, is that of the Markov processes, i.e. those which satisfy the property (known as the Markov property) of independence of the future with respect to the past conditionally to the knowledge of the present. The article *Über die analytischen Methoden in der Wahrscheinlichkeitsrechnung* [Kol31] sets definitely the analytical bases of the theory of Markov processes.

In this fundamental article published in 1931, the entire study of the process is focused around the function:

$$P(s, x, t, A)$$

which represents the probability such that at time t the random phenomenon is in one of the states of the set A, if it is in the state x at time s, prior to t ($0 \leq s < t$). As the necessary measurability assumptions are supposed to be satisfied, this function must verify the integral equation

$$P(s, x, t, A) = \int_E P(s, x, u, dy) \, P(u, y, t, A), \quad \text{for all } u \in]s, t[, \qquad (3.5)$$

where E stands for the set of all the possible states of the process. Equation (3.5), commonly called today Chapman-Kolmogorov's equation (Chapman had indeed noted it in a report [Cha28] on Brownian Motion in 1928), is the analytic translation of Markov's property and the measures $P(s, x, t, dy)$ represent the transition probabilities of the process: if we denote by $(X_t)_{t \geq 0}$ this process, then, for all measurable A (for the measure dy):

$$P(s, x, t, A) = P(X_t \in A \,|\, X_s = x).$$

However, as Kolmogorov's article is purely analytical, as we can easily see from its title, it doesn't mention any pathwise realizations of the random process.

Equation (3.5) can't be entirely solved in an explicit way in the too global framework in which it is posed. Thus, Kolmogorov seeks conditions of regularity on the probabilities $P(s, x, t, dy)$ which would make it possible to obtain a more accessible form. Eager to use the new techniques of analysis related to Lebesgue's integral, he naturally focuses on the case where $P(s, x, t, dy)$ is absolutely continuous, with density $f(s, x, t, y) \geq 0$, according to Lebesgue's measure.

Equation (3.5), which is satisfied by the transition probabilities $P(s, x, t, dy)$, translates for their densities as:

$$\int_{-\infty}^{\infty} f(s, x, t, z) \, dz = 1$$

$$\int_{-\infty}^{\infty} f(s, x, u, z) f(u, z, t, y) \, dz = f(s, x, t, y),$$

for all $u \in]s, t[$ and all $y \in \mathbb{R}$. Therefore, to obtain local conditions starting from (3.5), the natural idea is to realize a Taylor development of f, which needs regularity conditions on f and assumptions on the moments.

Kolmogorov asks that for all s, t, y, $f(s, x, t, y)$ admits third order derivatives on x and on y which are uniformly bounded on s and t, on any set of the type $\{s, t : s - t > k\}$, $k > 0$. Besides, under the following assumptions for the moments:

$$\text{for all } t \geq 0, \ \lim_{\Delta \to 0} \int_{-\infty}^{\infty} |y - x|^i f(t, x, t + \Delta, y) \, dy = 0, \quad i = 1, 2, 3, \ (3.6)$$

$$\lim_{\Delta \to 0} \frac{\int_{-\infty}^{\infty} |y - x|^3 f(t, x, t + \Delta, y) \, dy}{\int_{-\infty}^{\infty} |y - x|^2 f(t, x, t + \Delta, y) \, dy} = 0, \tag{3.7}$$

he shows the existence of the limits

$$A(s, x) = \lim_{\Delta \downarrow 0} \frac{1}{\Delta} \int_{-\infty}^{\infty} (y - x) f(s, x, s + \Delta, y) \, dy, \tag{3.8}$$

$$B^2(s, x) = \lim_{\Delta \downarrow 0} \frac{1}{\Delta} \int_{-\infty}^{\infty} (y - x)^2 f(s, x, s + \Delta, y) \, dy, \tag{3.9}$$

which he calls respectively the infinitesimal mean and the infinitesimal variance of the process and which will be known in the future, in the diffusions case, as the drift coefficient and diffusion coefficient. Thus, from the (3.5), the existence of the limits (3.8) and (3.9) and under the differentiability assumption of f mentioned previously, Kolmogorov obtains the two following partial differential equations:

$$\frac{\partial}{\partial s} f(s, x, t, y) = -A(s, x) \frac{\partial}{\partial x} f(s, x, t, y) - B^2(s, x) \frac{\partial^2}{\partial x^2} f(s, x, t, y), \tag{3.10}$$

$$\frac{\partial}{\partial t} f(s, x, t, y) = -\frac{\partial}{\partial y} [A(t, y) f(s, x, t, y)] + \frac{\partial^2}{\partial y^2} [B^2(t, y) f(s, x, t, y)]. \tag{3.11}$$

The importance of these equations is such that one can consider them as being at the origin of the modern theory of stochastic processes. Let us give e.g. the main arguments of the proof of the first equation, which the author calls *first fundamental differential equation* and which is now known as the *backward equation* (the second being the *forward equation*).

Proof of equation (3.10). If we apply Taylor-Lagrange's formula to the 3rd order in the variable z on the function $f(s + \Delta, z, t, y)$ and at the points x and z, we obtain for $s < s + \Delta < t$,

$$f(s, x, t, y) = \int_{-\infty}^{\infty} f(s, x, s + \Delta, z) f(s + \Delta, z, t, y) \, dz$$

$$= \int_{-\infty}^{\infty} f(s, x, s + \Delta, z) [f(s + \Delta, x, t, y)$$

$$+\frac{\partial}{\partial x}f(s+\Delta,x,t,y)(z-x)+\frac{\partial^2}{\partial x^2}f(s+\Delta,x,t,y)\frac{(z-x)^2}{2}$$
$$+\frac{\partial^3}{\partial x^3}f(s+\Delta,\alpha,t,y)\frac{(z-x)^3}{6}]\,dz\,,$$

with $\alpha=x+c(z-x)$, in the case where c is such that $0<c<1$. Thus, if we apply the notations

$$a(s,x,\Delta)=\int_{-\infty}^{\infty}(y-x)f(s,x,s+\Delta,y)\,dy\,,$$

$$b^2(s,x,\Delta)=\int_{-\infty}^{\infty}(y-x)^2f(s,x,s+\Delta,y)\,dy\,,$$

$$c(s,x,\Delta)=\int_{-\infty}^{\infty}|y-x|^3f(s,x,s+\Delta,y)\,dy\,,$$

we can write, under the boundedness assumption on the third order derivative, that for a value C independent from Δ and for θ such that $|\theta|<C$,

$$f(s,x,t,y)=f(s+\Delta,x,t,y)+\frac{\partial}{\partial x}f(s+\Delta,x,t,y)a(s,x,\Delta)$$
$$+\frac{\partial^2}{\partial x^2}f(s+\Delta,x,t,y)\frac{b^2(s,x,\Delta)}{2}+\theta\frac{c(s,x,\Delta)}{6}\,,$$

which brings us immediately to the finite difference formula

$$\frac{f(s+\Delta,x,t,y)-f(s,x,t,y)}{\Delta}=-\frac{\partial}{\partial x}f(s+\Delta,x,t,y)\frac{a(s,x,\Delta)}{\Delta}$$
$$-\frac{\partial^2}{\partial x^2}f(s+\Delta,x,t,y)\frac{b^2(s,x,\Delta)}{2\Delta}-\theta\frac{c(s,x,\Delta)}{6\Delta}\,.$$

To conclude, let us note that under the above assumptions, the ratio $c(s,x,\Delta)/\Delta$ tends towards 0 when Δ tends towards 0.   □

As we already pointed out, the study of " random movements" whose law is governed by the (3.5) had already been outlined by Chapman in 1928 ([Cha28]) in a context of theoretical physics. The name of "Chapman-Kolmogorov" equation should not let one believe however that it was the only occasion, before the article of 1931, when this equation appeared. Kolmogorov himself, in this article, mentions a particular case studied by Louis Bachelier in 1900 ([Bac00]). He underlines, in a section that he devotes to the work of Bachelier, that the equation (3.11) had been written in his work of 1900 without however having been proven, in the case where the process is homogeneous in space, i.e. when the densities $f(s,x,t,y)$ only depend on s, t and on the difference $y-x$. The equation appears also in the works of Marian Smoluchovski about the Brownian motion during the years 1910.

Some continuations of the article *Über die analytischen Methoden in der Wahrscheinlichkeitsrechnung* [Kol31] appeared shortly after its publication on behalf of other probabilists like Bernstein, to whom Kolmogorov's equations inspired his theory of the stochastic differential equations in 1932. However, this theory is based on the discrete model and only allows to obtain weak solutions in the continuous case. Another important work was that of Wolfgang Doeblin, carried out in 1940. Doeblin sent it whilst at war (where he died) to the *Académie des Sciences de Paris*, in a sealed envelope which wasn't opened until 2000 ([Doe40])[10]. In this manuscript, Doeblin, very much ahead of his time, considers the pathwise aspects of the stochastic processes. More exactly, he establishes links between the strictly analytical point of view of Kolmogorov and that of Lévy who concentrates primarily on the paths construction and the fine properties of the processes, and especially of the Brownian motion, by purely probabilistic methods which often left his contemporaries perplexed. Doeblin builds equations very close to Itô's stochastic differential equations established some ten or fifteen years later and whose solutions are Brownian motions with a modified temporal variable: if the law of (X_t) satisfies the (3.5), then

$$X_t = x + \beta_{H(t)} + \int_0^t A(s, X_s)\, ds\,,$$

where β is a real-valued Brownian motion and H the time change

$$H(t) = \int_0^t B^2(s, X_s)\, ds.$$

This pathwise vision of processes offers then quite more than the analytical *forward* and *backward* (3.10) and (3.11). It allows Doeblin to establish results on the regularity of trajectories, the comparison of solutions, the properties of iterated logarithm, the functional central limit theorems and especially a preliminary version of the formula of change of variable that Itô will obtain a few years later ([Itô44]) and which will inaugurate the era of stochastic calculus itself.

   To establish his formula, Doeblin considers a function $\varphi(t, x)$ of class $\mathcal{C}^{1,2}$ (i.e. of class $\mathcal{C}^1$ with respect to t and $\mathcal{C}^2$ with respect to x) and increasing with respect to x, which ensures quite easily that the law of the process $Y_t = \varphi(t, X_t)$ is a solution of Kolmogorov's equation as soon as the law of (X_t) is also a solution. Thus, he proves that (Y_t) satisfies

$$Y_t = \varphi(0, x) + \gamma_{\overline{H}(t)} + \int_0^t \overline{A}(s, X_s)\, ds\,,$$

where γ is a real-valued Brownian motion and

$$\overline{H}_t = \int_0^t \overline{B}^2(s, X_s)\, ds\,, \quad \overline{B}(s, x) = \left(\frac{\partial}{\partial x} \varphi(s, x) \right) B(s, x)$$

$$\overline{A}(s, x) = \frac{\partial \varphi(s, x)}{\partial x} A(s, x) + \frac{\partial}{\partial s} \varphi(s, x) + \frac{1}{2} \left(\frac{\partial^2}{\partial x^2} \varphi(s, x) \right) B^2(s, x)\,.$$

---

[10] The readers who may be less interested in the technical aspects can content themselves by reading the article [BY03]

It was necessary to await the construction of Itô's stochastic integral during and after the second world war to see the solutions of the (3.10) and (3.11) under a new aspect. These new equations are satisfied[11] by the process itself and no longer only by its transition probabilities. If β stands for a real-valued Brownian motion and if $A(t, x)$ and $B(t, x)$ are the functions defined as presented at the beginning of this section, then the transition probabilities of the process X solution of the stochastic differential equation

$$dX_t = A(t, X_t)\, dt + B(t, X_t)\, d\beta_t\,, \qquad (3.12)$$

satisfy the (3.5), (3.10) and (3.11). An essential tool for the obtention of this result was Itô's formula mentioned above. In its most common current form, this fundamental formula is stated as follows: if $\varphi(t, x)$ is a function of class $\mathcal{C}^{1,2}$ then

$$
\begin{aligned}
d\varphi(t, X_t) =& B(t, X_t)\frac{\partial}{\partial x}\varphi(t, X_t)\, d\beta_t \\
&+ \left(\frac{\partial}{\partial t}\varphi(t, X_t) + A(t, X_t)\frac{\partial}{\partial x}\varphi(t, X_t) + \frac{1}{2}B^2(t, X_t)\frac{\partial^2}{\partial x^2}\varphi(t, X_t) \right) dt\,,
\end{aligned}
$$
$$(3.13)$$

where X solves the stochastic differential equation (3.12). These are the bases of the stochastic calculus which was going to encounter several developments during all the second half of the twentieth century.

From the years 1950 onwards, Doob's martingale theory [Doo90], developed afterwards by P.A. Meyer and his school in Strasbourg, was going to make it possible to weaken the conditions imposed until then on the functions $A(t, x)$ and $B(t, x)$ to ensure the construction of stochastic processes. An essential remark in this direction was the observation that under natural assumptions of local boundedness and lipschitzianity, the (3.12) admits a single solution in law, in the sense that if β' is another Brownian motion (eventually defined on another probability space), a solution X' of:

$$dX'_t = A(t, X'_t)\, dt + B(t, X'_t)\, d\beta'_t$$

follows the same law as X. This made it possible in the years 1970 to define the concept of weak solution for the (3.12) which isn't related to the specific choice of a particular Brownian motion any longer. A famous theorem of Yamada and Watanabe [YW71] asserts that pathwise uniqueness (the one which corresponds to an equation directed by a fixed Brownian motion) implies the unicity in law of the weak solutions. A powerful formulation was proposed by Stroock and Varadhan [SV79] in terms of *martingale problems*. The associated

---

[11] In the very particular way of stochastic differential equations which necessitate the concept of Itô's stochastic integral

generator to the Markovian process (X_t) solution of (3.12) is the operator L on $\mathcal{C}^{1,2}$ defined by:

$$Lf(t,x) = \frac{1}{2}B^2(t,x)\frac{\partial^2}{\partial x^2}f(t,x) + A(t,x)\frac{\partial}{\partial x}f(t,x) + \frac{\partial}{\partial t}f(t,x).$$

Itô's formula allows to express this definition by saying that for all $f \in \mathcal{C}^{1,2}$,

$$(M_t) = \left(f(t,X_t) - \int_0^t Lf(s,X_s)ds\right)_{t\geq 0} \qquad (3.14)$$

is a local martingale[12].

Then, it becomes natural to define a solution for (3.12) as follows. Let $\mathcal{C} = \mathcal{C}(\mathbb{R}^+,\mathbb{R})$ the set of continuous functions from $\mathbb{R}^+$ to $\mathbb{R}$. We define the *canonical projections* on $\mathcal{C}$ by $X_t(\omega) = \omega(t)$ for $\omega \in \mathcal{C}$ and the *canonical filtration* by $\mathcal{C}_t = \sigma(X_s, s \leq t)$, $t \geq 0$. Thus, a solution of (3.12) is a probability P on $(\mathcal{C},(\mathcal{C}_t))$ such that under P the processes defined by (3.14) are local martingales.

The most interesting aspect of the work of Stroock and Varadhan is that under very weak conditions (approximately, the continuity of the functions A and B), they showed that the preceding martingales problem admits a solution P. Under this probability, the canonical process satisfies the Markovian properties which were at the origin of Kolmogorov's studies. The reader interested by these subjects can consult with interest the important treatise of Jacod [Jac79].

3.4.2 Processes with independent and stationary increments

Among the processes whose laws verify Chapman-Kolmogorov's equation (3.5), there is a very important family that Lévy started to study at the beginning of the years 1930: those where functions $f(s,x,t,y)$ are homogeneous in time and space, i.e. depend only on the differences $t - s$ and $y - x$. In other terms, we are talking about the *processes with independent and stationary increments* for which Kolmogorov attempts to characterize the law in an article edited in two parts in 1932: *Sulla forma generale di un processo stocastico omogeneo* [Kol32a] and *Ancora sulla forma generale di un processo omogeneo* [Kol32b]. He simply considers a "random time function" $X(\lambda)$, where $\lambda \geq 0$ represents the time variable, such that for all λ_1 and λ_2, $(\lambda_2 \geq \lambda_1)$ the difference $X(\lambda_2) - X(\lambda_1)$ is independent from $(X(\lambda), \lambda \leq \lambda_1)$ and the law of which only depends on $\lambda_2 - \lambda_1$, i.e. if we note $\Delta = \lambda_2 - \lambda_1$,

$$\Phi_\Delta(x) = P(X(\lambda_2) - X(\lambda_1) < x) = P(X(\lambda_2 - \lambda_1) < x), \text{ for all } x \in \mathbb{R}.$$

---

[12] I.e. there exists a sequence of a.s. finite stopping times (T_n), increasing towards $+\infty$, such that for all n, the process $(M_{t\wedge T_n})$ is a martingale

Then, he discovers that the relation

$$\Phi_{\Delta_1 + \Delta_2}(x) = \int_{-\infty}^{\infty} \Phi_{\Delta_1}(x - y)\, d\Phi_{\Delta_2}(y)\,,$$

is a particular case of the (3.5). Nevertheless, we can notice that at that time, these processes aren't related to Markov processes, which are the subject of the study mentioned above. In fact, this formalization will only appear in the years 1950.

The aim of the article of 1932 is, according to Kolmogorov himself, to generalize some results given by Bruno de Finetti [Fin30] in the case where the laws given by the repartition functions Φ_Δ admit second order moments, $\int x^2\, d\Phi_\Delta(x) < \infty$. We will use the following notations:

$$m_\Delta = \int x\, d\Phi_\Delta(x)\,, \quad \text{and} \quad \sigma_\Delta^2 = \int (x - m_\Delta)^2\, d\Phi_\Delta(x)\,.$$

Thus Kolmogorov obtains a particular case of the famous Lévy-Khinchin's formula: if $\psi_\Delta(t) = \int e^{itx}\, d\Phi_\Delta(x)$ then $\psi_\Delta(t) = [\psi_1(t)]^\Delta$ and

$$\log \psi_1(t) = itm_1 - \frac{\sigma_0^2}{2} t^2 + \int_{-\infty}^{\infty} \left(e^{itx} - 1 - \frac{itx}{1 + x^2} \right) \frac{1 + x^2}{x^2}\, G(dx)\,, \quad (3.15)$$

where G is a finite measure such that $G(\{0\}) = 0$ and where $m_1 \in \mathbb{R}$, $\sigma_0^2 \geq 0$. This formula is commonly attributed to Lévy and Khinchin who obtained its final version in 1934 ([Lév34]) and 1937 ([Khi37]), although, as we have seen, de Finetti and Kolmogorov had already established it in particular cases. More precisely, the result given by Kolmogorov is the following.

Theorem 4. *When the law given by the repartition function Φ_Δ has a second order moment, we have*

$$\log \psi_1(t) = itm_1 - \frac{\sigma_0^2}{2} t^2 + \int_{-\infty}^{\infty} \pi(x, t)\, dF(x)\,, \quad (3.16)$$

where $m_1 \in \mathbb{R}$, $\sigma_0^2 \geq 0$, $\pi(x, t) = (e^{itx} - 1 - itx)/x^2$ and where the measure $dF(x)$ is defined by an increasing and bounded function F.

The previous formulae only concern unidimensional laws of the random function $(X(\lambda))$; consequently, it would be better to talk about the characterization of *indefinitely divisible laws*[13] rather than of a result on processes with independent and stationary increments. Nevertheless, let us note that at that time, this terminology didn't exist.

---

[13] A law of probability P is said to be *indefinitely divisible* if, for all n, it is the n-th convolution power of a law μ_n. That amounts to saying that P is the law of the sum of n real-valued independent random variables with the same law μ_n

Kolmogorov's proof: First of all, we verify that ψ_Δ is continuous with respect to Δ. Indeed, for $\Delta \leq 1/n$, we have $\sigma_\Delta^2 = \sigma_{1/n}^2 - \sigma_{1/n-\Delta}^2 \leq \sigma_{1/n}^2 = (1/n)\sigma_1^2$, and so $\sigma_\Delta^2 \to 0$ when $\Delta \to 0$. Consequently, $\psi_\Delta(t) \to 1$ when $\Delta \to 0$. We conclude thanks to the equality $\psi_{\Delta_1+\Delta_2}(t) = \psi_{\Delta_1}(t)\psi_{\Delta_2}(t)$. The continuity of ψ_Δ in Δ allows to show that the equality

$$\psi_\Delta(t) = [\psi_1(t)]^\Delta,$$

which is true for all rationals Δ, is also verified for all reals Δ. The author deduces then (using de Finetti's proof) that

$$\log \psi_1(t) = \lim_{\Delta \downarrow 0} \frac{1}{\Delta}[\psi_\Delta(t) - 1].$$

Moreover, we have

$$\frac{1}{\Delta}[\psi_\Delta(t) - 1] = itm_1 + \frac{1}{\Delta}\int_{-\infty}^{\infty}(e^{itx} - 1 - itx)\,d\Phi_\Delta(x).$$

Let us note $F_\Delta(x) = \frac{1}{\Delta}\int_{-\infty}^{x} y^2\,d\Phi_\Delta(y)$, so that

$$\frac{1}{\Delta}\int_{-\infty}^{\infty}(e^{itx} - 1 - itx)\,d\Phi_\Delta(x) = \int_{-\infty}^{\infty}\pi(x,t)\,dF_\Delta(x).$$

A classical argument allows to justify that for any sequence Δ_n decreasing towards 0, there exists a subsequence Δ_{n_k} such that the sequence $F_{\Delta_{n_k}}(x)$ converges when k tends towards $+\infty$, towards the function $F(x)$ in all points x where the latter is continuous. Let us remark that F is an increasing function such that:

$$0 = \lim_{\Delta \downarrow 0} F_\Delta(-\infty) \leq F(-\infty) \leq F(\infty) \leq \lim_{\Delta \downarrow 0} F_\Delta(\infty) = \sigma_1^2$$

and taking into account that at fixed t, $\pi(x,t) \to 0$ when $x \to \pm\infty$, we obtain

$$\lim_{k \to \infty}\int_{-\infty}^{\infty}\pi(x,t)\,dF_{\Delta_{n_k}}(x) = \int_{-\infty}^{\infty}\pi(x,t)\,dF(x),$$

which implies that

$$\log \psi_1(t) = itm_1 + \int_{-\infty}^{\infty}\pi(x,t)\,dF(x).$$

But as σ_1^2 is finite, we have $\log \psi_1(t) = itm_1 - \frac{\sigma_1^2}{2}t^2 + o(t^2)$ $(t \to 0)$, and according to the previous computation

$$\log \psi_1(t) = itm_1 - \frac{t^2}{2}(F(\infty) - F(-\infty)) + o(t^2) \quad (t \to 0).$$

This implies in particular that $F(\infty) - F(-\infty) = \sigma_1^2$. Finally we deduce from the above that $F(-\infty) = 0$ and $F(\infty) = \sigma_1^2$; thus F is entirely determined.

Now, let us prove that for all increasing (and left continuous) functions F, with values between $F(-\infty) = 0$ and $F(+\infty) = \sigma_1^2 < +\infty$, the function ψ_1 given by $\log \psi_1(t) = itm_1 + \int_{-\infty}^{\infty} \pi(x, t)\, dF(x)$ is the characteristic function of an indefinitely divisible law. For this purpose, let us consider a step function T with steps:

$$\omega_k = T(x_k+) - T(x_k),$$

for a finite number of reals $x_1, x_2, \ldots, x_n$. We also suppose that T doesn't jump at point 0. We note $\sigma_1^2 = \omega_1 + \omega_2 + \cdots + \omega_n$ the sum of these jumps. Let us also note $\log \overline{\psi}_1(t) = itm_1 + \int_{-\infty}^{\infty} \pi(x, t)\, dT(x)$, $\eta = m_1 - \sum_k p_k$, $p_k = \omega_k / x_k^2$ and $\chi_k(t) = \exp(itx_k)$. Then we can easily verify the identities

$$\int_{-\infty}^{\infty} \pi(x, t)\, dT(x) = \sum_k p_k(e^{itx_k} - 1 - itx_k),$$

$$\log \overline{\psi}_1(t) = it\eta + \sum_k p_k(\chi_k(t) - 1).$$

Thus, we deduce that

$$\overline{\psi}_\Delta(t) = (\overline{\psi}_1(t))^\Delta = \exp(it\eta\Delta) + \sum \Delta p_k(\chi_k(t) - 1)).$$

Let us note that $\chi_k(t)$ and $\exp(it\eta\Delta)$ are characteristic functions and consequently $\psi_\Delta(t)$, as a product of characteristic functions, is a characteristic function itself. To conclude, let us note that we can find a sequence of step functions T_n such that $\int_{-\infty}^{\infty} \pi(x, t)\, dT_n(x)$ converges towards $\int_{-\infty}^{\infty} \pi(x, t)\, dF(x)$ uniformly on any bounded interval. Thus, the corresponding characteristic functions $\overline{\psi}_\Delta^{(n)}(t)$ converge towards $\psi_\Delta(t)$. This implies that ψ_Δ is a characteristic function, on one hand, and that this function verifies $\psi_\Delta(t) = (\psi_1(t))^\Delta$, on the other hand.   □

Kolmogorov completes this description of the law of processes with independent increments by observing that if $P_1(x) = \int_x^{\infty} y^{-2}\, dF(y)$ et $P_2(x) = \int_{-\infty}^{x} y^{-2}\, dF(y)$ then $P_1(x)\, d\lambda$ (respectively $P_2(x)\, d\lambda$) is the probability such that the process $X(\lambda)$ has a positive jump (respectively negative), of height bigger than (respectively smaller than) or equal to x, during the time increment $d\lambda$. The measures P_1 and P_2 are respectively the restrictions to $\mathbb{R}_+$ and $\mathbb{R}_-$ of what will be called afterwards the Lévy measure of the process $X(\lambda)$.

Following the works of Lévy, a number of authors during the years 1960 to 1980 such as Skorokhod, Zolotarev, Blumenthal, Getoor, Ray, Taylor, Fristedt, Bingham, Pitman, Jacod,... studied the fine properties of the trajectories of the processes with independent and stationary increments (now called Lévy processes). Then at the beginning of the years 1990, the synthesis works of Bertoin [Ber96] and Sato [Sat99] caused, among the probabilistic community, a renewed interest for the study of Lévy processes. This was again reinforced with the discovery of new grounds of applications, like financial mathematics, where the models using Lévy processes allow to compensate for the defects of the model known as the Black-Scholes model involving the geometrical Brownian motion.

3.4.3 Continuity and relative compactness criteria

It is remarkable that Kolmogorov also developed a certain number of tools for the study of the pathwise properties of random processes. Among those, a very effective criterion guaranteeing the pathwise continuity of the processes bears his name. It was found by the Soviet mathematician in 1934 and presented the same year during the Seminar at the University of Moscow. It was not the subject however of any of his publications and it was Slutsky who stated it and provided the first proof published in the *Giornale dell'Istituto Italiano dei Attuari* in 1937 [Slu37], allotting its paternity to Kolmogorov. Twenty years after this, Kolmogorov pointed out again to Chentsov an extension of this criterion for discontinuous processes which he published in 1956 [Che56] and which allows to conclude that the process in question does not have these discontinuities of the second order[14].

Theorem 5. *If for a family of real-valued random variables $(X_t, 0 \le t \le 1)$ there exist three strictly positive constants γ, c and ε such that:*

$$E(|X_t - X_s|^\gamma) \le c|t - s|^{1+\varepsilon} , \qquad (3.4.3.1)$$

then there exists a modification of X which is a.s. continuous.

Let us recall that a modification of a process X is a process $\tilde{X}$ such that for all fixed t, $\tilde{X}_t = X_t$, a.s. The following proof takes as a starting point the proof of [RY91], p. 26; see also [DM75], Vol. V, Chap. XXIII, p. 332.

Proof: Let D_m the set of reals on [0,1] of the form $2^{-m}i$, where $i = 0, 1, \ldots, 2^m$ and $D = \cup_m D_m$ the set of diadic numbers on [0,1]. Let Δ_m the set of pairs (s,t) of D_m^2 such that $|t - s| = 2^{-m}$. We note

$$Y_i = \sup_{(s,t)\in\Delta_i} |X_s - X_t|.$$

Since for all $s, t \in \Delta_i$, $E(|X_s - X_t|^\gamma) \le c\,(2^{-i})^{1+\varepsilon}$ and $\#\Delta_i = 2^i$, we have for all i,

$$E(Y_i^\gamma) \le \sum_{(s,t)\in\Delta_i} E(|X_s - X_t|^\gamma) \le c\,2^i(2^{-i})^{1+\varepsilon} \le c\,2^{-i\varepsilon} .$$

---

[14] In a nearby order of ideas, let us also underline the contribution of Kolmogorov to Skorokhod's topology for the space of right continuous and left limited trajectories [Kol56]

Now, let $s, t \in D$, $s \neq t$ and m the integer such that $2^{-m-1} < |t - s| \leq 2^{-m}$; thus there exist finite sequences $s_m, s_{m+1}, \ldots, s_p = s$ and $t_m, t_{m+1}, \ldots, t_k = t$ such that for all $i = m, m + 1, \ldots, p$ and $j = m, m + 1, \ldots, k$, we have $(s_i, s_{i+1}) \in \Delta_{i+1}$, $(t_j, t_{j+1}) \in \Delta_{j+1}$, $(s_m, t_m) \in \Delta_m$ and

$$X_s - X_t = \sum_{i=m}^{p-1} X_{s_{i+1}} - X_{s_i} + X_{s_m} - X_{t_m} + \sum_{j=m}^{k-1} X_{t_{j+1}} - X_{t_j}.$$

Thus, we deduce the inequalities

$$|X_s - X_t| \leq Y_m + 2 \sum_{i=m+1}^{\infty} Y_i \leq 2 \sum_{i=m}^{\infty} Y_i.$$

Let $\alpha \in [0, \varepsilon/\gamma[$. Let us define the random variable

$$M_\alpha = \sup\{|X_t - X_s|/|t - s|^\alpha : s, t \in D, s \neq t\}.$$

Then, we deduce from the above that:

$$M_\alpha \leq \sup_{m \in \mathbb{N}} \{2^{(m+1)\alpha} \sup_{2^{-m-1} < |t-s| \leq 2^m} |X_t - X_s| : s, t \in D, s \neq t\}$$

$$\leq \sup_{m \in \mathbb{N}} \{2 \cdot 2^{(m+1)\alpha} (\sum_{i=m}^{\infty} Y_i)\} \leq 2^{\alpha+1} \sum_{i=0}^{\infty} 2^{i\alpha} Y_i.$$

In the case where $\gamma \geq 1$, this inequality implies that

$$E(M_\alpha^\gamma)^{1/\gamma} \leq 2^{\alpha+1} \sum_{i=0}^{\infty} 2^{i\alpha} E(Y_i^\gamma)^{1/\gamma} \leq 2^{\alpha+1} \sum_{i=0}^{\infty} 2^{i\alpha} 2^{-i\varepsilon/\gamma} < \infty$$

and when $\gamma < 1$, the same inequality applies to $E(M_\alpha^\gamma)$. In particular, the variable M_α is finite and we have, for all $s, t \in D$, $|X_t(\omega) - X_s(\omega)| < K(\omega)|t - s|^\alpha$, where $K(\omega)$ is a constant which does not depend on s and t. Then, we deduce that X is uniformly continuous on D (and even hölderian of order α). The process $X_{|_D}$ (i.e. X restricted to D) extends then in a unique manner to a continuous process on $[0, 1]$:

$$\tilde{X}_t(\omega) = \lim_{s \to t, \, s \in D} X_s(\omega), \quad t \in [0, 1]$$

which is naturally also hölderian of order α. Finally, according to the assumption, for all $t \in [0, 1]$, $\lim_{s \to t} X_s = X_t$ in L^1, thus $\tilde{X}_t = X_t$, a.s. for all $t \in [0, 1]$. $\square$

As we can see, we have in fact proven a stronger result: *under the previous assumptions, there exists a modification of X with a.s. hölderian trajectories of order α for all $\alpha \in [0, \varepsilon/\gamma[$.* In particular, the Brownian motion is (a.s.) locally hölderian of order α for all $\alpha < 1/2$.

Theorem 5 admits some extensions such as the one of Chentsov mentioned previously. There also exists a criterion of the same type for processes with several indexes $(X_t, t \in [0,1]^d)$; then, in this case, we write the assumption (3.4.3.1) as follows:

$$E(|X_t - X_s|^\gamma) \le c|t - s|^{d+\varepsilon}.$$

The proof of this result is very close to the one we just gave and also applies to processes with values in general Banach spaces.

In 1970, Garsia, Rodemich and Rumsey suggested another generalization of this continuity criterion which is based on an entirely deterministic argument (see [SV79], p. 47 or [DM75], Vol. V, Chap. XXIII, p. 336). Let us briefly expose their result:

Let f denote a Borelian function defined on a ball B of $\mathbb{R}^d$, satisfying the integral condition:

$$\int_{B \times B} \Psi \left(\frac{|f(v) - f(u)|}{\varphi(|v - u|)} \right) du\, dv \stackrel{(def)}{=} A < \infty,$$

where Ψ and φ are two continuous functions on $\mathbb{R}_+$, null in 0, strictly increasing and unbounded, such that the integrals

$$h(x,t) = \int_0^t \Psi^{-1}(x/s^{2d})\, d\varphi(s)$$

converge for all $x > 0$. Then the function f admits a continuous version $\tilde{f}$ (i.e. $f(u) = \tilde{f}(u)$, a.e. in the sense of Lebesgue's measure) which possesses on the same ball B the modulus of continuity:

$$|\tilde{f}(v) - \tilde{f}(u)| \le 8h(k_d A, 2|v - u|),$$

where the constant k_d only depends on the dimension d.

It was then proven that this result is in fact a particular case of Sobolev's embedding theorems. The reader who may be interested in this topic can refer to [DM75], Vol. V, Chap. XXIII, p. 334. It was possible to apply the explicit form of this elaborated version of Kolmogorov's criterion to obtain fine continuity results for the local times of Lévy processes, and more generally of semi-martingales, see [Bar85] and [BY81]. One can compare this form of Kolmogorov's criterion with the *majorizing measures* technique of Fernique, Marcus, Talagrand,...

As one can easily imagine, the applications of Theorem 5 are numerous and diverse. One of the most important is doubtlessly the relative compactness criterion. Let us recall that a sequence of probability measures is said to be (weakly) relatively compact if one can extract from any subsequence, a new subsequence which converges weakly.

According to Prohorov's theorem, *a sequence of probability measures (P_n) on a separable and complete space is relatively compact if and only if it is tight,* i.e. if for any $\varepsilon > 0$, there exists a compact K such that $P_n(K) \ge 1 - \varepsilon$, for all n. This theorem presents a great interest, in particular when the relevant probability space is the set of continuous paths on the positive half-line. Then, let us denote by $\mathcal{C}$ the space of continuous functions on the positive half-line and with values in $\mathbb{R}^d$, which is fitted with the uniform convergence topology and the Borelian σ-algebra. In order to prove that a sequence of measures (P_n) on $\mathcal{C}$ converges weakly towards a measure P, we know that it suffices to verify that it is relatively compact and that its finite dimensional marginals converge towards the corresponding marginals of P. As $\mathcal{C}$ is separable and complete, relative compactness is equivalent to the tension property. In general, the convergence of the finite dimensional marginals can easily be stated. Relative compactness is often a problem. A very useful criterion is obtained as a direct consequence of Kolmogorov's theorem exposed in Theorem 5. One can refer to [Bil95] for a proof.

Corollary 4. *If a sequence of continuous processes* $(X_n(t), t \geq 0)^{15}$ *satisfies*

(i) The sequence of r.v.'s $(X_n(0))$ *is tight,*
(ii) There exist three strictly positive constants γ, c *and* ε *such that for any* n:

$$E(|X_n(t) - X_n(s)|^\gamma) \leq c|t - s|^{d+\varepsilon}, \qquad (3.4.3.2)$$

then, the sequence of laws of the processes (X_n) *is relatively compact.*

References

All the articles written by A. N. Kolmogorov listed below can be found in [KolII].

[Bac00]   Bachelier, L.: Théorie de la spéculation. Ann. Sci. École Norm. Sup., **17**, 21–86 (1900)
[Bar85]   Barlow, M.T.: Continuity of local times for Lévy processes. Z. Wahrsch. Verw. Gebiete, **69**, 23–35 (1985)
[Ber96]   Bertoin, J.: Lévy Processes. Cambridge University Press, Cambridge (1996)
[Bil95]   Billingsley, P.: Probability and Measure. Wiley, 3rd edition (1995)
[BMP01]   Baldi, P., Mazliak, L., Priouret, P.: Martingales et Chaînes de Markov. Hermann, paris (2001)
[BY81]   Barlow, M.T., Yor, M.: Semimartingale inequalities via the Garsia-Rodemich-Rumsey Lemma and application to local times. J. Funct. Analysis, **49**, 198–229 (1982)
[BY03]   Bru, B., Yor, M.: Wolfgang Doeblin, l'équation de Kolmogoroff. *La Recherche* (june 2003)
[Cha28]   Chapman, S.: On the brownian displacement and thermal diffusion of grains suspended in a non-uniform fluid. Proc. Roy. Soc. London, Ser. A, **119**, 34–54 (1928)

---

[15] This is equivalent to consider a sequence of random variables (X_n) with values in $\mathcal{C}$

[Che56]   Chentsov, N.N.: Weak convergence of stochastic processes whose trajec-
          tories have no discontinuities of second kind and the "heuristic" approach
          to the Kolmogorov-Smirnov tests. Teor. Ver, **1**, 140–144 (1956)

[Cra37]   Cramér, H.: Random Variables and Probability distributions. Cambridge
          (1937)

[DM75]   Dellacherie, C., Meyer, P.A.: Probabilités et Potentiel, Hermann, Paris
          (1975, 1992)

[Doe40]   Doeblin, W.: Sur l'équation de Kolmogoroff (sealed envelope, 1940),
          edited and commented by B. Bru and M. Yor. C.R. Acad. Sci. Paris
          Sér. I Math., **331**, 1033–1187 (2000)

[Doo90]   Doob, J.L.: Stochastic processes. Wiley Classics Library. John Wiley &
          Sons, Inc., New York (1990)

[Fin30]   de Finetti, B.: Le funzioni caratteristiche di legge istantanea. Rend. Acc.
          Lincei, **12**, 278–282 (1930)

[GS80]   Guikhman, I., Skorokhod, A.: Introduction à la théorie des processus
          aléatoires. (Fr. trans. D. Embarek). Éditions MIR, Moscow (1980)

[Itô44]   Itô, K.: Stochastic Integrals. Proc. Imp. Acad. Tokyo, **20**, 519–524 (1944)

[Itô86]   Itô, K.: Selected papers, edited by Stroock, D.W. and Varadhan, S.R.S.
          Springer (1986)

[Jac79]   Jacod, J.: Calcul stochastique et problèmes de martingales. Lecture Notes
          in Mathematics, 714. Springer, Berlin (1979)

[Khi24]   Khinchin, A.Y.: Über einen Satz der Wahrscheinlichkeitrechnung. Fund.
          Mat., **6**, 9–20 (1924)

[Khi37]   Khinchin, A.Y.: Déduction nouvelle d'une formule de M. Paul Lévy. Bull.
          Univ. Moscow Math-Mec., **1**, 1–5 (1937)

[KK25]   Khinchin, A.Y., Kolmogorov, A.N.: Über Konvergenz von Reihen, deren
          Glieder durch den Zufall bestimmt werden. Mat. Sb., **32**, 668–677 (1925)

[Kol29]   Kolmogorov, A.N.: Über das Gesetz des iterierten Logarithmus. Math.
          Ann., **101**, 126–135 (1929)

[Kol30]   Kolmogorov, A.N.: Sur la loi forte des grands nombres. CRAS Paris, **191**,
          910–912 (1930)

[Kol31]   Kolmogorov, A.N.: Über die analytischen Methoden in der Wahrschein-
          lichkeitrechnung. Math. Ann., **104**, 415–458 (1931)

[Kol32a] Kolmogorov, A.N.: Sulla forma generale di un processo stocastico omo-
          geneo. Atti Accad. Naz. Lincei Rend., **15**, 805–808 (1932)

[Kol32b] Kolmogorov, A.N.: Ancora sulla forma generale di un processo stocastico
          omogeneo, Atti Accad. Naz. Lincei Rend., **15**, 866–869 (1932).

[Kol33]   Kolmogorov, A.N.: Grundbegriffe der Wahrscheinlichkeitrechnung.
          Springer (1933)

[Kol56]   Kolmogorov, A.N.: On the Skorohod convergence. Teor. Veroyatnost. i
          Primenen., **1**, 239–247 (1956)

[KolII]   Shiryaev, A.N. (ed): Selected works of A.N. Kolmogorov, Vol. II (trans-
          lated by G. Lindquist). Kluwer Academic Publications (1992)

[Lév34]   Lévy, P.: Sur les intégrales dont les éléments sont des variables aléatoires
          indépendantes. Ann. Sc. Norm. Pisa, **3**, 337–366 (1934)

[Lév37]   Lévy, P.: Théorie de l'Addition des Variables aléatoires. Gauthier-Villars,
          Paris (1937)

[Maz03]   Mazliak, L.: Andrei Nikolaevitch Kolmogorov (1903–1987). Un aperçu
          de l'homme et de l'œuvre probabiliste. Cahiers du CAMS-EHESS,

Paris (2003). Available at http://www.proba.jussieu.fr/users/lma/recherche.html

[RY91]   Revuz, D., Yor, M.: Continuous martingales and Brownian motion. Springer (1991). 3rd edition, Springer (1999)

[Sat99]   Sato, K.I.: Lévy processes and infinitely divisible distributions. Translated from the 1990 Japanese original. Cambridge Studies in Advanced Mathematics, 68. University Press, Cambridge (1999)

[Shi89]   Shiryaev, A.N.: Kolmogorov: Life and creative activities. Ann. Prob., **17**, 866–944 (1989)

[Slu37]   Slutsky, E.B.: Qualche proposizione relativa alla teoria delle funzioni aleatorie. Gio. Ist. It. Att., **8**, 183–199 (1937)

[Str03]   Stroock, D.W.: Markov processes from Itô's perspective. Princeton University Press (2003)

[SV06]   Shafer, G., Vovk, V.: The origins of Kolmogorov's Grundbegriffe. Statistical Science, **21**:1, 70–98 (2006)

[SV79]   Stroock, D.W., Varadhan, S.R.S.: Multidimensional diffusion processes. Springer (1979)

[Szé86]   Székely, G.J.: Paradoxes in probability theory and mathematical statistics. Mathematics and its Applications (East European Series), **15**. D. Reidel Publishing Co., Dordrecht (1986)

[vonP94] von Plato, J.: Creating Modern Probability. Its mathematics, physics and philosophy in historical perspective. Cambridge University Press, Cambridge (1994)

[Wie23]   Wiener, N.: Differential space. Jour. Math. and Phys., **58**, 131–174 (1923)

[YW71]   Yamada, T., Watanabe, S.: On the uniqueness of solutions of stochastic differential equations II. J. Math. Kyoto Univ., **11**, 553–563 (1971)

4

Infinite-dimensional Kolmogorov equations

Giuseppe Da Prato

Scuola Normale Superiore, Pisa, Italy
http://www.sns.it/scienze/menunews/docentiscienze/daprato
daprato@sns.it

In [Kol31] (1931), Kolmogorov introduced very important partial differential equations. In the recent past years there was an increasing interest, due to applications in physics and in particular in statistical mechanics, to consider Kolmogorov equations in the extended context of infinite-dimensional Hilbert spaces. The aim of this chapter is to describe the "status of art" in this domain. We present different methods which have been used to solve these equations, as well as applications to some relevant stochastic partial differential equations.

4.1 Introduction and setting of the problem

4.1.1 The Kolmogorov setting

In his basic paper "Analytic methods in probability theory" see [Kol31], A. N. Kolmogorov studied the transition probability $P(s, x; t, I)$ that an event I (a Borel subset of $\mathbb{R}$) is realized at a time $t > s$ when the state at time s is x, under the basic assumption that $P(s, x; t, I)$ satisfies the following *Chapman-Kolmogorov equation*,

$$P(s, x; t, I) = \int_{\mathbb{R}} P(s, x; u, dy) P(u, y; t, I), \quad 0 \le s < u < t. \qquad (4.1)$$

It is known that this identity allows to construct a Markov process (X_t), $t \ge 0$ such that

$$\mathbb{P}(X_t \in I | X_s = x) = P(s, x; t, I).$$

In his paper [Kol31] Kolmogorov did not try to construct the process (X_t) but he derived from (4.1) some important partial differential equations for the transition probability. Let us briefly recall his argument.

Assume that the probability $P(s, x; t, dy)$ has a density $f(s, x; t, y)$. Then assume that the limits

$$A(s, x) := \lim_{\delta \downarrow 0} \frac{1}{\delta} \int_{-\infty}^{+\infty} (y - x) f(s, x; s + \delta, y) dy,$$

and

$$B^2(s, x) := \lim_{\delta \downarrow 0} \frac{1}{\delta} \int_{-\infty}^{+\infty} (y - x)^2 f(s, x; s + \delta, y) dy,$$

exist. Then, for $t > s$, and under some regularity conditions, f fulfills the *backward* equation,

$$-\frac{\partial}{\partial s} f(s, x; t, y) = A(s, x) \frac{\partial}{\partial x} f(s, x; t, y) + \frac{1}{2} B^2(s, x) \frac{\partial^2}{\partial x^2} f(s, x; t, y), \quad (4.2)$$

and the *forward* or Fokker-Planck equation

$$\frac{\partial}{\partial t} f(s, x; t, y) = -\frac{\partial}{\partial y} [A(t, y) f(s, x; t, y)] + \frac{1}{2} \frac{\partial^2}{\partial y^2} [B^2(t, y) f(s, x; t, y)]. \quad (4.3)$$

These equations can be easily generalized to a finite dimensional space $\mathbb{R}^n$. In this way Kolmogorov created a powerful analytic method for studying probabilistic properties of Markov diffusion processes in $\mathbb{R}^n$.

4.1.2 Infinite-dimensional Kolmogorov equations

We shall concentrate on the infinite-dimensional generalization of the backward Kolmogorov equation (4.2), where however we change the sign of the time to obtain a more familiar initial value problem. So, we shall be concerned with the following equation,

$$\begin{cases} D_t u(t, x) = K_0 u(t, x), \quad t > 0, \ x \in H, \\ u(0, x) = \varphi(x), \quad x \in H, \end{cases} \quad (4.4)$$

on a separable real Hilbert space H, where K_0 is the differential operator

$$K_0 \varphi(x) = \frac{1}{2} \operatorname{Tr} [C D^2 \varphi(x)] + \langle Ax + F(x), D\varphi(x) \rangle, \quad x \in D(A) \cap D(F). \quad (4.5)$$

Here $A \colon D(A) \subset H \to H$ is the infinitesimal generator of a strongly continuous semigroup e^{tA} in H, $C \colon H \to H$ is a symmetric nonnegative bounded linear operator in H (possibly the identity operator I), and $F \colon D(F) \subset H \to H$ is nonlinear. Moreover, φ belongs to a suitable functional space, D_t represents the derivative with respect to t, D the derivative with respect to x and Tr the trace. The operator K_0 is said to be *strictly elliptic* when C is continuously invertible, *elliptic (degenerate)* otherwise.

Obviously K_0 is not the more general second order operator (one could take A and C dependent on x and t or add a potential term as $V(x)u(t, x)$). However, we shall confine to this situation both for the sake of simplicity and also because the most part of available results concerns this case. If the space H is finite dimensional, K_0 is a usual elliptic operator (possibly degenerate).

Besides the parabolic (4.4) we shall also consider the elliptic equation

$$\lambda\psi - K_0\psi = f, \tag{4.6}$$

where $\lambda > 0$ and f are given.

There is a huge literature on Kolmogorov equations in finite dimensional spaces, which we shall not treat in this paper. Several results are classical (essentially when $A = 0$ and F is bounded), but the research is still very active in this field, see e.g. [BR95], [BKR01], [DL95], [DL04], [Ebe99], [Sta99].

First attempts to build a theory of partial differential equations on Hilbert spaces were made by P. Lévy, see [Lév51]. A different approach was initiated by L. Gross [Gro67] and Yu. Daleckij [Dal96] and [DF91], using functional analysis, Gaussian measures and stochastic differential equations, in the seventies. This approach was extensively developed recently, see the monograph [DZ02] and references therein.

One of the main motivations for studying (4.4) and (4.6), is that they are closely related to the following differential stochastic equation (in $X_t(x) = X(t, x)$):

$$\begin{cases} dX(t, x) = (AX(t, x) + F\left(X(t, x)\right)) dt + C^{1/2}dW(t), \quad t > 0, \ x \in H, \\[2mm] X(0, x) = x, \quad x \in H, \end{cases} \tag{4.7}$$

where $W(t)$ is a cylindrical Wiener process in H (see (4.12) below). The link is by the formal identities

$$u(t, x) = \mathbb{E}[\varphi(X(t, x))], \quad t \geq 0, \ x \in H, \tag{4.8}$$

and

$$\psi(x) = \int_0^{+\infty} e^{-\lambda t}\mathbb{E}[f(X(t, x))]dt, \quad x \in H, \tag{4.9}$$

respectively, where $\mathbb{E}$ denotes expectation.

In the applications (4.7) describes the evolution of an infinite-dimensional dynamical system perturbed by noise (the system being "isolated" when $F = 0$). Typical examples are reaction-diffusion, Burgers and Navier-Stokes equations. A represents often the Laplacian and the nonlinear mapping F describes the interaction.

4.1.3 Mild solutions

To take advantage of formulas (4.8) and (4.9), we should assume that (4.7) has a unique solution in some sense, *strong*, *mild* or in the sense of the *martingales*.

This is the situation taken into account in the first part of this introduction, where we shall deal with mild solutions. In the second part we shall consider a method for solving the Kolmogorov equation without using existence and uniqueness of problem (4.7).

Let us recall that, given $x \in H$, a *mild* solution of (4.7) is a mean square continuous stochastic process $X(\cdot, x)$ which is adapted to $W(t)$, such that $X(t, x) \in D(F)$ for any $t \geq 0$, and

$$X(t, x) = e^{tA}x + \int_0^t e^{(t-s)A}F(X(s, x))ds + W_A(t), \ t \geq 0, \tag{4.10}$$

where $W_A(t)$ is the *stochastic convolution* defined by

$$W_A(t) = \int_0^t e^{(t-s)A}C^{1/2}dW(s). \tag{4.11}$$

Let us precise the definition of $W_A(t)$. First we recall that a cylindrical process $W(t)$ can be written formally as

$$W(t) = \sum_{k=1}^{\infty} \beta_k(t)e_k, \quad t \geq 0, \tag{4.12}$$

where $\{e_k\}$ is a complete orthonormal basis in H and $\{\beta_k\}$ is a sequence of mutually independent standard Brownian motions in a probability space $(\Omega, \mathcal{F}, \mathbb{P})$. We notice that the series in (4.12) is a.s. divergent since $\mathbb{E}|W(t)|^2 = \sum_{k=1}^{\infty} t = +\infty$ (one can show that it is convergent in a space larger than H, see e.g. [DZ92]), however one can show that, under Hypothesis 4.1.1 below, $W_A(t)$ is meaningful. Then, expanding $W(t)$ in terms of the basis $\{e_k\}$ and using standard properties of the stochastic integral, we obtain the following identity

$$\mathbb{E}(|W_A(t)|^2) = \int_0^t \text{Tr} \left[e^{sA}Ce^{sA^*} \right]ds, \quad t \geq 0.$$

So the following assumption is natural (one can show that it is necessary):

Hypothesis 4.1.1
(i) A is the infinitesimal generator of a strongly continuous semigroup e^{tA} in H. There exists $M > 0$ and $\omega \in \mathbb{R}$ such that $\|e^{tA}\| \leq Me^{\omega t}$, $t \geq 0$.
(ii) $C \colon H \to H$ is a symmetric nonnegative bounded linear operator in H, and the operator Q_t defined by

$$Q_t x = \int_0^t e^{sA}Ce^{sA^*}xds, \quad x \in H, \tag{4.13}$$

is of trace class for any $t > 0$.

From now on we shall assume that Hypothesis 4.1.1 holds. It is easy to see that

$$\mathbb{E}\left(\langle W_A(t), x\rangle\langle W_A(t), y\rangle\right) = \langle Q_t x, y\rangle, \quad x, y \in H, \ t \geq 0,$$

so that $W_A(t)$ is a Gaussian random variable with mean 0 and covariance Q_t, and then $X(x, t)$ is a Gaussian random variable with law $N_{e^{tA}x, Q_t}$, that is with mean $e^{tA}x$ and covariance Q_t[1].

4.1.4 Transition semigroup and strong Feller property

Let us assume now that (4.7) has a unique mild solution $X(\cdot, x)$. Then we define the *transition semigroup* (P_t) setting

$$P_t \varphi(x) = \mathbb{E}[\varphi(X(t, x))], \quad \varphi \in B_b(H), \qquad (4.14)$$

where $B_b(H)$ is the Banach space of all Borel and bounded mappings $\varphi \colon H \to \mathbb{R}$, endowed with the norm

$$\|\varphi\|_0 = \sup_{x \in H} |\varphi(x)|, \quad \varphi \in B_b(H), \ t \geq 0.$$

The semigroup property

$$P_{t+s} = P_t P_s, \quad t, s \geq 0,$$

holds in view of the Markov property of the process $X(t, x)$.

(P_t) is a Markov semigroup, that is a one-parameter semigroup of linear operators on $B_b(H)$ such that $P_t \varphi \geq 0$ for all $\varphi \geq 0$ and $P_t 1 = 1$. Consequently, (P_t) is a contraction semigroup, that is $\|P_t \varphi\|_0 \leq \|\varphi\|_0$ for all $\varphi \in B_b(H)$.

Assume that the process $X(t, x)$ is mean square continuous in x and that $\varphi \in C_b(H)$, where $C_b(H)$ is the closed subspace of $B_b(H)$ of all uniformly continuous and bounded functions. It is easy to see that the function $u(t, x) = P_t \varphi(x)$ is continuous in (t, x). In particular, (P_t) is a *Feller* semigroup, that is $P_t \varphi \in C_b(H)$ for all $\varphi \in C_b(H)$. A stronger condition on (P_t) is that $P_t \varphi \in C_b(H)$ for all $t > 0$ and all $\varphi \in B_b(H)$. In this case (P_t) is called a *strong Feller* semigroup. This important property will be discussed later.

Notice that the restriction of (P_t) to $C_b(H)$ is not a strongly continuous semigroup in general. However, we shall define the *infinitesimal generator* K of (P_t) in $C_b(H)$ as follows, see [Cer94]. Consider the Laplace transform of P_t,

$$G(\lambda)f(x) := \int_0^{+\infty} e^{-\lambda t} P_t f(x) dt, \quad f \in C_b(H), \ \lambda > 0, \ x \in H.$$

This definition is meaningful since $P_t f(x)$ is continuous in (t, x) and $|P_t f(x)| \leq \|f\|_0$. One can check easily that $G(\lambda)f \in C_b(H)$ and the resolvent identity holds

$$G(\lambda) - G(\mu) = (\mu - \lambda)G(\lambda)G(\mu), \quad \lambda, \mu > 0.$$

---

[1] If $a \in H$ and $C \in L(H)$ is symmetric, nonnegative and of trace class, we shall denote by $N_{a,C}$ the Gaussian measure in H of mean a and covariance operator C. We shall set $N_{0,C} = N_C$

Since

$$\lim_{\lambda\to\infty} \lambda G(\lambda)f(x) = \lim_{\lambda\to\infty} \int_0^{+\infty} e^{-\tau} P_{\frac{\tau}{\lambda}} f(x)d\tau = f(x), \quad x \in H,$$

G is one-to-one so that it is a resolvent. Consequently, there exists a unique closed operator K in $C_b(H)$ such that $G(\lambda) = (\lambda - K)^{-1}$ for any $\lambda > 0$. K is called the infinitesimal generator of (P_t). It is clearly m-dissipative[2] in $C_b(H)$ since,

$$\|(\lambda - K)^{-1}f\|_0 \le \frac{1}{\lambda}\|f\|_0, \quad \lambda > 0, \ f \in C_b(H)^3.$$

It is an interesting problem (not easy) to understand the relationship between the abstract operator K and the differential operator K_0 defined by (4.5).

4.1.5 Invariant measures and irreducible semigroups

Let us come back to the Kolmogorov equation (4.4). Clearly, $u(t,x) = P_t\varphi(x)$ is the natural candidate to be a "solution" (in a sense to be made precise) of the Kolmogorov equation (4.4). In particular, it is natural to try to show, by a direct computation, that $u(t,x) = \mathbb{E}[\varphi(X(t,x))]$ is a strict solution to (4.4), at least for a regular φ. For this we have to justify the following formulas for any $h \in H$:

$$\langle Du(t,x), h\rangle = \mathbb{E}[\langle D\varphi(X(t,x)), DX(t,x)h\rangle], \quad t \ge 0, \ x, h \in H,$$
$$D^2 u(t,x)(h,h) = \mathbb{E}[D^2\varphi(X(t,x))(DX(t,x)h, DX(t,x)h)] \qquad (4.15)$$
$$+ \mathbb{E}[\langle D\varphi(X(t,x)), D^2 X(t,x)(h,h)\rangle]$$

and then to use the Itô formula. However, this procedure requires to know at least that $DX(t,x)$ and $D^2 X(t,x)$ do exist and are integrable (this is true for

---

[2] An operator A in a Banach space B is called *dissipative* if exists a function J from B to its dual such that $\forall x \in B$, $\langle J(x), x\rangle = |x|^2 = |J(x)|^2$ and that $\forall x \in D(A)$, $\mathrm{Re}\langle J(x), Ax\rangle \le 0$. A is said m-*dissipative* if, moreover, there exists $\lambda > 0$ such that $\lambda - A : D(A) \to B$ is one-to-one with a *bounded* inverse. An m-*dissipative* operator is always *maximal* dissipative (that is, it cannot be extended to a dissipative operator on a larger domain), hence the m in m-*dissipative*. (In Hilbert spaces, the converse is true: all maximal dissipative operators are m-dissipative.) The *raison d'être* of these notions, introduced in [LP61] (and somewhat before, by Phillips, in Hilbert spaces), is that a bounded operator A is the generator of a strongly continuous contraction semigroup *iff* it is dissipative and, more generally, that an (unbounded) operator A is the generator of a strongly continuous contraction semigroup *iff* it is m-dissipative with a dense domain (Lumer-Phillips theorem)
[3] In general the domain of K is not dense in $C_b(H)$ so that (P_t) is not strongly continuous in $C_b(H)$

instance when F is regular and C is of trace class, see [DZ02, Theorem 7.5.1]). For general stochastic partial differential equations this method can be very difficult to handle.

Another possibility, which has been extensively exploited recently, is to study Kolmogorov equations in the spaces $L^p(H, \nu), p \geq 1$ (we shall choose $p = 2$ for simplicity) where ν is an invariant measure for (P_t), that is ν is a Borel probability measure in H such that

$$\int_H P_t \varphi d\nu = \int_H \varphi d\nu \quad \text{for all } \varphi \in C_b(H). \tag{4.16}$$

Concerning existence and uniqueness of invariant measures, several results are available in the literature. An important case is when the operator A is self-adjoint negative with A^{-1} of trace class and $F = -DU$ where U is a suitable function (potential). In this case we say that (4.7) is a *gradient system* and, generalizing an idea of Kolmogorov [Kol31], one can show that the measure

$$\nu(dx) = \frac{e^{-2U(x)} \mu(dx)}{\int_H e^{-2U(y)} \mu(dy)}, \tag{4.17}$$

where μ is the Gaussian measure $\mu = N_{-\frac{1}{2} A^{-1}}$, is invariant. This situation is important in the applications since in this case P_t is symmetric in $L^2(H, \nu)$ and the Markov process $X(t, x)$ is reversible. However, there are situations, as for Burgers and Navier-Stokes equations, where the formula above does not hold and an explicit expression of the invariant measure for (P_t) is not known.

If H is finite dimensional, then classical results about existence and uniqueness of invariant measures are due to R. Z. Khas'minskii [Kha80]. These results have been considerably generalized requiring minimal regularity assumptions for the coefficients of K_0 in [BKR01], where some situation in infinite dimension have been studied.

Several papers have also been devoted to proving existence and uniqueness of invariant measures and their properties as ergodicity, mixing etc, for specific differential stocastic equations. We mention here: reaction-diffusion equations (see [DZ96], [Hai02]), Burgers equations (see [DDT94], [WKMS00]), wave equations [CDF97], [BD02a], Stefan problem [BD02b], porous media equations [DR02].

Moreover, much attention has been payed to Navier-Stokes equation. Here we mention [FM95], where existence and uniqueness of an invariant measure were established using irreducibility and strong Feller property of (P_t) when C is not "too degenerate". Later, in [Mat99], [Mat02], [KS01], [HM], the case of a degenerate C could be handled using a coupling argument.

A typical tool for the existence of an invariant measure is a bound (independent of t) for $\mathbb{E}(|X(t, x)|_Y^2)$ (or some other positive increasing function of $X(t, x)$) where Y is a Banach space compactly embedded[4] in X. Then the

---

[4] That is, the identity $Y \hookrightarrow X$ is a compact operator. Recall that an operator is *compact* if it maps the unit ball into a compact set

Krylov-Bogoliubov theorem will imply the existence. Concerning uniqueness the more used approaches are either to show that (P_t) is irreducible and strong Feller or to use a coupling argument.

We recall that (P_t) is *irreducible* if $P_t\chi_I(x) > 0$ for all $x \in H$ and all open set I where χ_I is the characteristic function of I. It is important to notice that when (P_t) is irreducible, any invariant measure ν is *full*, that is $\nu(I) > 0$ for any open set $I \subset H$. The following result, due to Doob and Khas'minskii, is important, see e.g. [DZ96, Theorem 4.2.1, Proposition 4.1.1].

Theorem 4.1.2 *Assume that (P_t) is irreducible, strong Feller and that it possesses an invariant measure ν. Then ν is unique, ergodic and strongly mixing, that is for all $x \in X$ and $\varphi \in C_b(H)$ we have*

$$\lim_{t\to+\infty} P_t\varphi(x) = \int_H \varphi(y)\nu(dy). \tag{4.18}$$

In several situations, irreducibility holds when the deterministic controlled system

$$y'(t) = Ay(t) + F(y(t)) + C^{1/2}\,u(t), \quad x \in H,$$

is *approximatively controllable* in any time $T > 0$. This means that for any $x, z \in H$, $T > 0$ and $\varepsilon > 0$ there exists a control $u \in L^2(0, T; H)$ such that $|y(T, x) - z| < \varepsilon$, see e.g. [DZ96, §7.2, §7.3]

To study the strong Feller property, a formula due to J. M. Bismut [Bis84] and D. Elworthy [Elw92] is a very useful tool, see [PZ95] for an infinite-dimensional generalization. This formula reads (formally) as follows (see e.g. [DZ96, §7.1]):

$$\langle DP_t\varphi(x), h \rangle = \frac{1}{t}\,\mathbb{E}\left[\varphi(X(t,x))\int_0^t \langle C^{-1/2}DX(s,x)\cdot h, dW(s)\rangle\right] \tag{4.19}$$

Assume now that there exists an invariant measure ν for (P_t), that is that (4.16) holds. Since, by the Hölder inequality, we have that

$$(P_t\varphi(x))^2 \le P_t(\varphi^2(x)), \quad \varphi \in C_b(H), \ x \in H,$$

we deduce that

$$\int_H (P_t\varphi)^2 d\nu \le \int_H P_t(\varphi^2) d\nu = \int_H \varphi^2 d\nu, \quad \varphi \in C_b(H),$$

thanks to the invariance of ν. Therefore, one can extend uniquely (P_t) to a strongly continuous semigroup of contractions in $L^2(H, \nu)$ (still denoted by (P_t)). We shall denote by K_2 its infinitesimal generator. Then we have to face the problem to see the relationship between the abstract operator K_2 and the differential operator K_0 defined by (4.5). For this, it is convenient to define arealization of K_0 in a space of functions depending only on a finite number of variables. We shall choose an algebra of *exponential functions*:

$$\mathcal{E}_A(H) = \text{Span}\{\varphi_h, \quad h \in D(A^*)\}, \tag{4.20}$$

where

$$\varphi_h : H \ni x \mapsto \varphi_h(x) = e^{i\langle h, x\rangle} \tag{4.21}$$

and A^* is the adjoint of A. This space is dense in $L^2(H, \nu)$. If $\varphi_h(x) = e^{i\langle h, x\rangle}$ we have

$$K_0\varphi_h(x) = -\left(\frac{1}{2}|C^{1/2}h|^2 + i\langle x, A^*h\rangle + i\langle F(x), h\rangle\right)\varphi_h(x), \quad x \in H.$$

So, $K_0\varphi_h$ belongs to $L^2(H, \nu)$ provided

$$x \mapsto \langle x, A^*h\rangle \in L^2(H, \nu), \quad \text{and} \quad x \mapsto \langle F(x), h\rangle \in L^2(H, \nu). \tag{4.22}$$

From now on we shall assume that (4.22) holds; we shall see in §4 below that it is fulfilled in several applications.

It is not difficult, using the Itô formula, to show that K_2 is an extension of K_0. More difficult (in some case, an open problem) is to show that K_2 is the closure of K_0 or, equivalently, that $\mathcal{E}_A(H)$ is a core[5] for K_2. Assume that this is the case. Then we can prove existence and uniqueness of a *strong* solution (in the sense of Friedrichs) to (4.6). That is, for any $\lambda > 0$ and any $f \in L^2(H, \nu)$ there exists a sequence $\{\varphi_n\} \subset \mathcal{E}_A(H)$ such that

$$\varphi_n \to \varphi, \quad \lambda\varphi_n - K_0\varphi_n \to f \quad \text{in } L^2(H, \nu).$$

Moreover, we can extend to K_2 several properties of K_0. Perhaps the most important is the following integration by parts formula (in French the "égalité du carré du champ").

$$\int_H K_2\varphi\,\varphi\,d\nu = -\frac{1}{2}\int_H |C^{1/2}\,D\varphi|^2 d\nu, \quad \varphi \in D(K_2). \tag{4.23}$$

We notice that identity (4.23) holds, by a straightforward proof, when $\varphi \in \mathcal{E}_A(H)$. In fact in this case one can check, by a direct computation, that the following identity holds

$$K_0(\varphi^2) = 2K_0\varphi\,\varphi + |C^{1/2}D\varphi|^2.$$

Now, since ν is invariant, we have that $\int_H K_0(\varphi^2)d\nu = 0$, and so (4.23) follows. Since $\mathcal{E}_A(H)$ is a core, then it holds for any $\varphi \in D(K_2)$.

Identity (4.23) has several interesting consequences as we shall show in §4 below devoted to applications.

---

[5] A core of a closed operator A is a subspace D of $D(A)$ such that $\forall x \in D(A)$ there exists a sequence (x_n) in D such that $x_n \to x$ and $Ax_n \to Ax$ (strongly). Equivalently, it is a dense subspace of the Banach space $D(A)$. If T denotes a closable operator, and $\overline{T}$ its closed extension, $D(T)$ is a core for $\overline{T}$

4.1.6 When the problem is not necessarily well-posed

We finally consider the situation when problem (4.7) is not necessarily well-posed, as for instance for the Navier-Stokes equation in 3 dimensions, where the existence of a martingale solution is known but the uniqueness is not known. However, even if several solutions exist it could happen that their law be the same. If one were able to prove existence and uniqueness of a regular solution of the Kolmogorov equation (4.4) then, following the ideas in [SV79], one could prove the uniqueness in law.

This program has been performed for some Kolmogorov equations with regular (continuous or Hölder) coefficients, see [GG94], [Zam00].

For the $3D$ Navier-Stokes equation existence of a regular solution was proved in [DD03]. If the Maximum Principle were applicable, this would imply uniqueness of the solution and therefore uniqueness in law of $3D$ Navier-Stokes equation. Unfortunately, the Maximum Principle for Kolmogorov equation (4.4) is known (when H is infinite-dimensional) only for regular functions F, see [DZ02, Proposition 5.2.2] and so the problem remains open.

Coming back to the situation when problem (4.7) is not well posed, another possibility is to solve directly (4.6) on a space $L^2(H, \nu)$ where ν is a probability measure, infinitesimally invariant for K_0, that is is such that

$$\int_H K_0\varphi(x)\nu(dx) = 0 \quad \text{for all } \varphi \in \mathcal{E}_A(H).$$

In the case of gradient systems this measure is given by (4.17). Otherwise, it is useful to consider a sequence of approximating operators

$$K_n\varphi(x) = \frac{1}{2} \operatorname{Tr} [CD^2\varphi(x)] + \langle Ax + F_n(x), D\varphi(x)\rangle, \quad \varphi \in \mathcal{E}_A(H),$$

where F_n are regular approximations of F such that K_n has an invariant measure ν_n. Then one can try to show that the sequence $\{\nu_n\}$ is tight[6] and that a cluster point ν is an (infinitesimally) invariant measure for K_0.

Assume now that an infinitesimally invariant measure ν for K_0 is given. Then K_0 is dissipative[7] in $L^2(H, \nu)$ and the goal is to show that its closure K_2 is m-dissipative. This method has been used in several situations: the stochastic quantization equation in [LR98] and [DT00], reaction-diffusion equations in [DR02], porous media equations in [DR04], Cahn-Hilliard equations in [DDT94].

Assume that we know that K_2 is m-dissipative and let e^{tK_2} be the semigroup generated by K_2, in $L^2(H, \nu)$. Now, an interesting problem is to solve (in a suitable sense) the stochastic differential equation (4.7). The idea is to

---

[6] That means that for all $\varepsilon > 0$, there exists a compact subset K_ε in H s.t. $\forall n$, $\nu_n(K_\varepsilon) \geq 1 - \varepsilon$. The interest of this notion rests on the *Prohorov's theorem*: from any tight sequence one can extract a weakly convergent subsequence

[7] See the definition in footnote 2, p. 72

construct a Markov process (X_t) such that for its transition semigroup p_t defined by

$$p_t\varphi(x) := \mathbb{E}\left(\varphi\left(X(t,x)\right)\right), \quad t \geq 0, \ x \in H, \ \varphi \in B_b(H),$$

we have that $p_t\varphi = e^{tK_2}\varphi$, ν-a.e.

To solve this problem one can use the theory of Dirichlet forms developped in [FOT94], [MR92] and generalized in [Sta99] (here no symmetry or sectoriality of the underlying operators is required). In this theory the following two main ingredients are needed:

(a) K_2 has a core which is an algebra.
(b) The capacity[8] determined by K_2 is tight.

If (a) and (b) are fulfilled, there exists a Borel subset $\overline{H}$ of H such that $\nu(\overline{H}) = 1$, and for all $x \in \overline{H}$

$$\mathbb{P}[X(t,x) \in \overline{H} \quad \forall\, t \geq 0] = 1,$$

and for all probability measures ν on $(H, \mathcal{B}(H))$ with $\nu(\overline{H}) = 1$

$$\varphi(X_t) - \int_0^t K_0\varphi(X_s)ds, \quad t \geq 0,$$

is an $(\mathcal{F}_t)$-martingale, where $\mathcal{F}_t$ is the completion of the σ-algebra generated by the X_s, $s \leq t$.

4.1.7 Outline of contents

Let us outline the contents of the chapter. In §2 we shall consider the important special case when $F = 0$ which corresponds to a *linear* deterministic system perturbed by noise. In this case the transition semigroup is called the *Ornstein-Uhlenbeck semigroup*. It plays an important role in describing systems in absence of interaction, the *free* fields.

§3 is devoted to the case when the nonlinear operator F is regular. Though this case is not very interesting for applications, however it is important, as a first step, to study approximating versions of more complicated problems. Finally, in §4 we shall discuss some recent results on Kolmogorov equations arising in the applications; we shall restrict for brevity to reaction-diffusion, Burgers and 2D Navier-Stokes equations.

---

[8] Roughly speaking: K_2 allows to define an *energy* (of the type of Dirichlet's) for L^2 functions whose gradient (in the sense of distributions) is also in L^2; by minimizing energy of the functions whose absolute value is ≥ 1 on an open set, one obtains the so-called *capacity* (in Choquet's sense) of this open set; the capacity of an arbitrary set is the infimum of capacities for all open supsets. A capacity is not a measure (it is only subadditive), but it has a lot of properties in common with measures

4.2 The Ornstein-Uhlenbeck semigroup

4.2.1 Definition and assumptions

Here we assume, besides Hypothesis 4.1.1, that $F = 0$. The Kolmogorov equation (4.4) reduces in this case to

$$
\begin{cases}
D_t u(t, x) = \dfrac{1}{2} \operatorname{Tr} [CD^2 u(t, x)] + \langle Ax, Du(t, x) \rangle, & t > 0, \ x \in D(A), \\[2mm]
u(0, x) = \varphi(x), & x \in H.
\end{cases}
\tag{4.24}
$$

We shall set

$$
L_0 \varphi(x) = \dfrac{1}{2} \operatorname{Tr} [CD^2 \varphi(x)] + \langle x, A^* D\varphi(x) \rangle, \quad x \in H, \ \varphi \in \mathcal{E}_A(H). \tag{4.25}
$$

Equation (4.7) reads in this case as follows,

$$
\begin{cases}
dX(t, x) = AX(t, x)dt + C^{1/2}dW(t), & t > 0, \ x \in H, \\[2mm]
X(0, x) = x, & x \in H,
\end{cases}
\tag{4.26}
$$

whose mild solution is given by

$$
X(t, x) = e^{tA}x + W_A(t),
$$

where $W_A(t)$ is defined by (4.11). Therefore, a candidate for the solution of (4.24) is given by (4.8). Since $X(t, x)$ is a Gaussian random variable of mean $e^{tA}x$ and covariance Q_t[9], we have

$$
u(t, x) = \int_H \varphi(e^{tA}x + y)N_{Q_t}(dy), \quad t \geq 0, \ x \in H. \tag{4.27}
$$

Let us set

$$
R_t \varphi(x) = \int_H \varphi(e^{tA}x + y)N_{Q_t}(dy), \quad t \geq 0, \ x \in H, \varphi \in B_b(H). \tag{4.28}
$$

Now, we can check that $u(t, x) := R_t \varphi$ is a strict solution of (4.24) for all $\varphi \in \mathcal{E}_A(H)$. Notice that, though $\mathcal{E}_A(H)$ is not dense in $C_b(H)$, for any $\varphi \in C_b(H)$ one can find a three index sequence φ_{n_1, n_2, n_3}[10] such that (see [DT01b], Lemma 2,4])

(i) $\displaystyle \lim_{n_1 \to \infty} \lim_{n_2 \to \infty} \lim_{n_3 \to \infty} \varphi_{n_1, n_2, n_3}(x) = \varphi(x), \quad x \in H,$

(ii) $\|\varphi_{n_1, n_2, n_3}\|_0 \leq \|\varphi\|_0.$

So, it is enough to prove several properties of $R_t \varphi$ for $\varphi \in \mathcal{E}_A(H)$ only.

---

[9] Recall the discussion after (4.11), p. 70

[10] Since convergences in (i) and (ii) below are only pointwise, we cannot substitute φ_{n_1, n_2, n_3} with a sequence depending on one index by a diagonal extraction procedure

Notice that $\mathcal{E}_A(H)$ is stable for R_t. Indeed, let $\varphi_h(x) = e^{i\langle x,h\rangle}$, with $h \in D(A^*)$. Then

$$R_t\varphi_h(x) = e^{i\langle e^{tA^*}h,x\rangle} \int_H e^{i\langle y,h\rangle} N_{Q_t}(dy), \quad x \in H, \ t > 0.$$

Recalling that the Fourier transform of N_{Q_t} is given by $e^{-\frac{1}{2}\langle Q_t h,h\rangle}$, we find that

$$R_t\varphi_h = \varphi_{e^{tA^*}h}, \quad h \in H. \tag{4.29}$$

By simple computations we see that $u(t,x) = R_t\varphi(x)$ is a strict solution of (4.24) and that the semigroup law

$$R_{t+s}\varphi = R_t R_s\varphi, \quad \varphi \in \mathcal{E}_A(H),$$

holds. Consequently it holds for any $\varphi \in C_b(H)$. R_t is called the *Ornstein-Uhlenbeck semigroup*. It is clear that $R_t\varphi$ is defined, through formula (4.28), for any Borel function φ such that

$$|\varphi(x)| \le a e^{\kappa|x|}, \quad x \in H.$$

By (4.29) we see that the semigroup R_t is not strongly continuous in $C_b(H)$, unless $A = 0$. We can define its infinitesimal generator L, proceeding as in §1, through its resolvent

$$(\lambda - L)^{-1}f(x) = \int_0^{+\infty} e^{-\lambda t} R_t f(x)dt, \quad f \in C_b(H), \ t \ge 0, x \in H.$$

We denote the domain of L by $D(L)$ or $D(L, C_b(H))$. Notice that $\mathcal{E}_A(H)$ is not included in $D(L, C_b(H))$. We have in fact

$$\lim_{h\to 0} \frac{1}{h}\left(R_t\varphi_h(x) - \varphi_h(x)\right) = -\frac{1}{2}|C^{1/2}h|^2 - i\langle x, A^*h\rangle$$

and the function in the right hand side is not bounded but has linear growth. For this reason it is useful to consider R_t acting in the space $C_{b,n}(H)$, the Banach space of all continuous functions in H uniformly continuous on bounded subsets of H and such that

$$\|\varphi\|_{b,n} := \sup_{x\in H} \frac{|\varphi(x)|}{1 + |x|^{2n}} < +\infty, \quad n \in \mathbb{N}. \tag{4.30}$$

One can define, as before, the infinitesimal generator L of R_t in $C_{b,n}(H)$ whose domain we shall denote by $D(L, C_{b,n}(H))$ see [Cer95]. One check easily that if $n \ge 1$ one has $\mathcal{E}_A(H) \subset D(L, C_{b,n}(H))$ and that if $\varphi \in D(L, C_{b,n}(H))$ we have $L\varphi = L_0\varphi$, where L_0 is given by (4.25).

Remark 4.2.1 (The Gross Laplacian) When $A = 0$ the semigroup R_t is called the *heat* semigroup. In this case R_t is strongly continuous in $C_b(H)$ and its infinitesimal generator is called the *Gross Laplacian*. Notice that in this case we have $Q_t = tC$, $t > 0$. Therefore C must be of trace class and so the Gross Laplacian is always degenerate in infinite dimension.

4.2.2 Ellipticity, hypoellipticity, smoothing

Let us assume that the operator L_0 is strictly elliptic, that is $C^{-1} \in L(H)$. Then, for Hypothesis 4.1.1 to be verified, A^{-1} must be a Hilbert-Schmidt operator. When L_0 is strictly elliptic it inherits some typical properties of elliptic operators in finite dimension. In particular R_t maps Borel bounded functions into C^∞ functions, that is[11]

$$\varphi \in B_b(H),\ t > 0 \Longrightarrow P_t\varphi \in C_b^\infty(H).$$

Let us give an idea of this fact.

Proposition 4.2.2 *Assume, besides Hypotheses 4.1.1, that C is invertible and $C^{-1} \in L(H)$. Then for all $\varphi \in B_b(H)$ and $t > 0$ we have $R_t\varphi \in C_b^\infty(H)$ and*

$$\langle DR_t\varphi, h \rangle = \int_H \langle \Gamma(t)h, Q_t^{-1/2}y \rangle \varphi(e^{tA}x + y) N_{Q_t}(dy) \qquad (4.31)$$

for all $h \in H$, where

$$\Gamma(t) := Q_t^{-1/2}e^{tA}, \quad t > 0, \qquad (4.32)$$

Moreover,

$$\|DR_t\varphi\|_0 \le Ct^{-1/2}e^{\omega t}\|\varphi\|_0. \qquad (4.33)$$

Before giving a sketch of the proof let us stress the fact that (4.32) means that

$$e^{tA}(H) \subset Q_t^{1/2}(H), \quad t \ge 0, \qquad (4.34)$$

and so, the operator $\Gamma(t)$ is a bounded operator in H by the closed graph theorem. Condition (4.32) will be discussed in Lemma 4.2.3 below.

Sketch of the proof. By a change of variables we have

$$R_t\varphi(x) = \int_H \varphi(y) N_{e^{tA}x, Q_t}(dy), \quad t \ge 0,\ x \in H, \varphi \in B_b(H).$$

In order to differentiate with respect to x we use the fact that $N_{e^{tA}x, Q_t} \ll N_{Q_t}$.[12] In fact, by (4.34) it follows that we can apply the Cameron-Martin formula[13] and conclude that

$$\frac{dN_{e^{tA}x, Q_t}}{dN_{Q_t}}(y) = e^{-\frac{1}{2}|\Gamma(t)x|^2 + \langle \Gamma(t)x, Q_t^{-1/2}y \rangle}, \quad y \in H.$$

---

[11] This is not true for the heat semigroup introduced in Remark 4.2.1. As shown in [Gro67] here the smoothing property only holds in the directions of the Cameron-Martin space

[12] Recall that this notation means that $N_{e^{tA}x, Q_t}$ is *absolutely continuous* with respect to N_{Q_t}, i.e. that for all $E \subset H$ s.t. $N_{Q_t}(E) = 0$, one has $N_{e^{tA}x, Q_t}(E) = 0$

[13] Let $Q \in L^+(H)$ and $a \in Q^{1/2}(H)$. Then $N_{a,Q}$ and N_Q are equivalent measures (that is, each one is absolutely continuous with respect to the other one) and $\frac{dN_{a,Q}}{dN_Q}(y) = \exp\{-\frac{1}{2}|Q^{-1/2}a|^2 + \langle Q^{-1/2}a, Q^{-1/2}y \rangle\},\ y \in H$

Now, we can write

$$R_t\varphi(x) = \int_H \varphi(y) N_{e^{tA}x,Q_t}(dy) = \int_H e^{-\frac{1}{2}|\Gamma(t)x|^2+\langle\Gamma(t)x,Q_t^{-1/2}y\rangle}\varphi(y)N_{Q_t}(dy)$$

and differentiating with respect to x the conclusion follows. $\square$

Let us explain the meaning of the operator $\Gamma(t)$. For this it is convenient to consider the following linear system

$$y'(t) = Ay(t) + C^{1/2}\,u(t), \quad y(0) = x, \qquad (4.35)$$

where y is the *state* and u is the *control*. Moreover the quantity

$$E(u) := \int_0^T |u(s)|^2 ds$$

is called the *energy* of u. System (4.35) is said to be *null controllable* if for any $T > 0$ there exists $u \in L^2(0,T;H)$ such that $y(T) = 0$.

Lemma 4.2.3 *Assume that $C = I$. Then we have*

$$e^{tA}(H) \subset Q_t^{1/2}(H), \quad t \geq 0. \qquad (4.36)$$

Moreover, there exists $c > 0$ such that[14]

$$\|\Gamma(t)\| \leq ct^{-1/2}e^{\omega t}, \quad t > 0. \qquad (4.37)$$

Sketch of the proof. It is well known, see [Zab81], that condition (4.36) is equivalent to the null controllability of system (4.35). Moreover we have

$$|\Gamma(t)x|^2 = \min\left\{\int_0^T |u(s)|^2 ds : u \in L^2(0,T;H),\ y(T) = 0\right\}. \qquad (4.38)$$

Now, taking $u(t) = -\frac{1}{T}\,e^{tA}x$, we find $y(T) = 0$ so that system (4.35) is null controllable and so (4.36) holds. Finally (4.37) follows from (4.38). $\square$

Remark 4.2.4 Condition (4.36) may be fulfilled even if L is not strictly elliptic. In this case we say that L is *hypoelliptic* because when H is finite dimensional, condition (4.36) reduces precisely to the Hörmander's hypoellipticity condition for the operator L. In this case if $\varphi \in C_b(H)$ and $t > 0$ we still have that $P_t\varphi \in C_b^\infty(H)$. Condition (4.36) is necessary and sufficient in order that R_t be strong Feller. In the following we shall confine to the elliptic case for brevity.

We notice also that the semigroup R_t is irreducible, provided $\text{Ker}\,C = \{0\}$.

---

[14] Recall Hypothesis 4.1.1-(i), p. 70

4.2.3 Invariant measure, ergodicity, mixing

We assume here that the semigroup e^{tA} is of negative type, that is the constant ω in Hypothesis 4.1.1-(i) is negative. In this case, it is easy to see that the linear operator

$$Q_\infty x := \int_0^{+\infty} e^{tA} C e^{tA^*} x, \quad x \in H, \tag{4.39}$$

is well defined and of trace class. Moreover, one can easily check that

$$\lim_{t \to +\infty} R_t \varphi(x) = \int_H \varphi d\mu := \overline{\varphi}, \quad \varphi \in C_b(H). \tag{4.40}$$

Consequently, the measure $\mu := N_{Q_\infty}$ is the unique invariant measure for R_t and it is *ergodic* and *strongly mixing*.

Let now consider the unique extension of R_t to $L^2(H, \mu)$, still denoted by R_t, and let L_2 be its infinitesimal generator. The expression of L_2 on $\mathcal{E}_A(H)$ can be easily computed by the very definition of the infinitesimal generator. We find

$$L_2 \varphi = L_0 \varphi = \frac{1}{2} \operatorname{Tr}[CD^2 \varphi(x)] + \langle x, A^* D\varphi(x) \rangle, \quad \varphi \in \mathcal{E}_A(H). \tag{4.41}$$

In this case $\mathcal{E}_A(H)$ is a core for R_t since $\mathcal{E}_A(H)$ is invariant for R_t and it is dense in $L^2(H, \mu)$, see e.g. [Dav80]. Moreover, the integration by parts formula (4.23) becomes

$$\int_H L_2 \varphi \, \varphi \, d\mu = -\frac{1}{2} \int_H |C^{1/2} D\varphi|^2 d\mu. \tag{4.42}$$

Remark 4.2.5 It is important to know when R_t is symmetric. The necessary and sufficient condition is given by, see [CG96],

$$\langle Cx, A^* y \rangle = \langle A^* x, Cy \rangle, \quad x, y \in D(A^*). \tag{4.43}$$

In particular, this condition is fulfilled when A is self-adjoint and commutes with C.

4.2.4 Smoothing in $L^2(H, \mu)$

Let us recall the definition of the Sobolev space $W^{1,2}(H, \mu)$, where $\mu = N_{Q_\infty}$ is as before. Let D be the linear operator

$$D : \mathcal{E}_A(H) \subset L^2(H, \mu) \to L^2(H, \mu; H), \quad \varphi \to D\varphi.$$

One can show that D is closable, see e.g [DZ02, Proposition 9.2.4]. If φ belongs to the domain of the closure of D, which we shall still denote by D, we shall say that $D\varphi$ belongs to $L^2(H, \mu; H)$. Now $W^{1,2}(H, \mu)$ is the linear space of all functions $\varphi \in L^2(H, \mu)$ such that $D\varphi \in L^2(H, \mu; H)$. $W^{1,2}(H, \mu)$, endowed with the inner product

$$\langle \varphi, \psi \rangle_{W^{1,2}(H,\mu)} = \langle \varphi, \psi \rangle_{L^2(H,\mu)} + \int_H \langle D\varphi(x), D\psi(x) \rangle \mu(dx),$$

is a Hilbert space.

Proposition 4.2.6 *Let* $\varphi \in L^2(H, \mu)$. *Then for all* $t > 0$ *we have* $R_t\varphi \in W^{1,2}(H, \mu)$ *and*

$$\|DR_t\varphi\|_{L^2(H,\mu)} \le Ct^{-1/2} \|\varphi\|_{L^2(H,\mu)}. \tag{4.44}$$

Proof. Let $h \in H$. From (4.31) we find, thanks to the Hölder inequality,

$$|\langle DR_t\varphi, h \rangle|^2 \le \int_H |\langle \Gamma(t)h, Q_t^{-1/2}y \rangle|^2 N_{Q_t}(dy) \int_H \varphi^2(e^{tA}x + y) N_{Q_t}(dy).$$

The first integral in the right hand side is equal to $|\Gamma(t)h|^2$, whereas the second one is equal to $R_t(\varphi^2)(x)$. So, we have

$$|\langle DR_t\varphi, h \rangle|^2 \le |\Gamma(t)h|^2 R_t(\varphi^2)(x).$$

Taking the supremum on h, integrating with respect to μ over H and taking into account the invariance of μ, yields the conclusion.   □

Remark 4.2.7 Taking the Laplace transform of (4.44) with respect to t, we find

$$\|D(\lambda - L)^{-1}f\|_{L^2(H,\mu)} \le C\sqrt{\frac{\pi}{\lambda}} \|f\|_{L^2(H,\mu)}, \quad f \in L^2(H, \mu). \tag{4.45}$$

This implies

$$D(L_2) \subset W^{1,2}(H, \mu), \tag{4.46}$$

with continuous embedding.

4.2.5 Perturbations of the Ornstein Uhlenbeck operator

We shall only consider regular perturbations of strictly elliptic Ornstein Uhlenbeck operators. We shall assume that Hypothesis 4.1.1 holds with $\omega < 0$ and that $C^{-1} \in L(H)$. We are given a nonlinear operator $F \in C_b(H; H)$. We denote by $\mu = N_{Q_\infty}$ the unique invariant measure for the semigroup R_t.

Let us define a Kolmogorov operator K_2 in $L^2(H, \mu)$ with domain $D(L_2)$, setting

$$K_2\varphi = L_2\varphi + \langle F(x), D\varphi \rangle, \quad \varphi \in D(L_2). \tag{4.47}$$

This definition is meaningful thanks to (4.45). Obviously μ is not an invariant measure for K_2 in general, but we are going to show, following [DZ96], that there exists an invariant measure ν for K_2 which is absolutely continuous with respect to μ.

Proposition 4.2.8 K_2 *is the infinitesimal generator of a strongly continuous semigroup* e^{tK_2} *on* $L^2(H, \mu)$. *Moreover, for any* $\lambda > \pi \|F\|_0^{2\ 15}$ *we have*

$$(\lambda - K_2)^{-1} = (\lambda - L_2)^{-1}(1 - T_\lambda)^{-1}, \qquad (4.48)$$

where

$$T_\lambda \varphi = \langle F, D(\lambda - L_2)^{-1} \varphi \rangle, \qquad \varphi \in L^2(H, \mu).$$

Sketch of the proof. Given $f \in L^2(H, \mu)$ and $\lambda > 0$, consider the equation

$$\lambda \varphi - L\varphi - \langle F(x), D\varphi \rangle = f.$$

Setting $\lambda \varphi - L\varphi = \psi$, we obtain $\psi - T_\lambda \psi = f$. But by (4.45) we have that $\|T_\lambda \psi\|_0 \le \sqrt{\frac{\pi}{\lambda}} \|F\|_0$. Thus, the conclusion follows from the contraction principle. $\square$

We now consider the adjoint semigroup $e^{tK_2^*}$ of e^{tK_2}. We denote by Σ^* the set of all its stationary points:

$$\Sigma^* = \left\{ \varphi \in L^2(H, \mu) : \ e^{tK_2^*} \varphi = \varphi, \ t \ge 0 \right\}.$$

It is easy to see that Σ^* is a lattice. Now we can prove (see [DZ02]):

Proposition 4.2.9 *There exists an invariant measure* ν *for* e^{tK_2} *absolutely continuous with respect to* μ.

Sketch of the proof. Let $\lambda > 0$ be fixed and let φ_0 be the function identically equal to 1. Clearly $\varphi_0 \in D(K_2)$ and we have $K_2 \varphi_0 = 0$. Consequently $1/\lambda$ is an eigenvalue of $(\lambda - K_2)^{-1}$ since $(\lambda - K_2)^{-1} \varphi_0 = \frac{1}{\lambda} \varphi_0$. Moreover $1/\lambda$ is a simple eigenvalue because μ is ergodic (recall (4.40)). Since the embedding $W^{1,2}(H, \mu) \subset L^2(H, \mu)$ is compact, see [DMN92] and [Pes93], and $D(L_2) \subset W^{1,2}(H, \mu)$ by (4.46), it follows that $(\lambda - K_2)^{-1}$ is compact as well for any $\lambda > 0$. Therefore $(\lambda - K_2^*)^{-1}$ is compact and $1/\lambda$ is a simple eigenvalue for $(\lambda - K_2^*)^{-1}$. Consequently there exists $\rho \in L^2(H, \mu)$ such that

$$(\lambda - K_2^*)^{-1} \rho = \frac{1}{\lambda} \rho.$$

It follows that $\rho \in D(K_2^*)$ and $K_2^* \rho = 0$, then $\rho \in \Sigma^*$. Since Σ^* is a lattice, ρ can be chosen to be nonnegative and such that $\int_H \rho d\mu = 1$. Now $\nu(dx) = \rho(x) \mu(dx)$ is an invariant measure for e^{tK_2}. $\square$

---

[15] We set $\|F\|_0 = \sup_{x \in H} |F(x)|$

4.3 Regular nonlinearities

4.3.1 Introduction

We are here concerned with the Kolmogorov equation (4.4) under Hypothesis 4.1.1 where we take $M = 1$ so that $\|e^{tA}\| \leq e^{t\omega}$, $t \geq 0$. We suppose moreover that

Hypothesis 4.3.1 F is Lipschitz continuous and of class C^1.

Let us denote by κ the minimal number such that

$$\langle F(x) - F(y), x - y \rangle \leq \kappa |x - y|^2, \quad x, y \in H.$$

Clearly $|\kappa| \leq K$ where K is the Lipschitz constant of F.

Proposition 4.3.2 *Assume that Hypotheses 4.1.1 and 4.3.1 hold. Then for any $x \in H$ there exists a unique mild solution $X(\cdot, x)$ of (4.7) (that is a solution of the integral equation (4.10)). Moreover, $X(t, x)$ is differentiable with respect to x and for any $h \in H$ we have $DX(t, x) \cdot h = \eta^h(t, x)$, where $\eta^h(t, x)$ is the mild solution of the equation*

$$\begin{cases} \dfrac{d}{dt}\, \eta^h(t, x) = A\eta^h(t, x) + DF(X(t, x)) \cdot \eta^h(t, x), \\[2mm] \eta^h(0, x) \quad = h. \end{cases} \tag{4.49}$$

Finally,

$$|\eta^h(t, x)| \leq e^{(\omega + \kappa)t}|h|, \quad t \geq 0. \tag{4.50}$$

Sketch of the proof. The mild (4.10) can be considered as a family of deterministic integral equations with Lipschitz nonlinearities, indexed by the points of the probability space Ω. Thus, by standard fixed point arguments, one can show that it has a unique solution $X(t, x)$ and that $X(t, x)$ is differentiable with respect to x. Let us show (4.50). Recalling that $\langle Ax, x \rangle \leq \omega |x|^2$, $x \in D(A)$, by assumption and using Hypothesis 4.3.1, it follows that

$$\langle DF(y) \cdot x, x \rangle \leq (\omega + \kappa)|x|^2, \quad x \in D(A), \ y \in H.$$

Consequently, multiplying both sides of the first equation in (4.49) by $\eta^h(t, x)$ we find

$$\frac{1}{2}\frac{d}{dt}\, |\eta^h(t, x)|^2 \leq (\omega + \kappa)|\eta^h(t, x)|^2,$$

which, by a standard comparison result, yields inequality (4.50). $\square$

Now, we can define the transition semigroup (P_t) by (4.14); it is easy to see that (P_t) is Feller. Moreover if $C = I$, in view of the Bismut-Elworthy formula (4.19), it follows that (P_t) is strong Feller.

4.3.2 Invariant measures

Let us assume that the system is *dissipative*, that is $\omega + \kappa < 0$ and set $\omega_1 = -\kappa - \omega$. Then the following result holds, see [DZ96, Theorem 6.3.3].

Theorem 4.3.3 *The law $\mathcal{L}(X(t,x))$ of $X(t,x)$ is convergent as $t \to +\infty$ to the unique invariant measure ν of (P_t). Moreover ν is ergodic, and*

$$\lim_{t \to +\infty} P_t \varphi(x) = \int_H \varphi(y) \nu(dy), \quad x \in H, \ \varphi \in C_b(H). \tag{4.51}$$

Sketch of the proof. It is convenient to consider the solution $X(t,s,x)$ of (4.7) with initial time $s \in \mathbb{R}$,

$$\begin{cases} dX(t,x,s) = (AX(t,s,x) + F(X(t,s,x))) \, dt + C^{1/2} dW(t), \quad t > s, \ x \in H, \\ X(s,s,x) = x, \quad x \in H, \end{cases}$$
$$\tag{4.52}$$

Then, taking advantage of the dissipativity it is possible to show that there exists the limit (independent of t)

$$\lim_{s \to -\infty} X(t,s,x) := \eta \quad \text{in } L^2(\Omega, \mathcal{F}, \mathbb{P}).$$

Consequently, the law ν of η is the invariant measure and fulfills the required conditions.   □

4.4 Some Kolmogorov equations arising in the applications

In this section we shall consider some realizations of (4.4) for which there exists a unique mild solution $X(t,x)$ and an invariant measure ν. Then we shall consider the unique extension in $L^2(H,\nu)$ of the transition semigroup (P_t) defined by (4.14), whose infinitesimal generator we shall denote by K_2. Our goal is to show that the algebra of all exponential functions $\mathcal{E}_A(H)$ is a core for K_2 and then to deduce some qualitative properties of (P_t).

Let us explain the general idea of the proof. We consider an approximating equation

$$\lambda \varphi_n(x) - L\varphi_n(x) - \langle F_n(x), D\varphi_n(x) \rangle = f(x), \tag{4.53}$$

where $\lambda > 0$ and $f \in \mathcal{E}_A(H)$ are given. Here F_n are regular approximations of F, for instance the finite dimensional Galerkin approximations or the Yosida approximations (when F is dissipative)[16].

---

[16] The Yosida approximation of F is defined by $F_n = n(J_n - 1)$ where $J_n = (1 - \frac{1}{n}F)^{-1}$ see e.g. [Bré73]

Since F_n is regular, we can use the results proved in §3 and show that (4.53) has a unique solution φ_n. If F_n is bounded we have that $L\varphi_n$ is bounded as well and so $\varphi_n \in D(L, C_b(H)) \cap C_b^1(H)$, whereas if F_n is Lipschitz then $L\varphi_n$ belongs to $C_{b,1}(H)$ so that $\varphi_n \in D(L, C_{b,1}(H)) \cap C_b^1(H)$[17].

Then we proceed in the following steps:

(i) We prove, using the Itô formula, that K_2 is an extension of K_0. This implies that K_0 is dissipative in $L^2(H, \nu)$ and consequently it is closable; we call $\overline{K_0}$ its closure.

(ii) We prove that φ_n belongs to the domain of $\overline{K_0}$. This requires a suitable approximation of φ_n by functions $\varphi_{n,k}$ in $\mathcal{E}_A(H)$ such that $\varphi_{n,k}$ belongs to $D(L, C_b(H)) \cap C_b^1(H)$ (when F_n is bounded) or to $D(L, C_{b,1}(H)) \cap C_b^1(H)$ (when F_n is Lipschitz). This approximation is provided by [DT01b, Proposition 2.7].

As a consequence, we can write (4.53) in the form

$$\lambda\varphi_n - \overline{K_0}\varphi_n = f + \langle F_n(x) - F(x), D\varphi_n \rangle. \tag{4.54}$$

(iii) We show that

$$\lim_{\lambda \to 0} \langle F_n(x) - F(x), D\varphi_n(x) \rangle = 0 \quad \text{in } L^2(H, \nu). \tag{4.55}$$

This requires an estimate, uniform in n,

$$|\langle F_n(x), D\varphi_n \rangle|^2 \leq |G(x)|, \tag{4.56}$$

where G is integrable with respect to ν. Then (4.56) yields (4.55) by the dominated convergence theorem.

Finally, by (4.55) it follows that the closure of the range of $\lambda - \overline{K_0}$ includes $\mathcal{E}_A(H)$ and so it is dense in $L^2(H, \nu)$. Now, the Lumer-Phillips theorem (see [LP61] and footnote 2 above, p. 72), implies that $\overline{K_0} = K_2$.

To prove (4.56) we notice that the solution φ_n of (4.53) is given by

$$\varphi_n(x) = \int_0^{+\infty} e^{-\lambda t} \mathbb{E}\left[f\left(X_n(t, x) \right) \right] dt, \quad x \in H,$$

where $X_n(t, x)$ is the solution of (4.4) with F_n replacing F. Moreover, for any $h \in H$, we have

$$\langle D\varphi_n(x), h \rangle = \int_0^{+\infty} e^{-\lambda t} \mathbb{E}\left[\langle Df(X_n(t, x)), \eta_n^h(t, x) \rangle \right] dt, \quad x \in H,$$

where $\eta_n^h(t, x) = DX_n(t, x) \cdot h$, so that

$$|\langle D\varphi_n(x), h \rangle| \leq \|Df\|_0 \int_0^{+\infty} e^{-\lambda t} \mathbb{E}\left[|\eta_n^h(t, x)| \right] dt, \quad x \in H.$$

So, the key point is to find an estimate for $\mathbb{E}\left[|\eta_n^h(t, x)| \right]$.

In §4.1 we apply this procedure to reaction-diffusion equations, in §4.2 to the Burgers equation and in §4.3 to the Navier-Stokes equation in dimension 2.

---

[17] For the definition of $C_{b,1}(H)$ see the end of §2.1 and (4.30)

4.4.1 Reaction-diffusion equations

We are here concerned with the following stochastic reaction-diffusion equation in the domain $D = [0,1]^n$,

$$\begin{cases} dX(t) = (AX(t) + p\,(X(t)))\,dt + C^{1/2}\,dW(t), \\ \\ X(0) \;= x. \end{cases} \tag{4.57}$$

Here A denotes the realization of the Laplace operator Δ with Dirichlet boundary conditions, in the Hilbert space $H = L^2(D)$, that is

$$Ax = \Delta_\xi x, \quad D(A) = H^2(D) \cap H^1_0(D).$$

As well known, A is self-adjoint in $L^2(D)$, it possesses a complete set of eigenfunctions $\{e_k\}$ where

$$e_k(\xi) = (2/\pi)^{n/2}\,\sin(\pi k_1 \xi_1) \cdots (\sin \pi k_n \xi_n), \quad \xi \in D$$

and

$$Ae_k = -\pi^2 |k|^2\,e_k, \quad k = (k_1, \ldots, k_n) \in \mathbb{N}^n.$$

Moreover, p is a polynomial of odd degree and with negative leading coefficient; we notice that there exists $\lambda \in \mathbb{R}$ such that $p'(\xi) \geq -\lambda$. Finally, we choose $C = (-A)^{-\gamma/2}$ where γ is such that $\gamma > \frac{n}{2} - 1$. Thus, we can take $\gamma = 0$ and $C = I$ only for $n = 1$.

Notice that

$$\|e^{tA}\| \leq e^{-\pi^2 t}, \quad t \geq 0.$$

so that Hypothesis 4.1.1-(i) is fulfilled with $M = 1$ and $\omega = -\pi^2$. Moreover, since

$$\mathrm{Tr}\,[(-A)^{-(1+\gamma)}] = |\pi|^{-2(1+\gamma)} \sum_{k \in \mathbb{N}^n} |k|^{-2(1+\gamma)} < +\infty,$$

Hypothesis 4.1.1-(ii) is fulfilled as well.

We are going to solve (4.57) following [DZ96]. For more general reaction-diffusion equations (including systems), see the monograph [Cer01].

To solve (4.7) we introduce an approximating problem,

$$\begin{cases} dX_\alpha(t) = (AX_\alpha(t) + F_\alpha\,(X_\alpha(t)))\,dt + (-A)^{-\gamma/2}dW(t), \\ \\ X_\alpha(0) = x \in H, \end{cases} \tag{4.58}$$

where for any $\alpha > 0$, F_α is defined by $F_\alpha(x)(\xi) = p_\alpha(x(\xi))$ and p_α are the Yosida approximations of p (see footnote(9) with α replacing $1/n$).

F_α is Lipschitz continuous, with Lipschitz constant $\frac{2}{\alpha}$. Thus, in view of Proposition 4.3.2, for any $\alpha > 0$, and any $x \in H$, problem (4.58) has a unique solution $X_\alpha(\cdot, x)$. From which we deduce:

Proposition 4.4.1 *If $x \in L^{2d}(D)$, problem (4.57) has a unique mild solution $X(\cdot, x)$. Moreover for any $m \in \mathbb{N}$, there is $c_{m,p,T} > 0$ such that*

$$\mathbb{E}\left(|X(t,x)|_{L^{2d}(D)}^{2m} \right) \le c_{m,p,T}\left(1 + |x|_{L^{2d}(D)}^{2m} \right).$$

If $x \in L^2(D)$ there exists a sequence $\{x_n\} \subset L^{2d}(D)$ convergent to x in $L^2(D)$ and such that $\{X(\cdot, x_n)\}$ converges to $X(\cdot, x)$ in quadratic mean. In this case we call X a generalized solution.

Sketch of the proof. We reduce (4.58) to a family of deterministic integral equations. That is, setting $Y_\alpha(t) = X_\alpha(t) - W_A(t)$, we obtain

$$\begin{cases} Y'_\alpha(t) = AY_\alpha(t) + F_\alpha\left(Y_\alpha(t) + W_A(t) \right), \ t \in [0,T], \\ Y_\alpha(0) = x. \end{cases} \tag{4.59}$$

Using the monotonicity of p we find the estimate

$$|Y_\alpha(t)|_{L^{2d}(D)} \le e^{(\lambda - \pi^2)t}|x|_{L^{2d}(D)} + \int_0^t e^{(\lambda - \pi^2)(t-s)}|F(W_A(s)|_{L^{2d}(D)}ds$$

and, using some properties of the stochastic convolution, see [DZ96], we find that

$$\mathbb{E}\left(|X_\alpha(t,x)|_{L^{2d}(D)}^{2d} \right) \le c\left(|x|_{L^{2d}(D)}^{2d} + 1 \right), \quad t \in [0,T]. \tag{4.60}$$

Now we can prove that the sequence $\{X_\alpha\}$ is convergent. In fact, using again monotonicity, we find that for $\alpha > \beta > 0$,

$$\frac{1}{2}\frac{d}{dt}\mathbb{E}\left(|X_\alpha(t,x) - X_\beta(t,x)|^2 \right) \le C_1 \, \alpha \left(|x|_{L^{2d}(D)}^{2d} + 1 \right).$$

Then there exists the limit in $L^2(\Omega, \mathcal{F}, \mathbb{P}; H)$

$$X(t) = \lim_{\alpha \to 0} X_\alpha(t), \quad \text{uniformly in } t \in [0,T].$$

Using (4.60), we can take limits of both sides of the equation

$$X_\alpha(t) = e^{tA}x + \int_0^t e^{(t-s)A}F_\alpha(X_\alpha(s))ds + W_A(t), \quad t \ge 0,$$

and we find that X fulfills (4.8). □

Let now $\lambda \le 0$. Since

$$\langle A(x-y) + p(x) - p(y), x - y \rangle \le (-\pi^2 + \lambda)|x - y|^2, \quad x,y \in D(A) \cap L^{2d}(D),$$

system (4.57) is dissipative and, proceeding as in §3.2, we find that there is a unique invariant measure ν. If $\lambda > 0$, one can show again the existence of an invariant measure, see [DZ96]. Moreover, if $C = I$ one can prove that (P_t) is irreducible and strong Feller so that there is a unique ergodic invariant measure and (4.18) holds.

We now show (see [DZ02] and [DDG02]), that $\mathcal{E}_A(H)$ is a core of K_2. Following the program outlined at the beginning of this section, we need to prove estimate (4.56). This is not difficult in this case. Let us consider in fact $\eta_n^h(t, x) = DX_n(t, x) \cdot h$ which is the solution of (4.49) (with η_n^h replacing η^h). Moreover, arguing as in the proof of Proposition 4.3.2, we find that

$$|\eta_n^h(t, x)| \le e^{(\lambda - \pi^2)t}|h|, \quad h, x \in H, \ t \ge 0, \tag{4.61}$$

and so (4.56) follows. Now we can prove the following result. In its formulation we assume for simplicity that $C = I$ and $\lambda - \pi^2 < 0$ but the result holds when $C^{-1} \in L(H)$ and $\lambda \in \mathbb{R}$ if the degree of the polynomial is greater than 1, see [DDG02]. We set $\omega_1 = \pi^2 - \lambda$.

Theorem 4.4.2 *Assume that $C = I$ and $\omega_1 > 0$. Then K_2 is the closure of K_0. Moreover the following statements hold.*

(i) *The operator $D : \mathcal{E}_A(H) \subset L^2(H, \nu) \to L^2(H, \nu; H)$ is closable. We shall still denote by D its closure and by $W^{1,2}(H, \mu)$ its domain.*

(ii) *$D(K_2) \subset W^{1,2}(H, \nu)$; moreover for all $\varphi \in D(K_2)$ we have*

$$\int_H K_2\varphi \, \varphi \, d\nu = -\frac{1}{2} \int_H |D\varphi|^2 d\nu. \tag{4.62}$$

(iii) *(Poincaré's inequality) For any $\varphi \in W^{1,2}(H, \nu)$ we have*

$$\int_H |\varphi - \overline{\varphi}|^2 d\nu \le \frac{1}{2\omega_1} \int_H |D\varphi|^2 d\nu, \tag{4.63}$$

     where

$$\overline{\varphi} = \int_H \varphi d\nu.$$

(iv) *(Exponential convergence to equilibrium) For all $\varphi \in L^2(H, \nu)$ we have*

$$\int_H |P_t\varphi - \overline{\varphi}|^2 d\nu \le e^{-2\omega_1 t} \int_H |\varphi|^2 d\nu, \quad t \ge 0. \tag{4.64}$$

(iv) *(Log-Sobolev inequality) For all $\varphi \in L^2(H, \nu)$ we have*

$$\int_H \varphi^2 \log(\varphi^2) d\mu \le \frac{1}{\omega_1} \int_H |D\varphi|^2 d\mu + \|\varphi\|_2^2 \log(\|\varphi\|_2^2). \tag{4.65}$$

Sketch of the proof. The starting point is the identity (4.23) (which implies (4.62)) from which it follows that the operator D is closable, see [DDG02, Proposition 3.5]. Also, (4.62)) easily implies that

$$\int_H |P_t\varphi|^2 \, d\nu + \int_0^t ds \int_H |DP_s\varphi|^2 \, d\nu = \int_H |\varphi|^2 \, d\nu, \qquad (4.66)$$

for all $\varphi \in L^2(H,\nu)$. Letting $t \to \infty$ and recalling (4.18), we find that

$$(\overline{\varphi})^2 + \int_0^{+\infty} ds \int_H |DP_s\varphi|^2 \, d\nu = \int_H |\varphi|^2 \, d\nu,$$

which is equivalent to

$$\int_H |\varphi - \overline{\varphi}|^2 d\nu = \int_0^{+\infty} ds \int_H |DP_s\varphi|^2 \, d\nu. \qquad (4.67)$$

Now, by the identity

$$DP_t\varphi(x) \cdot h = \mathbb{E}[\langle D\varphi(X(t,x)), \eta^h(t,x)\rangle]$$

and (4.61), we see that

$$|DP_t\varphi(x)|^2 \le e^{-2\omega_1 t} P_t(|D\varphi|^2)(x), \quad x \in H, t \ge 0.$$

Finally, by (4.67) and the invariance of the measure ν, it follows that

$$\int_H |\varphi - \overline{\varphi}|^2 d\nu \le \int_0^{+\infty} ds e^{-\omega_1 s} \int_H |D\varphi|^2 \, d\nu,$$

which yields (4.64). The proof of (4.65) is similar. □

4.4.2 Burgers equation

We are here concerned with the stochastic Burgers equation in the interval $[0,1]$ with Dirichlet boundary conditions,

$$\begin{cases} dX(t) = (AX(t) + b\,(X(t)))\,dt + dW(t), \\ \\ X(0) \;= x, \end{cases} \qquad (4.68)$$

where

- $A = D_\xi^2$, $\quad D(A) = H^2(0,1) \cap H^1(0,1)$,
- $b(x) = D_\xi(x^2)$, $\quad D(b) = H_0^1(0,1)$,
- W is a cylindrical process on $H = L^2(0,1)$.

Problem (4.68) has a unique solution which we denote by $X(t, x)$, see [DDT94]. Moreover there exists a unique invariant measure, see [DDT94] and [DG].

To prove that $\mathcal{E}_A(H)$ is a core for K_2 we need again to find an estimate for $\mathbb{E}(|\eta_n^h(t, x)|)]$. In this case the estimate is tricky, see [DD06],

$$\mathbb{E}(|\eta_n^h(t, x)|_{H^1(D)}) \leq c(|x|_{L^4(D)} + 1)^7 e^{ct}(t^{-7/8} + 1).$$

Proceeding as before, we can conclude that $\mathcal{E}_A(H)$ is a core for K_2, we can define the Sobolev space $W^{1,2}(H, \nu)$ and prove identities (4.62), (4.66) and (4.67). It follows that $P_t \varphi \in W^{1,2}(H, \nu)$ for almost all t but we do not know if the Poincaré inequality holds and if there is a gap in the spectrum of K_2.

4.4.3 2D-Navier-Stokes equation

We are concerned with the problem

$$\begin{cases} dX(t) = (AX(t) + b(X(t))dt + C^{1/2}dW(t), & t > 0 \\ X(0) = x, \end{cases} \tag{4.69}$$

where $D = [0, 2\pi]^2$,

$$H = \{x \in (L^2(D))^2 : \text{ div } x = 0 \text{ in } D\}, \quad V = (H_\#^1(D))^2 \cap H,$$

and the subscript $\#$ means periodicity.

The operator $C \in L(H)$ is nonnegative, symmetric and of trace class and W is an H-valued Wiener process.

Moreover, A is the Stokes operator

$$A = \nu_0 P \Delta, \quad D(A) = V \cap (H_\#^2(D))^2, \quad V = \{y \in (H_\#^1(D))^2 : \text{ div } y = 0\},$$

where P is the projector on the divergence free vectors, and b is defined by

$$\langle b(x), y \rangle = \sum_{i,j=1}^{2} \int_D x_i D_i x_j y_j d\xi, \quad x \in H_\#^1(D), \ y \in L^2(D).$$

Also in this case the estimate of $|\eta_n^h(t, x)|$ is tricky. One can show, see [BDD04a], that there exists $\kappa > 0, \delta \in (0, 2)$ such that

$$|\eta_n^h(t, x)|^2 \leq e^{\kappa \int_0^t |AX(s,x)|^\delta ds} |h|^2, \quad x, h \in H \tag{4.70}$$

and moreover that if $\alpha \leq \frac{1}{\|C\|}$ there exists $\omega_\alpha > 0$ such that

$$\mathbb{E}\left(|\eta^h(t, x)|^2\right) \leq e^{2\omega_\alpha t} e^{\alpha \|x\|^2} |h|^2, \quad t \geq 0 \quad x, h \in H. \tag{4.71}$$

Then, we can proceed as before and prove identity (4.23).

Remark 4.4.3 Spectral gap, and exponential convergence to equilibrium has been proved in [WMS01], [HM].

References

[AR90]   Albeverio, S., Röckner, M.: New developments in the theory and appli-
            cations of Dirichlet forms in Stochastic processes, Physics and Geom-
            etry. In: Albeverio, S. et al. (eds), World Scientific, Singapore, 27–76
            (1990)
[AR91]   Albeverio, S., Röckner, M.: Stochastic differential equations in infinite
            dimensions: solutions via Dirichlet forms. Probab. Theory Relat. Fields,
            89, 347–386 (1991)
[BD02a]   Barbu, V., Da Prato, G.: The stochastic nonlinear damped wave equa-
            tion. Appl. Math. Optimiz., **46**, 125–141 (2002)
[BD02b]   Barbu, V., Da Prato, G.: The two phase stochastic Stefan problem.
            Probab. Theory Relat. Fields, **124**, 544–560 (2002)
[BDD04a]   Barbu, V., Da Prato, G., Debussche, A.: Essential m-dissipativity
            of Kolmogorov operators corresponding to periodic 2D-Navier Stokes
            equations. Atti Accad. Naz. Lincei Cl. Sci. Fis. Mat. Natur. Rend. Lin-
            cei (9) Mat. Appl., **15**, 29–38 (2004)
[BDD04b]   Barbu, V., Da Prato, G., Debussche, A.: The Kolmogorov equation as-
            sociated to the stochastic Navier-Stokes equations in 2D. Infin. Dimens.
            Anal. Quantum Probab. Relat. Top., **7**, 163–182 (2004)
[Bis84]   Bismut, J.M.: Large deviations and the Malliavin calculus, Birkhäuser,
            Boston, MA (1984)
[BKR01]   Bogachev, V.I., Krylov, N.V., Röckner, M.: On regularity of transition
            probabilities and invariant measures of singular diffusions under mini-
            mal conditions. Comm. Partial Diff. Equations, **26**, 2037–2080 (2001)
[BR95]   Bogachev, V.I., Röckner, M.: Regularity of invariant measures on finite
            and infinite-dimensional spaces and applications. J. Funct. Anal., **133**,
            168–223 (1995)
[BR01]   Bogachev, V.I., Röckner, M.: Elliptic equations for measures on infinite-
            dimensional spaces and applications. Probab. Theory Relat. Fields,
            120, 445–496 (2001)
[Bré73]   Brézis, H.: Opérateurs maximaux monotones et semi-groupes de con-
            tractions dans les espaces de Hilbert. North-Holland Mathematics
            Studies, No. 5. Notas de Matemtica (50). North-Holland Publishing
            Co., Amsterdam, London. Elsevier, New York (1973)
[CDF97]   Crauel, H., Debussche, A., Flandoli, F.: Random attractors. J. Dynam.
            Differential Equations, **9**, 307–341 (1997)
[Cer94]   Cerrai, S.: A Hille–Yosida theorem for weakly continuous semigroups.
            Semigroup Forum, **49**, 349–367 (1994)
[Cer95]   Cerrai, S.: Weakly continuous semigroups in the space of functions with
            polynomial growth. Dyn. Syst. Appl., **4**, 351–372 (1995)
[Cer01]   Cerrai, S.: Second order PDE's in finite and infinite dimensions: A prob-
            abilistic approach. Lecture Notes in Mathematics, vol. 1762, Springer,
            Berlin (2001)
[CG95]   Chojnowska-Michalik, A., Goldys, B.: Existence, uniqueness and in-
            variant measures for stochastic semilinear equations. Probab. Theory.
            Relat. Fields, **102**, 331–356 (1995)
[CG96]   Chojnowska-Michalik, A., Goldys, B.: On regularity properties of non-
            symmetric Ornstein-Uhlenbeck semigroup in L^p spaces. Stochastics and
            Stochastic Reports, **59**, 183–209 (1996)

[Dal96]   Daleckij, Yu.: Differential equations with functional derivatives and stochastic equations for generalized random processes. Dokl. Akad. Nauk SSSR, **166**, 1035–1038 (1996)

[Dav80]   Davies, E.B.: One parameter semigroups. Academic Press, London, New York (1980)

[DD03]   Da Prato, G, Debussche, A.: Ergodicity for the $3D$ stochastic Navier-Stokes equations. Journal Math. Pures Appl., **82**, 877–947 (2003)

[DD06]   Da Prato, G., Debussche, A.: m-dissipativity of Kolmogorov operators corresponding to Burgers equations with space-time white noise. Potential Analysis (to appear)

[DDG02]   Da Prato, G., Debussche, A., Goldys, B.: Invariant measures of non symmetric dissipative stochastic systems. Probab. Theory Relat. Fields, **123**, 355–380 (2002)

[DDT94]   Da Prato, G., Debussche, A., Temam, R.: Stochastic Burgers equation. Nonlinear Diff. Eq. Appl. **1**, 389–402 (1994)

[DDT04]   Da Prato, G., Debussche, A., Tubaro, L.: Irregular semi-convex gradient systems perturbed by noise and application to the stochastic Cahn-Hilliard equation. Ann. Inst. H. Poincaré Probab. Statist., **40**, 73–88 (2004)

[DF91]   Daleckij, Yu., Fomin, S.V.: Measures and differential equations in infinite-dimensional space, Kluwer (1991)

[DG]   Da Prato, G., Gątarek, D.: Stochastic Burgers equation with correlated noise. Stochastics and Stochastic Reports **52**, 29–41 (1995)

[DL95]   Da Prato, G., Lunardi, A.: On the Ornstein-Uhlenbeck operator in spaces of continuous functions. J. Funct. Anal., **131**, 94–114 (1995)

[DL04]   Da Prato, G., Lunardi, A.: Elliptic operators with unbounded drift coefficients and Neumann boundary condition. J. Differential Equations, **198**, 35–52 (2004)

[DMN92]   Da Prato, G., Malliavin, P., Nualart, D.: Compact families of Wiener functionals. C. R. Acad. Sci. Paris, **315**, 1287–1291 (1992)

[DR02]   Da Prato, G., Röckner, M.: Singular dissipative stochastic equations in Hilbert spaces. Probab. Theory Relat. Fields **124**, 261–303 (2002)

[DR04]   Da Prato, G., Röckner, M.: Invariant measures for a stochastic porous medium equation. Stochastic analysis and related topics in Kyoto, 13–29, Adv. Stud. Pure Math., **41**, Math. Soc. Japan, Tokyo (2004)

[DT00]   Da Prato, G., Tubaro, L.: A new method to prove self-adjointness of some infinite-dimensional Dirichlet operator. Probab. Theory Relat. Fields, **118**, 131–145 (2000)

[DT01a]   Da Prato, G., Tubaro, L.: On a class of gradient systems with irregular potentials. Infinite-Dimensional Analysis, Quantum Probability and related topics, **4**, 183–194 (2001)

[DT01b]   Da Prato, G., Tubaro, L.: Some results about dissipativity of Kolmogorov operators. Czechoslovak Mathematical Journal, **51(126)**, 685–699 (2001)

[DZ92]   Da Prato, G., Zabczyk, J.: Stochastic equations in infinite dimensions. Cambridge University Press, Cambridge, UK (1992)

[DZ96]   Da Prato, G., Zabczyk, J.: Ergodicity for infinite-dimensional systems. London Math. Soc. Lecture Notes **229**, Cambridge University Press, Cambridge, UK (1996)

[DZ02]   Da Prato, G., Zabczyk, J.: Second Order Partial Differential Equations
            in Hilbert Spaces. London Math. Soc. Lecture Notes **293**, Cambridge
            University Press, Cambridge, UK (2002)
[Ebe99]   Eberle, A.: Uniqueness and non-uniqueness of singular diffusion oper-
            ators. Lecture Notes in Mathematics **1718**, Springer (1999)
[Elw92]   Elworthy, K.D.: Stochastic flows on Riemannian manifolds. In: Pinsky,
            M.A., Wihstutz, V. (eds): Diffusion processes and related problems in
            analysis, Vol. II, 33–72, Birkhäuser (1992)
[Fla94]   Flandoli, F.: Dissipativity and invariant measures for stochastic Navier-
            Stokes equations. NoDEA **1**, 403–423 (1994)
[FM95]   Flandoli, F., Maslowski, B.: Ergodicity of the 2-D Navier-Stokes equa-
            tion under random perturbations. Commun. Math. Phys. **171**, 119–141
            (1995)
[FOT94]   Fukushima, M., Oshima, Y., Takeda, M.: Dirichlet forms and symmet-
            ric Markov processes. de Gruyter, Berlin (1994)
[GG94]   Gątarek, D., Goldys, B.: On weak solutions of stochastic equations in
            Hilbert spaces. Stochastics and Stochastic Reports, **46**, 41–51 (1994)
[Gro67]   Gross, L.: Potential theory on Hilbert spaces. J. Funct. Anal., **1**,
            123–181 (1967)
[Hai02]   Hairer, M.: Exponential mixing properties of stochastic PDEs through
            asymptotic coupling. Probab. Theory Relat. Fields, **124**, 345–380
            (2002)
[HM]   Hairer, M, Mattingly J.: Ergodicity of the 2D Navier-Stokes Equa-
            tions with Degenerate Stochastic Forcing. Annals of Mathematics,
            (to appear)
[Kha80]   Khas'minskii, R.Z.: Stochastic Stability of Differential Equations.
            Sijthoff and Noordhoff, Alphen aan den Rijn — Germantown,
            Md. (1980)
[Kol31]   Kolmogorov, A.N.: Über die analytischen Methoden in der Wahrschein-
            lichkeitsrechnung. Math. Ann., **104**, 415–458 (1931)
[KS01]   Kuksin, S., Shirikyan, A.: Ergodicity for the randomly forced 2D
            Navier-Stokes equations. Math. Phys. Anal. Geom., **4**, 147–195 (2001)
[Lév51]   Lévy, P.: Problèmes concrets d'analyse fonctionnelle. Gauthier-Villars,
            Paris (1951)
[LP61]   Lumer, G., Phillips, R.S.: Dissipative operators in a Banach space. Pac.
            J. Math., **11**, 679–698 (1961)
[LR98]   Liskevich, V., Röckner, M.: Strong uniqueness for a class of infinite-
            dimensional Dirichlet operators and application to stochastic quanti-
            zation. Ann. Scuola Norm. Sup. Pisa Cl. Sci. (4), **XXVII**, 69–91 (1998)
[Mat99]   Mattingly, J.C.: Ergodicty of 2D Navier-Stokes equations with random
            forcing and large viscosity. Commun. Math. Phys., **206**, 273–288 (1999)
[Mat02]   Mattingly, J.C.: Exponential convergence for the stochastically forced
            Navier-Stokes equations and other partially dissipative dynamics.
            Comm. Math. Phys. **230**, 421–462 (2002)
[MR92]   Ma, Z.M., Röckner, M.: Introduction to the Theory of (Non Symmetric)
            Dirichlet Forms. Springer (1992)
[Pes93]   Peszat, S.: On a Sobolev space of functions of infinite numbers of vari-
            ables. Bull. Pol. Acad. Sci., **40**, 55–60 (1993)
[Pri99]   Priola, E.: On a class of Markov type semigroups in spaces of uniformly
            continuous and bounded functions. Studia Math., **136**, 271–295 (1999)

[PZ95]   Peszat, S., Zabczyk, J.: Strong Feller property and irreducibility for diffusions on Hilbert spaces. Ann. Probab., **23**, 157–172 (1995)

[Sta99]   Stannat, W.: The theory of generalized Dirichlet forms and its applications in Analysis and Stochastics. Memoirs AMS **678**, Amer. Math. Soc. (1999)

[SV79]   Stroock, D.W., Varhadan, S.R.S: Multidimensional Diffusion Processes, Springer (1979)

[WKMS00]   Weinan, E., Khanin, K., Mazel, A., Sinai, Y.G.: *Invariant measures for Burgers equation with stochastic forcing.* Ann. of Math. (2), **151**, 877–960 (2000)

[WMS01]   Weinan, E., Mattingly, J.C., Sinai, Y.G.: Gibbsian dynamics and ergodicity for the stochastically forced Navier-Stokes equation. Commun. Math. Phys., **224**, 83–106 (2001)

[Zab81]   Zabczyk, J.: Linear stochastic systems in Hilbert spaces: spectral properties and limit behavior. Report **236**, Institute of Math., Polish Academy of Sci. (1981). Also in: Banach Center Publications, **41**, 591–609 (1985)

[Zam00]   Zambotti, L.: A new approach to existence and uniqueness for martingale problems in infinite dimensions. Probab. Theory Relat. Fields, **118**, 147–168 (2000)

5

From Kolmogorov's theorem on empirical distribution to number theory

Kevin Ford

Department of Mathematics, University of Illinois at Urbana-Champaign, USA
http://www.math.uiuc.edu/~ford
ford@math.uiuc.edu

We describe some new estimates for the probability that an empirical distribution function of uniform-[0,1] random variables stays on one side of a given line, and give applications to number theory.

5.1 Introduction

Let $X_1, \ldots, X_n$ be real-valued independent random variables, each with distribution function $F(t)$. Let

$$F_n(t) = \frac{1}{n} \#\{i : X_i \leq t\}$$

be the corresponding empirical distribution function. For n, t fixed, $F_n(t)$ is a random variable. Applying the strong law of large numbers to the Bernoulli variables

$$\mathbf{1}_{\{X_n \leq t\}} \quad (= 1 \text{ if } X_n \leq t, 0 \text{ otherwise}),$$

we see that $F_n(t) \xrightarrow[n \to \infty]{} F(t)$ almost surely. In 1933, Glivenko [Gli33] and (slightly later) Cantelli [Can33] proved that the convergence is uniform on the real line : $\sup | F_n(t) - F(t) | \xrightarrow[n \to \infty]{} 0$ almost surely. Immediately, in his seminal paper [Kol33], Kolmogorov made a careful study of the convergence of $F_n(t)$ to $F(t)$ as $n \to \infty$: he showed that if F is continuous, then for each $\lambda > 0$, the probability $\mathbf{P}(\sup |F_n(t) - F(t)| < \lambda/\sqrt{n})$ is independent of F, and that

$$\mathbf{P}(\sup |F_n(t) - F(t)| < \lambda/\sqrt{n}) \to \sum_{k=-\infty}^{\infty} (-1)^k e^{-2k^2\lambda^2} \quad (n \to \infty) \qquad (5.1)$$

uniformly in λ.[1]

---

[1] Notice that applying the Central Limit Theorem to the Bernoulli variables $\mathbf{1}_{\{X_n \leq t\}}$, we have only

The three papers of Glivenko, Kolmogorov and Cantelli appeared (in this order) in the same issue of the *Giornale dell Istituto Italiano degli Attuari*, all in Italian, and with almost the same title. The paper [Kol33] of Kolmogorov also appears in his Selected Works ([KolII], pp. 139–146; comments pp. 574–583).

Six years later, Smirnov [Smi39] studied the corresponding one-sided bounds, showing for $\lambda \geq 0$ that

$$\mathbf{P}(\sup(F_n(t) - F(t)) < \lambda/\sqrt{n}) \to 1 - e^{-2\lambda^2} \quad (n \to \infty). \qquad (5.2)$$

Together, (5.1) and (5.2) form the basis for the well-known *Kolmogorov-Smirnov* goodness-of-fit tests.

It is sometimes convenient to express probabilities of the above type in terms of the "order statistics" of $X_1, \ldots, X_n$, which is the increasing sequence $\xi_1 \leq \cdots \leq \xi_n$ obtained by ordering (each realization of) $X_1, \ldots, X_n$.

From now on, we will consider uniform distribution on $[0, 1]$, that is

$$F(t) = \begin{cases} 0 & t \leq 0 \\ t & 0 < t < 1 \\ 1 & t \geq 1. \end{cases} \qquad (5.3)$$

In this case, the numbers $\xi_1, \ldots, \xi_n$ are called *uniform order statistics*. In this note, we are interested in the behavior of

$$Q_n(u, v) = \mathbf{P}\left(\forall i \in \{1, \ldots, n\} : \xi_i \geq \frac{i - u}{v}\right).$$

In this notation, Smirnov's theorem reads[2] $Q_n(\lambda\sqrt{n}, n) \to 1 - e^{-2\lambda^2}$.

$$\mathbf{P}(|F_n(t) - F(t)| < \lambda/\sqrt{n}) \to \frac{1}{2\pi} \int_{-\lambda/\sigma(t)}^{\lambda/\sigma(t)} e^{-s^2/2} ds,$$

with $\sigma(t) = \sqrt{F(t)(1 - F(t))}$. In Kolmogorov's theorem, $|F_n(t) - F(t)|$ is replaced by its supremum over t, and the limit in the right-hand side is a universal (independent of F) function, of which Kolmogorov gave the first table of values

[2] Notice that

$$F_n(t) = \begin{cases} 0 & t \in (-\infty, \xi_1) \\ i/n & t \in [\xi_i, \xi_{i+1}) \quad (1 \leq i \leq n - 1) \\ 1 & t \in [\xi_n, +\infty) \end{cases}$$

thus we see (with (5.3)) that

$$\mathbf{P}(\sup(F_n(t) - F(t)) < \lambda/\sqrt{n}) = \mathbf{P}\left(\max_i(\frac{i}{n} - \xi_i) < \lambda/\sqrt{n}\right) = Q_n(\lambda\sqrt{n}, n)$$

Refinements to (5.2) were given later in the range $\lambda_0 \leq \lambda = O(n^{1/6})$ for a *fixed* positive λ_0 (e.g. Smirnov [Smi44], Lauwerier [Lau63]; see also Chap. 9 of [SW86]), in particular

$$Q_n(\lambda\sqrt{n}) = 1 - e^{-2\lambda^2}\left(1 - \frac{2\lambda}{3n^{1/2}} + O\left(\frac{\lambda^4 + 1}{n}\right)\right). \qquad (5.4)$$

Let $w = u + v - n$. Trivially $Q_n(u,v) = 0$ when $w \leq 0$ and $Q_n(u,v) = 1$ when $u \geq n$ (recall that $0 \leq X_i \leq 1$ from the choice of F). If $u \leq 1$ and $w > 0$, the exact formula $Q_n(u,v) = \dfrac{w}{v}(1 + u/v)^{n-1}$ was found by Daniels [Dan45]. Estimating $Q_n(u,v)$ when $u > 1$ is much more difficult, however there is an exact formula

$$\begin{aligned} Q_n(u,v) &= \frac{w}{v^n} \sum_{0 \leq j < u} \binom{n}{j}(w + n - j)^{n-j-1}(j-u)^j \\ &= 1 - \frac{w}{v^n} \sum_{u < j \leq n} \binom{n}{j}(w + n - j)^{n-j-1}(j-u)^j. \end{aligned} \qquad (5.5)$$

The special case $v = n$ of (5.5) is due to Smirnov [Smi44], and the general case is due to Pyke [Pyk59]. The equivalence of the two expressions for $Q_n(u,v)$ follows from one of Abel's identities ([Rio68], p. 18, (13a)). The first is more convenient when u is very small and fixed, while the second is more convenient for larger u because all summands are positive.

Smirnov [Smi44] estimated $Q_n(\lambda\sqrt{n}, n)$ using (5.5) and Stirling's formula for $k!$, and Csáki [Csá74] used similar methods to show

$$Q_n(\alpha\sqrt{n}, n + (\beta - \alpha)\sqrt{n}) \to 1 - e^{-2\alpha\beta} \qquad (n \to \infty).$$

for fixed $\alpha \geq 0$, $\beta \geq 0$. Lauwerier [Lau63] and Penkov [Pen76], by contrast, started with (5.5) and used complex analytic methods to approximate $Q_n(\lambda\sqrt{n})$. Yet another approach is based on what are called "almost sure invariance principles" or "strong approximation theorems" ([CR81], [Phi86]). The strong Komlós-Major-Tusnády theorem [KMT75] implies

$$|F_n(t) - t - n^{-1/2}B_n(t)| \ll \frac{\log n}{n} \qquad (0 \leq t \leq 1)$$

with probability $\geq 1 - O(1/n)$, where $B_n(t)$ is a Brownian bridge process. The order $\frac{\log n}{n}$ on the right side is also best possible [KMT75] (see also Chap. 4 of [CR81]). Since

$$\mathbf{P}\left(\sup_{0 \leq t \leq 1}(B_n(t) - (at + b)) \leq 0\right) = 1 - e^{-2b(a+b)},$$

the KMT theorem implies the uniform estimate

$$Q_n(u, v) = O\left(\frac{1}{n}\right) + 1 - e^{-\frac{2(u+O(\log n))(w+O(\log n))}{n}}$$

$$= 1 - e^{-2uw/n} + O\left(\frac{(u + w + \log n) \log n}{n}\right). \tag{5.6}$$

This gives an asymptotic for $Q_n(u, v)$ in a wide range of u and w, but requiring $\frac{u}{\log n} \to \infty$ and $\frac{w}{\log n} \to \infty$.

For the application to number theory in [For04a], we need sharper uniform bounds than (5.6). In particular, we need the bound $Q_n(u, v) = O(u/n)$ uniformly for $n \geq 1$, $w = O(1)$ and $1 \leq u \leq n$.

5.2 New estimates for uniform order statistics

Theorem 1. *Uniformly in $u > 0$, $w > 0$ and $n \geq 1$, we have*

$$Q_n(u, v) = 1 - e^{-2uw/n} + O\left(\frac{u + w}{n}\right),$$

i.e. $|O(\frac{u+w}{n})| \leq \mathrm{const}(\frac{u+w}{n})$ where the constant is independent of u, v, n.

In addition we have the following useful approximation.

Corollary 1. *Uniformly in $u > 0$, $w > 0$ and $n \geq 1$, we have*

$$Q_n(u, v) = \frac{2uw}{n}\left(1 + O\left(\frac{1}{u} + \frac{1}{w} + \frac{uw}{n}\right)\right).$$

In particular, when $uw/n \to 0$, $u \to \infty$ and $w \to \infty$ as $n \to \infty$, we see that $Q_n(u, v)$ is asymptotic to $2uw/n$. Starting with (5.5), a complicated modification of the complex analytic method of Lauwerier [Lau63] can be used to prove Theorem 1. This was carried out in the original version of [For04a], and a sketch of the argument appears in [For04b].

Here we outline a new method based on the theory of random walks, full details of which appear in [For06a]. Rather than work with (5.5), we reinterpret $Q_n(u, v)$ in terms of a random walk. Let $Y_1, \cdots, Y_{n+1}$ be independent random variables with exponential distribution, and let $W_k = Y_1 + \cdots + Y_k$ for $1 \leq k \leq n + 1$. By a well-known theorem of Rényi [Rén53], the vectors $(\xi_1, \ldots, \xi_n)$ and $(W_1/W_{n+1}, \ldots, W_n/W_{n+1})$ have identical distributions. Similarly, given that $W_{n+1} = v$, the probability density function of the vector $(W_1/v, \ldots, W_n/v)$ is identically $n!$ on the set $\{(x_1, \ldots, x_n) : 0 \leq x_1 \leq \cdots \leq x_n \leq 1\}$. Therefore,

$$Q_n(u, v) = \mathbf{P}[\min_{1 \leq i \leq n} (W_i - i) \geq -u \mid W_{n+1} = v].$$

Put $X_i = 1 - Y_i$, so that the X_i have mean 0, variance 1 and $X_i < 1$ for all i. Let

$$S_i = X_1 + \cdots + X_i, \quad T_i = \max(0, S_1, \ldots, S_i) \quad (i \geq 0).$$

The sequence $0, S_1, S_2, \ldots$ can be thought of as a recurrent random walk on the real line, with T_i measuring the farthest extent to the right that the walk has achieved during the first i steps. Setting

$$R_m(x, y) = \mathbf{P}[T_{m-1} < y \mid S_m = x],$$

we have

$$Q_n(u, v) = R_{n+1}(n + 1 - v, u). \tag{5.7}$$

If we label the point y as a barrier, then $R_m(x, y)$ is the probability of stopping after m steps at x without crossing the barrier.

In proving (5.1) in [Kol33], Kolmogorov used a relation similar to (5.7). Specifically, let $Y_1, Y_2, \ldots, Y_n$ be independent random variables with discrete distribution

$$\mathbf{P}[Y_j = r - 1] = \frac{e^{-1}}{r!} \quad (r = 0, 1, \ldots)$$

and let $Z_j = Y_1 + \cdots + Y_j$ for $j \geq 1$. The variables Y_i have mean 0 and variance 1. Kolmogorov proved that for integers $u \geq 1$,

$$\mathbf{P}(\sup |F_n(t) - F(t)| \leq u/n) = \frac{n! e^n}{n^n} \mathbf{P}\left(\max_{0 \leq j \leq n-1} |Z_j| < u, Z_n = 0\right)$$

$$= \mathbf{P}\left(\max_{0 \leq j \leq n-1} |Z_j| < u \mid Z_n = 0\right).$$

Small modifications to the proof yield, for *integers* $u \geq 1$ and for $n \geq 2$, that

$$Q_n(u, n) = \mathbf{P}\left(\max_{0 \leq j \leq n-1} Z_j < u \mid Z_n = 0\right).$$

Let f_m be the pdf (probability density function) for S_m $(m = 1, 2, \ldots)$. The Central Limit Theorem for densities (e.g. Theorem 1 in §46 of [GK68]) implies that for large m and $|x| \ll \sqrt{m}$, $f_m(x) \approx (2\pi m)^{-1/2} e^{-x^2/2m}$. However, there are asymmetries in the distribution for $|x| > \sqrt{m}$, which can be seen using the exact formula

$$f_m(x) = \begin{cases} \frac{(m-x)^{m-1}}{e^{m-x}(m-1)!} & x \leq m \\ 0 & x > m, \end{cases} \tag{5.8}$$

easily proved by induction on m.

Our principal tool for estimating $R_n(x, y)$ is a reccurrence formula based on the reflection principle for random walks. Suppose $y \geq 0$ and $y \geq x$. By reflecting about the point y that part of the walk beyond the first crossing of y, a recurrent random walk of n steps that crosses the point y and ends at the point x is about as likely as a random walk which ends at $2y - x$ after n steps. This of course is inexact, since the steps of a random walk may not be symmetric and the walk may not hit y exactly. The next Lemma 2 gives a precise measure of the accuracy of the reflection principle for our specific walk. For convenience, define

$$\widetilde{R}_n(x, y) = f_n(x) R_n(x, y) = \mathbf{D}[T_{n-1} < y, S_n = x],$$

where the last expression stands for $\frac{d}{dx}\mathbf{P}[T_{n-1} < y, S_n < x]$. From the reflection principle we expect that $\widetilde{R}_n(x, y) \approx f_n(x) - f_n(2y - x)$.

Lemma 1. *For a positive integer $n \geq 2$, real $y > 0$, real x, and real $a \geq 1$,*

$$\widetilde{R}_n(x, y) = f_n(x) - f_n(y + a) + \int_0^1 \sum_{k=1}^{n-1} \widetilde{R}_k(y + \xi, y) \tag{5.9}$$
$$(f_{n-k}(a - \xi) - f_{n-k}(x - y - \xi))\, d\xi.$$

Proof. Start with

$$\widetilde{R}_n(x, y) = f_n(x) - f_n(y + a) + f_n(y + a) - \mathbf{D}[T_{n-1} \geq y, S_n = x].$$

If $S_n = y + a$, then there is a unique k, $1 \leq k \leq n - 1$, so that $T_{k-1} < y$ and $S_k \geq y$. Thus,

$$f_n(y + a) = \sum_{k=1}^{n-1} \mathbf{D}[T_{k-1} < y, S_k \geq y, S_n = y + a]$$
$$= \sum_{k=1}^{n-1} \int_0^1 \mathbf{D}[T_{k-1} < y, S_k = y + \xi, S_n = y + a]\, d\xi$$
$$= \sum_{k=1}^{n-1} \int_0^1 \widetilde{R}_k(y + \xi, y) f_{n-k}(a - \xi)\, d\xi.$$

Similarly,

$$\mathbf{D}[T_{n-1} \geq y, S_n = x] = \sum_{k=1}^{n-1} \mathbf{D}[T_{k-1} < y, S_k \geq y, S_n = x]$$
$$= \sum_{k=1}^{n-1} \int_0^1 \widetilde{R}_k(y + \xi, y) f_{n-k}(x - y - \xi)\, d\xi.$$

In Lemma 1, we choose $a = y - x - b(n, y - x)$, where $b = b(n, z)$ is the unique solution of $f_n(-z) = f_n(z - b)$ ith $-2 \leq b \leq z - 1$ ($b(n, z)$ exists and is unique since $f_n(x)$ is unimodular with maximum at $x = 1$). This makes $|f_{n-k}(a - \xi) - f_{n-k}(x - y - \xi)|$ small, at least when k is small. Also, $\widetilde{R}_k(y + \xi, y)$ should be small, since it measures the likelihood of a walk staying to the left of y for $n-1$ steps and jumping over y on the n-th step. Suppose $n \geq 10$, $0 \leq y \leq \frac{n}{10}$, and $y \leq x \leq y+1$. We have $f_n(1+x) \leq f_n(1-x)$ for $x \geq 0$, thus when $0 \leq \xi \leq 1$ and $1 \leq j \leq n-1$, $f_j(5-\xi) \leq f_j(x-y-\xi)$. By Lemma 1 with $a = 5$,

$$\widetilde{R}_n(x,y) \leq f_n(x) - f_n(y+5) = \int_x^{y+5} \frac{t-1}{n-t} f_n(t)\, dt \ll \frac{(y+1)f_n(y)}{n}.$$

Together with estimates for $|f_{n-k}(a-\xi) - f_{n-k}(x-y-\xi)|$ obtained from (5.8), the integral-sum on the right of (5.9) can be shown to be small. We conclude that, with small error,

$$R_n(x,y) \approx 1 - \frac{f_n(2y-x-b)}{f_n(x)}.$$

The desired asymptotic for $Q_n(u,v)$ now follows from (5.8) and the asymptotic $b = b(n,z) = -2 + O(\frac{(z+1)^2}{n-1})$.

We note that when the steps in a recurrent random walk have an arbitrary continuous or lattice distribution, one can define a quantity analogous to $R_n(x,y)$. The same argument provides an analogous formula to (5.9) and an analog of Theorem 1, namely

$$R_m(y-z,y) = 1 - e^{-2yz/n} + O\left(\frac{y+z+1}{n}\right) \quad (0 \leq y \ll \sqrt{n}, 0 \leq z \ll \sqrt{n}),$$

can be shown to hold for a very general class of distributions (see [For06b]).

5.3 Number theory applications

Hardy and Ramanujan initiated the study of the statistical distribution of the prime factors of integers in their ground-breaking 1917 paper [HR17], and much work has been done on this topic since then. Write an arbitrary integer $n = p_1 p_2 \cdots p_k$, where the p_i are primes and $p_1 \leq \cdots \leq p_k$. Roughly speaking, the quantities $g_j = \log\log p_{j+1} - \log\log p_j$ behave like independent exponentially distributed random variables. Of course the g_j have discrete distributions, but the distributions approach the exponential distribution as $j \to \infty$. It is well-known that a typical integer n has about $\log\log b - \log\log a$ prime factors in an interval $(a,b]$ (see e.g. Chap. 1 of [HT88]), and the probability that n has at least one prime factor in $(a,b]$ is approximately[3]

$$1 - \prod_{a<p\leq b} (1-1/p) = 1 - \frac{\log a + O(1)}{\log b}.$$

---

[3] p will always denote a prime number; $\prod_{a<p\leq b}$ will be a product on primes, $\sum_{a<p\leq b}$ a sum on primes

One can also consider integers with a fixed number of prime factors and examine the statistics

$$(\xi_1, \cdots, \xi_m), \qquad \xi_i = \frac{\log\log p_{j+i} - \log\log p_j}{\log\log p_k - \log\log p_j}, \qquad m = k - 1 - j.$$

With k and j fixed, the numbers $\xi_1, \ldots, \xi_m$ behave much like uniform order statistics. This means that for "nice" functions $f : [0,1]^m \to \mathbb{R}$, the average of $f(\xi_1, \ldots, \xi_m)$ over n which are the product of k primes is about

$$m! \int_{0 \leq x_1 \leq \cdots \leq x_m \leq 1} f(x_1, \ldots, x_m)\, dx_1 \cdots dx_m.$$

The approximation gets better as $j \to \infty$.

These phenomena can be explained by considering the following "model" of the integers (known as the Kubilius model). Let $\{X_p : p \text{ prime}\}$ be independent Bernoulli random variables so that $\mathbf{P}(X_p = 0) = 1 - \frac{1}{p}$ and $\mathbf{P}(X_p = 1) = \frac{1}{p}$. Thus X_p models the event that a random integer is divisible by p. By an elementary estimate,

$$\sum_{a < p \leq b} \mathbf{E}(X_p) = \sum_{a < p \leq b} \frac{1}{p} = \log\log b - \log\log a + O(1/\log a).$$

(The $\log\log$, rather than $\log$, are due to the fact that we sum only on primes.) For more about probabilistic number theory, the reader may consult the excellent monographs of Elliott [Ell79].

Questions about the distribution of all divisors of integers are much more difficult, since the corresponding random variables $\{X_d : d \geq 1\}$ are not at all independent (e.g. $X_6 = 1 \implies X_3 = 1$). Consider the problem of estimating $\varepsilon(y, z)$, the probability that a random integer has a divisor d satisfying $y < d \leq z$. More precisely,

$$\varepsilon(y, z) = \lim_{x \to \infty} \frac{\#\{n \leq x : \exists\, d | n, y < d \leq z\}}{x}.$$

Similarly, let $\varepsilon_r(y, z)$ be the probability that a random integer has exactly r divisors in the interval $(y, z]$. Interest in bounding $\varepsilon(y, z)$ began in the 1930s with a paper by Besicovitch [Bes34], who proved that $\liminf_{y \to \infty} \varepsilon(y, 2y) = 0$. A year later, Erdős [Erd35] improved this to $\lim_{y \to \infty} \varepsilon(y, 2y) = 0$. Later work, especially by Erdős [Erd36], [Erd60] and Tenenbaum [Ten84], focused on determining the rate at which $\varepsilon(y, 2y) \to 0$ and on bounding $\varepsilon(y, z)$ for more general y, z. Chapter 2 of the book [HT88] contains a thorough exposition on such bounds and their applications. The main theorem of [For04a] is a determination of the order of magnitude of $\varepsilon(y, z)$ for all y, z; that is, bounding $\varepsilon(y, z)$

between two constant multiples of a smooth function of y, z. In particular, we show that for some positive constants c_1 and c_2,

$$\frac{c_1}{(\log y)^\delta (\log \log y)^{3/2}} \leq \varepsilon(y, 2y) \leq \frac{c_2}{(\log y)^\delta (\log \log y)^{3/2}} \qquad (5.10)$$

where $\delta = 1 - \dfrac{1 + \log \log 2}{\log 2} = 0.08607\ldots$ A relatively short, complete proof of this special case is given in [For06c].

Concerning the behavior of $\varepsilon_r(y, z)$, Erdős conjectured in [Erd60] that

$$\lim_{y \to \infty} \frac{\varepsilon_1(y, 2y)}{\varepsilon(y, 2y)} = 0.$$

The ratio $\dfrac{\varepsilon_r(y, z)}{\varepsilon(y, z)}$ can be considered as the conditional probability that a random integer contains exactly r divisors in $(y, z]$ given that it has at least one such divisor. In [Ten87] a lower bound $\dfrac{\varepsilon_r(y, 2y)}{\varepsilon(y, 2y)} \geq c_3 f(y)$ was given, where $f(y) \to 0$ very slowly as $y \to \infty$. Erdős conjecture is disproved in [For04a], where the order of $\varepsilon_r(y, z)$ is determined for a wide range of y, z. In particular, for any $r \geq 1$ and any constant $c > 1$,

$$\liminf_{y \to \infty} \frac{\varepsilon_r(y, cy)}{\varepsilon(y, cy)} > 0.$$

Also,

$$\frac{\varepsilon_r(y, z)}{\varepsilon(y, z)} \to 0 \qquad (z/y \to \infty),$$

confirming a conjecture of Tenenbaum [Ten87].

We now say a few words about the proofs. Let m be the product of the distinct prime factors of n which are $\leq y$. First, $\varepsilon(y, 2y)$ can be estimated in terms of

$$\sum_m \frac{L(m)}{m}, \qquad L(m) = \mu\{u : \exists\, d|m, e^u < d \leq 2e^u\},$$

where μ denotes Lebesgue measure. The quantity $L(m)$ is a kind of measure of the global distribution of the divisors of m. If $m = p_1 \cdots p_k$, then

$$L(m) \leq \min_{0 \leq h \leq k} 2^{k-h} \log(2p_1 \cdots p_h).$$

Most of the time, we expect $\log(2p_1 \cdots p_h) = O(\log p_h)$, so

$$L(m) \approx O\left(2^k \exp\{\min_{1 \leq h \leq k} (-h \log 2 + \log \log p_h)\}\right).$$

Putting $\xi_i = \dfrac{\log \log p_i}{\log \log y}$, then $\xi_1, \ldots, \xi_k$ behave much like uniform order statistics. Thus, upper bounds for averages of $L(m)$ depend on the size of $Q_k(u, v)$

with $v = \dfrac{\log\log y}{\log 2}$. Utilizing Theorem 1 (actually, the weaker bound $Q_n(u,v)$

$= O(\frac{(u+1)(w+1)^2}{n})$ proved in [For04a] suffices) leads to the upper bound in (5.10). Furthermore, the bulk of the contribution comes from numbers n with $k = \dfrac{\log\log y}{\log 2} + O(1)$. This implies that most integers which have a divisor in $(y, 2y]$ have about $\dfrac{\log\log y}{\log 2}$ prime factors $\leq y$. By contrast, most integers n have about $\log\log y$ prime factors $\leq y$.

Acknowledgement

This research was supported by National Science Foundation grants DMS-0301083 and DMS-0555367.

References

[Bes34]   Besicovitch, A.S.: On the density of certain sequences of integers. Math. Ann., **110**, 336–341 (1934)

[Can33]   Cantelli, F.G.: Sulla determinazione empirica delle leggi di probabilità. Giorn. Ist. Ital. Attuari, **4**, 421–424 (1933)

[CR81]   Csörgő, M., Révész, P.: Strong Approximations in probability and statistics. Academic Press (1981)

[Csá74]   Csáki, E.: On tests based on empirical distribution functions (Hungarian). Magyar Tud. Akad. Mat. Fiz. Oszt. Közl., **23**, 239–327 (1977). English translation in: Leifman, L.J. (ed) Selected translations in mathematical statistics and probability, **15**, Amer. Math. Soc., 229–317 (1981)

[Dan45]   Daniels, H.E.: The statistical theory of the strength of bundles of threads, I. Proc. Roy. Soc. London. Ser. A., **183**, 405–435 (1945)

[Ell79]   Elliott, P.D.T.A.: Probabilistic number theory I, II. Grund. Math. Wissen. 239, 240. Springer, Berlin Heidelberg New York (1979, 1980)

[Erd35]   Erdős, P.: Note on the sequences of integers no one of which is divisible by any other. J. London Math. Soc., **10**, 126–128 (1935)

[Erd36]   Erdős, P.: A generalization of a theorem of Besicovitch. J. London Math. Soc., **11**, 92–98 (1936)

[Erd60]   Erdős, P.: An asymptotic inequality in the theory of numbers (Russian). Vestnik Leningrad. Univ., **15**, 41–49 (1960)

[For04a] Ford, K.: The distribution of integers with a divisor in a given interval (2004), 62 pages. To appear in Annals of Math. (2007). Preprint available at:
          http://front.math.ucdavis.edu/math.NT/0401223

[For04b] Ford, K.: Du théorème de Kolmogorov sur les distributions empiriques à la théorie des nombres. In: L'héritage de Kolmogorov en mathématiques. Éditions Belin, Paris, 111–120 (2004)

[For06a]   Ford, K.: Sharp probability estimates for generalized Smirnov statistics (2006), 10 pages. preprint available at:
http://front.math.ucdavis.edu/math.PR/0609224

[For06b]   Ford, K.: Sharp probability estimates for random walks with barriers (2006), preprint available on the ArXiv: math.PR/0610450.

[For06c]   Ford, K.: Integers with a divisor in $(y, 2y]$, 18 pages. To appear in the Proceedings of the Anatomy of Integers Workshop, Montreal, March 2006 (2007). Preprint available at:
http://front.math.ucdavis.edu/math.NT/0607473

[GK68]   Gnedenko, B. V., Kolmogorov, A. N.: Limit distributions for sums of independent random variables. (Translated from the Russian, annotated, and revised by K. L. Chung. With appendices by J. L. Doob and P. L. Hsu. Revised edition), Addison-Wesley, Reading, Mass.-London-Don Mills., Ont. (1968)

[Gli33]   Glivenko, V.: Sulla determinazione empirica delle leggi di probabilità. Giorn. Ist. Ital. Attuari, **4**, 92–99 (1933)

[HR17]   Hardy, G.H., Ramanujan, S.: The normal number of prime factors of a number n. Quart. J. Math., **158**, 76–92 (1917)

[HT88]   Hall, R.R., Tenenbaum, G.: Divisors. Cambridge Tracts in Mathematics, **90**, Cambridge University Press, Cambridge, UK (1988)

[KMT75]   Komlós, J., Major, P., Tusnády, G.: An approximation of partial sums of independent RV's and the sample DF. I. Z. Wahrscheinlichkeitstheorie und Verw. Gebiete, **32**, 111–131 (1975)

[Kol33]   Kolmogorov, A.N.: Sulla determinazione empirica di una legge di distribuzione (On the empirical determination of a distribution law). Giorn. Ist. Ital. Attuar., **4**, 83–91 (1933)

[KolII]   Kolmogorov, A.N.: Selected works, vol. II: Probability theory and mathematical statistics (with a preface by Aleksandrov, P.S.; translated from the Russian by Lindquist, G.; translation edited by Shiryayev, A.N.). Kluwer Academic Publishers Group, Dordrecht (1992)

[Lau63]   Lauwerier, H.A.: The asymptotic expansion of the statistical distribution of N. V. Smirnov (German). Z. Wahrscheinlichkeitstheorie und Verw. Gebiete, **2**, 61–68 (1963)

[Pen76]   Penkov, B.I.: Asymptotic distribution of Pyke's statistics (Russian). Teor. Verojatnost. i Primenen., **21**, 378–383 (1976). English translation in: Theory of probability and its applications, **21**, 370–374 (1976)

[Per39]   Perron, O.: Über Bruwiersche Reihen. Math. Z., **45**, 127–141 (1939)

[Phi86]   Philipp, W.: Invariance principles for independent and weakly dependent random variables, in *Dependence in Probability and Statistics (Oberwolfach, 1985)*, Progr. Probab. Statist. **11**, Birkhäuser Boston, Boston, MA. 225–268 (1986)

[Pyk59]   Pyke, R.: The supremum and infimum of the Poisson process. Ann. Math. Statist., **30**, 568–576 (1959)

[Rén53]   Rényi, A.: On the theory of order statistics. Acta Math. Acad. Sci. Hung., **4**, 191–232 (1953)

[Rio68]   J. Riordan: Combinatorial identities. Wiley, New York (1968)

[Smi39]   Smirnov, N.V.: Sur les écarts de la courbe de distribution empirique (Russian, French summary). Rec. Math. Moscou (Mat. Sbornik), **6**, 3–26 (1939)

[Smi44]   Smirnov, N.V.: Approximate laws of distribution of random variables from empirical data (Russian). Uspekhi Matem. Nauk, **10**, 179–206 (1944)

[SW86]   Shorack, G.R., Wellner, J.A.: Empirical processes with applications to statistics. Wiley Series in Probability and Mathematical Statistics. Wiley, New York (1986)

[Ten84]   Tenenbaum, G.: Sur la probabilité qu'un entier possède un diviseur dans un intervalle donné. Compositio Math., **51**, 243–263 (1984)

[Ten87]   Tenenbaum, G.: Un problème de probabilité conditionnelle en arithmétique. Acta Arith., **49**, 165–187 (1987)

Kolmogorov's ε-entropy and the problem of statistical estimation

Mikhail Nikouline[1] and Valentin Solev[2]

[1] Statistique Mathématique et ses Applications, University Victor Segalen
(Bordeaux II), France
http://www.sm.u-bordeaux2.fr/stat
nikouline@sm.u-bordeaux2.fr
[2] Laboratory of Statistical Methods, Steklov Institute of Mathematics, RAS, Saint
Petersburg, Russia
http://www.pdmi.ras.ru/staff/solev.html
solev@pdmi.ras.ru

Translated from the French by Francis Brown

In 1933, Andreï Kolmogorov published the fundamental result of *non-parametric statistics*, which is a theorem on the convergence of empirical measure towards theoretical measure, upon which depends the famous *Kolmogorov test*. Since then, research on this topic has expanded considerably — see, for example, Dudley (1976, 1997), Van der Vaart (1989, 2000), Deheuvels and Mason (1992), Van der Vaart and Wellner (1996).

At the beginning of the 1950's, Kolmogorov turned his attention to information theory and its relation with complexity theory, the theory of functions, and statistical estimation. It was at this time that he introduced the very deep notion of ε-*entropy*, which has since played a fundamental rôle in non-parametric and semi-parametric statistics. We will present all these concepts, and give a survey of the current state of knowledge.

In this article, we will only assume a basic knowledge of the theory of functions and functional analysis, probability theory and mathematical statistics. It is intended for researchers and master's degree students who wish to gain familiarity with the general problems of the theory of non-parametric estimation, and to get an idea of the influence of A. N. Kolmogorov's ideas on the development of statistical estimation.

6.1 Overview of the problem: parametric and non-parametric statistics

Every statistician comes across the following general situation: one observes a certain random variable X, which takes random values in a known set $\mathcal{X}$

following a certain distribution (or probability law) $P \in \mathcal{P}$. The distribution P is unknown and the problem is to estimate it; all information available to the statistician lies in the observed value of X and in the set of all *a priori* possible laws $\mathcal{P}$.

Let us consider the situation where a statistician observes the independent random variables $X_1, \ldots, X_n$ which obey the same law $P \in \mathcal{P}$. One can often assume that the set $\mathcal{P}$ is embedded in a certain metric space. The case which is the most well-understood is when $\mathcal{P}$ is a "smooth" set in a Euclidean space $\mathbb{R}^s$ of dimension s: this case is referred to as the *parametric* problem — see, for example, Halmos (1950), Kolmogorov (1950, 1951), Lehmann (1983), Voinov and Nikulin (1993, 1996), Le Cam and Yang (1999), Bosq and Lecoutre (1987), etc.

Unfortunately, in practice, one rarely meets statistical problems in which the *a priori* knowledge about the unknown law can be interpreted in this manner, and in fact, the most interesting case occurs when the set $\mathcal{P}$ is infinite. This is known as the *non-parametric* problem.

In order to obtain a reasonable method of statistical estimation in this case, it is necessary to approximate the given problem with a different problem in which the set $\mathcal{P}$ is *finite*. The accuracy of this approximation is expressed in terms of Kolmogorov's ε-*entropy* and the ε-*capacity* of the parametric set $\mathcal{P}$. Here, we will focus on the influence of Kolmogorov's concept of ε-entropy on the development of functional estimation theory.

Roughly speaking (details will be given presently), in order to define the Kolmogorov ε-entropy of the parametric set $\mathcal{P}$, which we denote $\mathcal{H}(\varepsilon, \mathcal{P})$, we assume that $\mathcal{P}$ is equipped with a metric (a distance), which is suited to the problem under consideration: $\mathcal{H}(\varepsilon, \mathcal{P})$ is by definition the logarithm of the minimal number of balls of radius ε needed to cover $\mathcal{P}$. If one takes successive values $\varepsilon = \varepsilon_n$ in such a way that[3]

$$\varepsilon_n \asymp \sqrt{\frac{\mathcal{H}(\varepsilon_n, \mathcal{P})}{n}}, \tag{6.1}$$

then the rate at which ε_n tends to 0 in (6.1) will determine, in a certain sense, the accuracy of the statistical estimation. (This is a delicate problem.)

There has been much research on this subject. Functional estimation is now a well-developed theory and has its origins in the work of Le Cam (1973, 1986), Khas'minskii (1978), Ibragimov and Khas'minskii (1981), Birgé (1983), which gave rise to several important results. For general results on this question, we refer the reader to Assouad (1983), Birgé and Massart (1998), Bretagnolle and Huber (1979), Ceci and Mazliak (2004), Dudley (1976, 1997), Hall, Huber, Owen and Coventry (1994), Huber (1997), Le Cam and Yang (1999), Nikulin and Solev (2002), Van der Geer (1993, 1995), Van der Vaart

---

[3] $u_n \asymp v_n$ means that u_n/v_n and v_n/u_n are bounded for all n greater than a certain n_0

(2000), Wong and Shen (1995), Shen (1997), Yatrakos (1985), Yu (1997), etc.
(See the bibliography at the end of this chapter.)

6.2 Notations and definitions

6.2.1 Hidden laws and estimators

Let $\mathcal{P} = \{P_\theta : \theta \in \Theta\}$ denote a family of probability measures P_θ on a measurable space $\{\mathcal{X}, \mathcal{B}\}$, which are dominated by a σ-finite measure μ.[4] We denote by

$$f_\theta = \frac{d\,P_\theta}{d\,\mu} \quad \text{the density of } P_\theta \text{ with respect to } \mu,$$

and

$$\mathcal{F} = \left\{ f : f = \frac{d\,P}{d\,\mu}, P \in \mathcal{P} \right\} \tag{6.2}$$

the set of densities f_θ. Suppose that the map $\theta : \theta(P_\theta) = \theta$ is a bijection from $\mathcal{P}$ to Θ. We thus have

$$\Theta = \{\theta : \theta = \theta(P), P \in \mathcal{P}\}.$$

We shall use the same notation for the function $\theta : \theta(f_\theta) = \theta$, defined on $\mathcal{F}$. (Any ambiguity should be clear from the context.)

On the probability space $\{\mathcal{X}, \mathcal{B}, \mathcal{P}\}$, we consider a sequence of independent random variables $X_1, \ldots, X_n, \ldots$ which obey the same probability law $P \in \mathcal{P}$, with density $f \in \mathcal{F}$. Let $\hat{f}_n = \hat{f}_n(X_1, \ldots, X_n)$ be an estimator of the unknown density $f \in \mathcal{F}$, which is constructed by means of the observations $X_1, \ldots, X_n$. We denote by $l(\hat{f}_n, f)$ the *loss function*[5] of the estimator $\hat{f}_n$, and we define the risk function:

$$R(\hat{f}_n) = \sup_{f \in \mathcal{F}} \mathbf{E}_f \, l(\hat{f}_n, f)$$

---

[4] Let us recall some definitions. A measure μ on $\{\mathcal{X}, \mathcal{B}\}$ is said to be σ-finite if $\mathcal{X}$ is a countable union $\mathcal{X} = \cup_n A_n$ where, for all n, $A_n \in \mathcal{B}$ and $\mu(A_n) < +\infty$. The family $\{P_\theta : \theta \in \Theta\}$ is said to be dominated by μ if, for all $\theta \in \Theta$, P_θ is absolutely continuous with respect to μ (in other words, if for all $A \in \mathcal{B}$ such that $\mu(A) = 0$, we have $P_\theta(A) = 0$): which is denoted $P_\theta \ll \mu$. The theorem of Lebesgue-Radon-Nikodym guarantees the existence of a function f_θ, which is integrable with respect to the mesure μ, such that $dP_\theta = f_\theta d\mu$. This function is denoted by $f_\theta = \frac{d\,P_\theta}{d\,\mu}$, and is called the Lebesgue-Radon-Nikodym derivative − or density − of P_θ with respect to μ

[5] Let $\hat{f}_n = \hat{f}_n(X_1, \ldots, X_n) : \mathbb{R}^n \to \mathcal{F}$ denote a point-wise estimator of the parameter f ($f \in \mathcal{F}$). Every positive function $l(\cdot, \cdot) : \mathcal{F} \times \mathcal{F} \to \mathbb{R}^1_+$ is called a *loss function* of the estimator $\hat{f}_n$. Loss functions enable us to measure the quality of estimators. This assumes that the observed value $l(\hat{f}_n, f)$ of $l(\cdot, \cdot)$ represents for every f the *loss* which results from using $\hat{f}_n$ instead of f. It is natural to assume that $l(f, f) = 0$

of the estimator $\hat{f}_n$ with respect to the loss function $l(\cdot,\cdot)$. Here, $\mathbf{E}_f$ denotes the expectation which is calculated when the X_i obey the density law f. The problem is to construct *optimal (minimax) operators* which minimize the risk $R(\hat{f}_n)$:

$$R_n(\mathcal{F}) = \inf_{\hat{f}_n} R(\hat{f}_n),$$

or *asymptotically optimal (asymptotically minimax) estimators*, which satisfy the following relation:

$$\overline{\lim}_{n\to\infty} \frac{R(\hat{f}_n)}{R_n(\mathcal{F})} = 1.$$

In practice it is perfectly good enough to have estimators which are asymptotically minimax, or even estimators which satisfy the following condition:

$$\overline{\lim}_{n\to\infty} \frac{R(\hat{f}_n)}{R_n(\mathcal{F})} < \infty.$$

For more detailed background on statistical concepts, we refer the reader to Lehmann (1983), Voinov and Nikulin (1993), or Van der Vaart (2000), for example.

Let P_n denote the *empirical measure*, based on the samples $X_1,\ldots,X_n$:

$$P_n(A) = \frac{1}{n} \sum_{j=1}^{n} \delta_{X_j}(A), \quad A \in \mathcal{B},$$

where

$$\delta_X(A) = \begin{cases} 1 & \text{if} \quad X \in A, \\ 0 & \text{if} \quad X \notin A, \end{cases} \tag{6.3}$$

from which it follows that for all φ:

$$\int \varphi \, dP_n = \frac{1}{n} \sum_{j=1}^{n} \varphi(X_j).$$

In the following, we shall use the notations:

$$P_n\varphi = \int \varphi \, dP_n, \quad P\varphi = \int \varphi \, dP, \quad \nu_n(\varphi) = P_n\varphi - P\varphi$$

for all $\varphi \in L^1(dP)$. If the function φ is bounded: $a \le \varphi \le b$, then the accuracy of the approximation of $P\varphi$ by the estimator $P_n\varphi$ is given by *Hoeffding's inequality*:

$$P\{\nu_n(\varphi) > y\} \le \exp\left\{ -\frac{2ny^2}{(b-a)^2} \right\},$$

whose bilateral version has the following form:

$$P\{|\nu_n(\varphi)| > y\} \le 2\exp\left\{ -\frac{2ny^2}{(b-a)^2} \right\}. \tag{6.4}$$

In particular, if we take $\varphi(x) = \delta_x(A)$, where A is fixed, and $a = 0$, $b = 1$, we have:

$$P\{|P_n(A) - P(A)| > y\} \leq 2\exp\{-2ny^2\}. \tag{6.5}$$

We now state the problem, which is to construct the estimator $\hat{P}_n \in \mathcal{P}$ of the unknown law P, and hence of the estimator $\hat{\theta}_n = \theta(\hat{P}_n)$ of the unknown parameter θ, using the fact that we know *a priori* that $P \in \mathcal{P}$. One should expect that:

1. the method of constructing of a reasonable estimator depends on the parametrising set $\mathcal{P}$ and the way that the quality of the estimator is measured;
2. in certain cases, the estimator $\hat{P}_n$ may be better than the empirical estimator P_n;
3. the quality of the best estimator of this kind depends on the "richness" of the family $\mathcal{P}$.

One can measure the "richness" of the family $\mathcal{P}$ using Kolmogorov's ε-entropy, which we define presently.

6.2.2 ε-entropy, and ε-capacity

Let $(\mathcal{Y}, \rho)$ denote a metric space, equipped with a distance function ρ. A subset $B \subset \mathcal{Y}$ is said to be *totally bounded*, if for all $\varepsilon > 0$ there exists a *finite* set $T_\varepsilon \subset \mathcal{Y}$, which is called an ε-*covering* of B, such that for all $y \in B$,

$$\min_{x \in T_\varepsilon} \rho(y, x) \leq \varepsilon.$$

(In other words, B is covered by the closed balls of radius ε whose centers are points of T_ε, which are finite in number.)

Definition 1. *Let B denote a totally bounded subset of the metric space $(\mathcal{Y}, \rho)$ and let $N(\varepsilon, B) = N(\varepsilon, \rho, B)$ denote the smallest number of closed balls of radius ε which cover B. The quantity*

$$\mathcal{H}(\varepsilon, B) = \mathcal{H}(\varepsilon, \rho, B) = \ln N(\varepsilon, \rho, B)$$

is called the Kolmogorov ε-entropy of the set B.

Definition 2. *A set of points $x_1, \ldots, x_m$ of $\mathcal{Y}$ is called ε-distinguishable if for all i, j such that $i \neq j$, $\rho(x_i, x_j) > \varepsilon$. For $B \subset \mathcal{Y}$, let $\mathcal{N}(\varepsilon, B) = \mathcal{N}(\varepsilon, \rho, B)$ denote the largest number m for which there exists a set of ε-distinguishable points $x_1, \ldots, x_m$ in B. The quantity*

$$\mathcal{C}(\varepsilon, B) = \mathcal{C}(\varepsilon, \rho, B) = \ln \mathcal{N}(\varepsilon, B)$$

is called the ε-capacity of the set B.

It is obvious that

$$\mathcal{N}(2\varepsilon, B) \leq N(\varepsilon, B) \leq \mathcal{N}(\varepsilon, B), \quad \text{which gives} \quad \mathcal{C}(2\varepsilon, B) \leq \mathcal{H}(\varepsilon, B) \leq \mathcal{C}(\varepsilon, B).$$
(6.6)

The properties of ε-entropy and ε-capacity are well presented in the book by Kolmogorov and Tikhomirov (1959).

The function $\mathcal{H}(\varepsilon, \rho, \mathcal{P})$ describes in an adequate fashion (for statistical applications) the "richness" of the parametrising set $\mathcal{P}$ (on condition that there is a good choice of distance function ρ).

6.3 The Kullback-Leibler distance and the maximum likelihood estimator

Let P and G be two probability measures, whose densities are respectively f and g with respect to the measure μ. We consider the *Kullback-Leibler distance* between P and G (or between f and g):

$$K(P,G) = K(f,g) = \int_{f>0} \ln\left(\frac{f}{g}\right) f \, d\mu = \int_{f>0} \ln\left(f/g\right) dP. \qquad (6.7)$$

(One can verify that $K(P,G) \in [0, +\infty]$. One does not have $K(P,G) = K(G,P)$, but the tradition is to call it a distance nonetheless.) We wish to estimate the unknown density f of the probability law $P \in \mathcal{P}$, using the independent observations $X_1, \ldots, X_n, \ldots$ which obey this law $P \in \mathcal{P}$, and using the fact that we already know that $f \in \mathcal{F}$, where $\mathcal{F}$ is a class of densities:

$$\mathcal{F} = \left\{ f : f = \frac{dP}{d\mu}, P \in \mathcal{P} \right\}. \qquad (6.8)$$

We denote

$$K_n(g,f) = \int_{f>0} \ln\left(f/g\right) dP_n. \qquad (6.9)$$

The approach suggested by R. Fisher for finding the estimator $\hat{f}_n$, is to minimize the functional $K(f, \cdot)$ given the empirical data. More precisely, the estimator $\hat{f}_n$ is the point in $\mathcal{F}$ where the functional $K_n(f, \cdot)$ reaches its minimum:

$$\int_{f>0} \ln\left(f/\hat{f}_n\right) dP_n \leq \int_{f>0} \ln\left(f/g\right) dP_n \quad \text{for all } g \in \mathcal{F}. \qquad (6.10)$$

In this way, we have chosen a method for measuring the average risk of using the density g when the real density is equal to f (in the present case the average risk is given by $K(f,g)$), and then we have taken the density $\hat{f}_n \in \mathcal{F}$

which minimizes the average risk given the empirical data, i.e. the quantity $K_n(f, \hat{f}_n)$.

The estimator $\hat{f}_n$, given by (6.10), is called the *maximum likelihood estimator* of f because it maximizes, over $\mathcal{F}$, the *likelihood function* $\mathcal{L}$ (as a function of g):

$$\mathcal{L}(g; X_1, \ldots, X_n) = \prod_{j=1}^{n} g(X_j), \quad \max_{g \in \mathcal{F}} \mathcal{L}(g; X_1, \ldots, X_n) = \mathcal{L}(\hat{f}_n; X_1, \ldots, X_n).$$

(6.11)

The *Hellinger distance* $H(P, P_*)$ between two probability measures P and P_* is defined by the formula

$$H^2(P, P_*) = \frac{1}{2} \int \left(\sqrt{\frac{d\,P}{d\,\mu}} - \sqrt{\frac{d\,P_*}{d\,\mu}} \right)^2 d\mu, \tag{6.12}$$

where μ is a measure dominating the measures P and P_*. The quantity $H(P, P_*)$ does not depend on the choice of the measure μ. If the measure μ is fixed, we will write (whenever it is convenient) $h^2(f, f_*)$ instead of $H^2(P, P_*)$, where f, f_* denote the densities of the measures P, P_*. We observe that

$$h^2(f, f_*) = h^2_\mu(f, f_*) = 1 - \int \sqrt{f_*(x)/f(x)}\, f(x)\, d\mu. \tag{6.13}$$

6.4 The entropy of a partition and Fano's inequality

6.4.1 The entropy of a partition

Let $\tau = \{A_1, \ldots, A_N\}$ be a measurable partition of the probability space $\{\Omega, \mathcal{B}, P\}$:

$$\bigcup_{j=1}^{N} A_j = \Omega, \quad A_i \cap A_j = \emptyset \ (i \neq j), \quad A_j \in \mathcal{B}.$$

The quantity

$$H(\tau) = - \sum_{j=1}^{N} P(A_j) \ln P(A_j)$$

is called the *entropy of the partition* τ, the convention being that the terms in the sum for which $P(A_j) = 0$ are zero. It is clear[6] that

$$0 < - \sum_{j=1}^{N} P(A_j) \ln P(A_j) \leq \ln N. \tag{6.14}$$

---

[6] Maximize $- \sum\limits_{j=1}^{N} x_j \ln x_j$ subject to the constraint $\sum\limits_{j=1}^{N} x_j = 1$

Suppose that $\tau_* = \{A_1^*, \ldots, A_{N*}^*\}$ is a *finer* partition than the partition τ: in other words, each A_i^* is included in an A_j. We will denote this relation: $\tau_* \succ \tau$. Let, for example, $A_k = \underset{j \in J_k}{\cup} A_j^*$.

Let us recall that if we are given a *convex* function φ, a probability law P on a measurable space $\{\mathcal{X}, \mathcal{B}\}$, and a measurable function ξ, then *Jensen's inequality* holds:

$$\int_{\mathcal{X}} (\varphi \circ \xi) \, dP \geq \varphi \left(\int_{\mathcal{X}} \xi \, dP \right).$$

One deduces, in particular, that

$$P(A_k) \ln P(A_k) \geq \sum_{j \in J_k} P(A_j^*) \ln P(A_j^*),$$

which shows that the entropy of a partition finer than τ is greater than the entropy of τ:

$$H(\tau) \leq H(\tau_*) \qquad \text{if } \tau_* \succeq \tau. \tag{6.15}$$

6.4.2 Conditional entropy

If $P(B) > 0$, the quantity

$$H_B(\tau) = -\sum_{j=1}^{N} P(A_j|B) \ln P(A_j|B)$$

is called the *conditional entropy* (on observing B).

If $\tau = \{A_1, \ldots, A_N\}$, let $\sigma(\tau)$ denote the system of subsets:

$$B = \underset{j \in J}{\cup} A_j, \quad J \subset \{1, \ldots, N\}. \tag{6.16}$$

6.4.3 The entropy of a random variable

It is easiest to imagine the case when τ is defined by a random element Y and by a partition $\tau_* = \{A_1^*, \ldots, A_N^*\}$ of the space of possible values of Y, i.e.

$$A_i = \{Y \in A_i^*\}.$$

In this case $\sigma(\tau) \subset \sigma(Y)$, where $\sigma(Y)$ is the smallest σ-algebra with respect to which the function Y is measurable. If Y takes values in a finite or countable set, and if τ is the finer of the above partitions, then by definition

$$H(Y) = H(\tau).$$

6.4.4 The quantity of information

Suppose that we have two partitions:

$$\tau_1 = \{B_1, \ldots, B_N\} \quad \text{and} \quad \tau_2 = \{D_1, \ldots, D_r\},$$

and let

$$\tau = \tau_1 \vee \tau_2 = \left\{ A_{i,j} = B_i \cap D_j \ \text{s.t.} \ P(A_{i,j}) > 0 \ (1 \le i \le N, \ 1 \le j \le r) \right\}.$$

We shall write $H(\tau_1, \tau_2)$ instead of $H(\tau_1 \vee \tau_2)$. The *quantity of information of the partition* τ_1 *with respect to* τ_2 is denoted by

$$I(\tau_1, \tau_2) = \sum_{i,j} P(A_{ij}) \ln \frac{P(A_{ij})}{P(B_i)P(D_j)}. \tag{6.17}$$

Observe that

$$I(\tau_1, \tau_2) = H(\tau_1) + H(\tau_2) - H(\tau_1, \tau_2) = I(\tau_2, \tau_1). \tag{6.18}$$

By Jensen's inequality

$$I(\tau_1, \tau_2) \le I(\tau_1^*, \tau_2^*), \ \text{if} \ \tau_1^* \succeq \tau_1 \ \text{and} \ \tau_2^* \succeq \tau_2.$$

Let X, Y denote two random elements defined on a probability space. Let

$$I(X, Y) = \sup_{\tau_1, \tau_2} I(\tau_1, \tau_2) \tag{6.19}$$

denote the quantity of information contained in X with respect to Y, where the supremum is calculated over the set of all partitions τ_1, τ_2 satisfying the conditions: $\sigma(\tau_1) \subset \sigma(X)$, $\sigma(\tau_2) \subset \sigma(Y)$. The supremum $I(X, Y)$ is equal to

$$I(X, Y) = \begin{cases} K\left(P_{X,Y}, P_X \times P_Y\right) & \text{if } P_{X,Y} \ll P_X \times P_Y \\ \infty & \text{otherwise} \end{cases} \tag{6.20}$$

where $P_{X,Y}$ is the probability law of the vector (X, Y), P_X, P_Y the probability laws of the random elements X, Y (where we recall that the relation $\mu \ll \nu$ means that the measure μ is absolutely continuous with respect to the measure ν). We observe that (6.20), unlike (6.18), correctly defines the quantity $I(X, Y)$ even in the case where the quantities $H(X), H(Y)$ are infinite.

6.4.5 Connection between the quantity of information and the Kullback-Leibler distance

We will need to know certain properties of the quantity $I(X, Y)$ in the particular case when the probability law P_Y of the random variable Y is uniform over the finite set $\mathcal{Y} = \{1, ..., N\}$. Let us denote this law by ν.

Let $\{f_y, y \in \mathcal{Y}\}$ be the family of densities (with respect to a measure μ on $\mathcal{X}$) of probability laws P_{f_y} on a measurable space $\{\mathcal{X}, \mathcal{B}\}$. Suppose that (Y, X) is a random element chosen in $\Gamma_N = \mathcal{Y} \times \mathcal{X}$ according to the law:

$$P_{Y,X}\{J \times A\} = P\{Y \in J, X \in A\} = \frac{1}{N} \sum_{y \in J} P_{f_y}\{X \in A\}.$$

Let us calculate the density of the probability law $P_{Y,X}$ with respect to the measure $\nu \times \mu$ on Γ_N:

$$\frac{d\,P_{Y,X}}{d\,(\nu \times \mu)}\,(y,x) = \frac{1}{N} f_y(x).$$

Therefore

$$I(Y, X) = \frac{1}{N} \sum_{j=1}^{N} \int f_j(x) \ln \left(\frac{f_j(x)}{\frac{1}{N} \sum\limits_{i=1}^{N} f_i(x)} \right) d\mu(x) = \frac{1}{N} \sum_{j=1}^{N} K\left(f_j, \overline{f}\right),$$

$$(6.21)$$

where

$$\overline{f} = \frac{1}{N} \sum_{i=1}^{N} f_i(x).$$

If g is the density (with respect to the measure μ) of a probability law, we thus have

$$\frac{1}{N} \sum_{j=1}^{N} \int f_j(x) \ln \left(\frac{g(x)}{\frac{1}{N} \sum\limits_{i=1}^{N} f_i(x)} \right) d\mu(x) = \int \overline{f}(x) \ln \left(\frac{g(x)}{\overline{f}(x)} \right) d\mu(x) \tag{6.22}$$

$$= -K\left(\overline{f}, g\right) \le 0,$$

and we therefore obtain

$$I(Y, X) \le I(Y, X) + K\left(\overline{f}, g\right) = \frac{1}{N} \sum_{j=1}^{N} \int f_j(x) \ln \left(\frac{f_j(x)}{g(x)} \right) d\mu(x)$$

$$(6.23)$$

$$= \frac{1}{N} \sum_{j=1}^{N} K\left(f_j, g\right).$$

It follows from (6.21) and (6.23) that

$$I(Y, X) = \inf_{g} \frac{1}{N} \sum_{j=1}^{N} K\left(f_j, g\right). \tag{6.24}$$

Let us now consider the case where $X = (X_1, \ldots, X_n)$ is a vector whose coordinates X_i take values in a measurable space $\{\mathcal{X}_*, \mathcal{B}_*\}$, and are conditionally independent given $Y = y$:

$$f_y(x) = \prod_{j=1}^n g_y(x_j) = g_y^{(n)}(x), \quad x = (x_1, \ldots, x_n), \quad x_j \in \mathcal{X}_*,$$

$$x \in \mathcal{X} = \mathcal{X}_*^n, \quad d\mu(x) = d\mu_*(x_1) \times \ldots \times d\mu_*(x_n). \tag{6.25}$$

Here, μ_* is a measure on $\{\mathcal{X}_*, \mathcal{B}_*\}$, and $\{g_y, \ y \in \mathcal{Y}\}$ is the family of probability densities (with respect to the measure μ_*) on the measurable space $\{\mathcal{X}_*, \mathcal{B}_*\}$.

By (6.24), for each probability law Q with density q with respect to μ_* we have:

$$I(Y, X) \le \frac{1}{N} \sum_{j=1}^N K\left(g_j^{(n)}, q^{(n)}\right) = \frac{n}{N} \sum_{j=1}^N K\left(g_j, q\right).$$

By choosing

$$q(x) = \bar{g}(x) = \frac{1}{N} \sum_{i=1}^N g_i(x),$$

we obtain

$$I\left(Y, (X_1, \ldots, X_n)\right) \le n I(Y, X_1) = n \inf_q \frac{1}{N} \sum_{j=1}^N K\left(g_j, q\right). \tag{6.26}$$

Let q be the density of a probability law. We will call a K-ball with center q and radius r the set of densities (with respect to the measure μ):

$$U(q, r) = \{f : K(f, q) \le r\}. \tag{6.27}$$

Let $U(q, r)$ be a K-ball of minimal radius which contains $\mathcal{F}$. We denote by $\tau(\mathcal{F})$ its radius. Relation (6.26) gives the following estimate:

$$I(Y, X) \le n \tau(\mathcal{F}). \tag{6.28}$$

If X, Y are random elements with discrete probability laws we will use the following notation:

$$H(Y|X) = \sum_x H_{X=x}(Y) P\{X = x\}. \tag{6.29}$$

It is clear that
$$I(X, Y) = H(Y) - H(Y|X). \tag{6.30}$$

6.4.6 Fano's inequality

Let θ be a random element, which takes its values in a finite set $\Theta = \{\theta_1, \ldots, \theta_N\}$, and let $\hat{\theta} = \psi(X) \in \Theta$ be the estimator of θ given the observation X.

Lemma 1 (Fano's inequality). *The following inequality holds:*

$$P\{\theta \neq \hat{\theta}\} \geq \frac{H(\theta)}{\ln(N-1)} - \frac{I(\theta, X) + \ln 2}{\ln(N-1)}. \tag{6.31}$$

Proof. Let

$$B = \{\theta \neq \psi\}, \ \psi = \psi(X).$$

Then

$$H_{\psi=\theta_k}(\theta) = -\sum_{j \neq k} P\{\theta = \theta_j | \psi = \theta_k\} \ln P\{\theta = \theta_j | \psi = \theta_k\} -$$

$$-P\{\theta = \theta_k | \psi = \theta_k\} \ln P\{\theta = \theta_k | \psi = \theta_k\} =$$

$$= -P\{B | \psi = \theta_k\} \sum_{j \neq k} \frac{P\{\theta = \theta_j | \psi = \theta_k\}}{P\{B | \psi = \theta_k\}} \ln \frac{P\{\theta = \theta_j | \psi = \theta_k\}}{P\{B | \psi = \theta_k\}} -$$

$$-P\{B | \psi = \theta_k\} \ln P\{B | \psi = \theta_k\} - P\{\overline{B} | \psi = \theta_k\} \ln P\{\overline{B} | \psi = \theta_k\}.$$

Since (see (6.14))

$$-P\{B | \psi = \theta_k\} \ln P\{B | \psi = \theta_k\} - P\{\overline{B} | \psi = \theta_k\} \ln P\{\overline{B} | \psi = \theta_k\} \leq \ln 2$$

and

$$-\sum_{j \neq k} \frac{P\{\theta = \theta_j | \psi = \theta_k\}}{P\{B | \psi = \theta_k\}} \ln \frac{P\{\theta = \theta_j | \psi = \theta_k\}}{P\{B | \psi = \theta_k\}} \leq \ln(N-1),$$

then

$$P(B | \psi = \theta_j) \geq \frac{H_{\psi(X)=\theta_j}(\theta) - \ln 2}{\ln(N-1)},$$

which implies that

$$P(B) = \sum_{j=1}^{N} P(B | \psi = \theta_j) P\{\psi = \theta_j\} \geq \frac{H(\theta | \psi) - \ln 2}{\ln(N-1)}.$$

If we replace $H(\theta | \psi(X))$ by $H(\theta) - I(\theta, \psi(X))$ (see (6.30)), we get

$$P(B) \geq \frac{H(\theta)}{\ln(N-1)} - \frac{I(\theta, \psi(X)) + \ln 2}{\ln(N-1)},$$

from which we obtain (6.31), after applying the inequality

$$I(\theta, \psi(X)) \leq I(\theta, X).$$

□

6.5 The lower bound for the minimax risk

Consider independent random variables $X_1, \ldots, X_n$ taking values in a measurable space $\{\mathcal{X}, \mathcal{B}\}$, with the same probability law $P \in \mathcal{P}$ with density f with respect to the measure μ. Let X denote the random vector $X = (X_1, \ldots, X_n)$. Suppose that we have chosen a metric $d(P, Q)$ to measure the proximity of the laws P and Q. If f and q denote the densities of P and Q with respect to the fixed measure μ, we will also write $d(f, q) = d(P, Q)$.

6.5.1 The case where the set of densities is finite

First of all let us suppose that $f \in \mathcal{F}$ where $\mathcal{F}$ is *finite*: $\mathcal{F} = \{f_1, ..., f_N\}$, where the points f_i are ε-distinguishable: $d(f_i, f_j) > \varepsilon$ $(i \neq j)$.

The problem is to construct an estimator $\hat{f}_n$ for the unknown density f. For $\theta \in \{1, \ldots, N\}$, let $\theta(f)$ denote the index of the unknown density $f : f = f_\theta$. Let $\hat{f}_n = \hat{f}_n(X_1, \ldots, X_n)$ be an estimator of the unknown density f, taking values in $\mathcal{F}$, and let $\hat{\theta}_n = \theta(\hat{f}_n)$ be its index. Suppose that $\hat{\theta}_n = \psi(X)$ is an estimator for $\theta = \theta(f)$.

We define the following loss function:

$$l_*(\hat{f}_n, f) = \begin{cases} 0 & \text{if } \hat{f}_n = f, \\ 1 & \text{if } \hat{f}_n \neq f. \end{cases}$$

The accuracy of the estimate is therefore measured by the risk function

$$R_*(\hat{f}_n) = R_*(\hat{\theta}_n) = \max_{f \in \mathcal{F}} P_f\{\theta(f) \neq \hat{\theta}_n\}.$$

Suppose that (θ, X) is an element in $\Gamma_n = \{1, ..., N\} \times \mathcal{X}$ chosen randomly according to the probability law $P_{\theta, X}$:

$$P_{\theta, X}\{J \times A\} = P\{\theta \in J, X \in A\} = \frac{1}{N} \sum_{\theta \in J} P_{f_\theta}\{X \in A\}.$$

It is clear that

$$R_*(\hat{f}_n) = R_*(\hat{\theta}_n) = \max_{f \in \mathcal{F}} P_f\{\theta(f) \neq \hat{\theta}_n\} \geq \frac{1}{N} \sum_\theta P_{f_\theta}\{\theta \neq \hat{\theta}_n\}$$

$$= P_{\theta, X}\{\theta \neq \hat{\theta}_n\},$$

which, by Fano's inequality, gives:

$$R_*(\hat{\theta}_n) \geq \frac{H(\theta)}{\ln(N-1)} - \frac{I(\theta, X) + \ln 2}{\ln(N-1)} = \frac{\ln N}{\ln(N-1)} - \frac{I(\theta, X) + \ln 2}{\ln(N-1)}$$

$$\geq 1 - \frac{I(\theta, X) + \ln 2}{\ln(N-1)}.$$

(6.32)

It therefore follows from (6.28) that

$$R_*(\hat{f}_n) = R_*(\hat{\theta}_n) \geq 1 - \frac{n\tau(\mathcal{F}) + \ln 2}{\ln(\mathcal{N} - 1)}. \tag{6.33}$$

Now let us consider another loss function generated by a function l which is a positive increasing function on $[0, \infty[$. The losses caused by using the estimator $\hat{f}_n$, when f is the true density, will be measured by $l\left(d(\hat{f}_n, f)\right)$. Therefore, since

$$R(\hat{f}_n) = \sup_{f \in \mathcal{F}} \mathbf{E}_f \, l\left(d(\hat{f}_n, f)\right) \geq l(\varepsilon/2) \sup_{f \in \mathcal{F}} P_f\{\hat{f}_n \neq f\} = l(\varepsilon/2) R_*(\hat{f}_n),$$

we obtain the following lower bound for the minimax risk $R(\hat{f}_n)$:

$$R(\hat{f}_n) \geq l(\varepsilon/2) \left(1 - \frac{n\tau(\mathcal{F})}{\ln(\mathcal{N} - 1)} - \frac{\ln 2}{\ln(\mathcal{N} - 1)}\right). \tag{6.34}$$

6.5.2 The general case

Now let us consider the general case: we no longer assume that the set $\mathcal{F}$ is *finite* but *totally bounded* with respect to the metric d. Recall that this means that for every $\varepsilon > 0$, there exists a finite set of densities, T_ε, such that

$$\mathcal{F} \subset \bigcup_{g \in T_\varepsilon} V(g, \varepsilon)$$

where $V(g, \varepsilon)$ is the ball with center g and radius ε for the metric d.

We consider a finite subset $\mathcal{F}_* = \{f_1, \ldots, f_{\mathcal{N}}\} \subset \mathcal{F}$, which only contains points which are ε-distinguishable (i.e. $d(f_i, f_j) > \varepsilon$ when $i \neq j$), and chosen in such a way that for every point $q \in \mathcal{F} \setminus \mathcal{F}_*$, we have:

$$\min_{f \in \mathcal{F}_*} d(f, q) \leq \varepsilon. \tag{6.35}$$

We consider an estimator $\hat{f}_n = \hat{f}_n(X_1, \ldots, X_n) \in \mathcal{F}$ with unknown density f. Using $\hat{f}_n$, we will construct a new estimator $\hat{f}_n^* \in \mathcal{F}_*$ such that

$$d(\hat{f}_n^*, \hat{f}_n) \leq d(f, \hat{f}_n), \ f \in \mathcal{F}_*. \tag{6.36}$$

So,

$$\text{if } f \in \mathcal{F}_* \text{ and } d(f, \hat{f}_n) \geq \varepsilon/2, \text{ then } f \neq \hat{f}_n^*.$$

It is clear that

$$R(\hat{f}_n) = R(\hat{f}_n, \mathcal{F}) = \sup_{f \in \mathcal{F}} \mathbf{E}_f \, l\left(d(\hat{f}_n, f)\right) \geq \sup_{f \in \mathcal{F}_*} \mathbf{E}_f \, l\left(d(\hat{f}_n, f)\right)$$

$$\geq l(\varepsilon/2) \sup_{f \in \mathcal{F}_*} P_f\left(d(\hat{f}_n, f) \geq \varepsilon/2\right). \tag{6.37}$$

For every $f \in \mathcal{F}_*$, the event $\{d(\hat{f}_n, f) \geq \varepsilon/2\}$ coincides with the event $\{\hat{f}_n^* \neq f\}$. Therefore, by applying (6.37) and (6.34) (to the finite set $\mathcal{F}_*$) it follows that

$$R(\hat{f}_n) \geq l(\varepsilon/2) \left(1 - \frac{n\tau(\mathcal{F}_*)}{\ln(\mathcal{N} - 1)} - \frac{\ln 2}{\ln(\mathcal{N} - 1)} \right), \tag{6.38}$$

and since $\tau(\mathcal{F}_*) \leq \tau(\mathcal{F})$, we conclude that relation (6.34) is also valid in the general case (without necessarily assuming that the set $\mathcal{F}$ is finite).

6.5.3 A lower bound for the risk

In order to use inequality (6.38), a finite set $\mathcal{F}_* = \{f_1, \ldots, f_{\mathcal{N}}\} \subset \mathcal{F}$, ε and n must be chosen suitably. More precisely, one must choose ε in such a way that

$$\frac{n\tau(\mathcal{F}_*)}{\ln(\mathcal{N}(\varepsilon) - 1)} < C < 1.$$

In this case, (6.38) implies that

$$R(\hat{f}_n) \geq C_* \, l(\varepsilon/2), \text{ for } \mathcal{N} = \mathcal{N}(\varepsilon) \text{ sufficiently large and } C_* < 1 - C.$$

So, let $\mathcal{F}_N \subset \mathcal{F}$ denote a finite set, consisting of N ε-distinguishable points. It is clear that $R(\hat{f}_n, \mathcal{F}) \geq R(\hat{f}_n, \mathcal{F}_N)$. One can make the following

Remark 6.5.1 If

$$\frac{n\tau(\mathcal{F}_N)}{\ln(N - 1)} < C < 1, \tag{6.39}$$

then for n sufficiently large and $0 < C_* < 1 - C$:

$$R(\hat{f}_n, \mathcal{F}) \geq C_* \, l(\varepsilon/2). \tag{6.40}$$

6.5.4 The quantity of ε-distinguishable points in a discrete set

We will show, on an example, how to deduce a lower bound on the risk $R(\hat{f}_n)$ from the inequality (6.40). In order to do this, we will need an upper bound for the number of ε-distinguishable points in a discrete set. Let A be the set of vectors $\mathbf{a} = (a_1, \ldots, a_N)$ with coordinates $a_j \in \{-1, 1\}$. We therefore have $\text{card } A = 2^N$. Consider the Hamming distance

$$v(\mathbf{a}, \mathbf{b}) = \text{card} \, \{j : a_j \neq b_j\} \geq r.$$

Given an integer $r > 1$, we will say that a subset $B(r) \subset A$ consists of r-*distinguishable* points if for every pair $(\mathbf{a}, \mathbf{b})$ of points on $B(r)$, with $\mathbf{a} = (a_1, \ldots, a_N)$, $\mathbf{b} = (b_1, \ldots, b_N)$, we have

$$v(\mathbf{a}, \mathbf{b}) \geq r.$$

Then, we shall assume that the set $B(r) \subset A$ is maximal, in the following sense:

$$v(\mathbf{a}, B(r)) = \min_{\mathbf{b} \in B(r)} v(\mathbf{a}, \mathbf{b}) \leq r, \quad \text{si} \quad \mathbf{a} \in A \setminus B(r).$$

Our aim is to estimate the quantity $M(N, r) = \text{card } B(r)$.

To each point $\mathbf{a} \in B$ there corresponds a ball $V(\mathbf{a})$ (for the "metric" v) of radius $(r - 1)$ centered at $\mathbf{a}$:

$$V(\mathbf{a}) = \{\mathbf{b} : \mathbf{b} \in A, \ v(\mathbf{a}, \mathbf{b}) \leq r - 1\}.$$

The number of elements card $V(\mathbf{a})$ of this ball is easy to calculate:

$$\text{card } V(\mathbf{a}) = \sum_{j=0}^{r-1} C_N^j = 2^N P\{\xi > N - r\},$$

where ξ is a random variable with binomial distribution $B(N, 1/2)$. Therefore, using Hoeffding's inequality for $r = N/4$ we deduce that

$$P\{\xi > N - r\} \leq \exp\{-N/8\},$$

and since $\bigcup_{\mathbf{a} \in B(r)} V(\mathbf{a}) = A$ and card $B(r) \times$ card $V(\mathbf{a}) \geq$ card A, we obtain the following lemma:

Lemma 2. *We have the inequality*

$$M(N, N/4) \geq \exp\{N/8\}$$

(where $M(N, r) = \text{card } B(r)$).

6.5.5 Example: An estimating problem for smooth density

Let $\mathcal{F}$ be the set of densities f which are defined on $\mathbb{R}$, which are r-differentiable, and whose rth derivative $f^{(r)}$ satisfies the following condition (for fixed $\alpha \in]0, 1[$):

$$\|f^{(r)}(\cdot + h) - f^{(r)}(\cdot)\|_p \leq C|h|^\alpha, \text{ where } \|g(\cdot)\|_p = \left(\int_{-\infty}^\infty |g(x)|^p \, dx\right)^{1/p}.$$

(6.41)

We write $\beta = r + \alpha$. Corresponding to the estimator $\hat{f}_n$, which was constructed using the observations $X_1, \ldots, X_n$, we write

$$R(\hat{f}_n) = \sup_{f \in \mathcal{F}} \|\hat{f}_n - f\|_p.$$

Suppose that $\varphi \in C^\infty$ with supp $\varphi \subset [1/4, 3/4]$ and

$$\int \varphi(x) \, dx = 0.$$

Let $N > 1$. We set

$$\psi_j(x) = \varphi\left(N(x - j/N)\right), \qquad j = 0, 1, \ldots, N - 1.$$

Let $f_* \in \mathcal{F}$ be a density satisfying (6.41) with the constant $C/2$ such that $f_*(x) \geq 1/2$ if $x \in [0, 1]$. We set

$$f(x) = f_\mathbf{a}(x) = f_*(x) + \psi(x), \quad \psi(x) = \psi^\mathbf{a}(x) = C(N) \sum_{j=0}^{N-1} a_j \psi_j(x). \quad (6.42)$$

Here $\mathbf{a} = (a_1, \ldots, a_N) \in B(N/4)$, and the normalization constant $C(N)$ will be chosen later in such a way that $C(N) \to 0$ when $N \to \infty$. Therefore the function f is positive when N is sufficiently large, and is indeed a probability density:

$$\int f(x)\, dx = 1.$$

Let $\mathcal{F}_N$ denote the set of all the densities f which can be written in the form (6.42). In this case, by Lemma 2, we have

$$\operatorname{card} \mathcal{F}_N \geq C_* \exp\{N/8\}. \quad (6.43)$$

We observe that

$$\|\psi_j^{(k)}\|_p^p = N^{kp-1}\|\varphi^{(k)}\|_p^p \qquad (k = 0, 1, \ldots)$$

and thus (the functions $\psi_j^{(k)}$ are supported on disjoint intervals):

$$\|\psi^{(k)}\|_p^p = C^p(N) \sum_{j=0}^{N-1} \|\psi_j^{(k)}\|_p^p = C^p(N) N^{kp}\|\varphi^{(k)}\|_p^p \quad (6.44)$$

and

$$\|\psi_j^{(r)}(\cdot + h) - \psi_j^{(r)}(\cdot)\|_p^p = N^{rp-1}\|\varphi^{(r)}(\cdot + h) - \varphi^{(r)}(\cdot)\|_p^p. \quad (6.45)$$

Let $\mathbf{a}$ and $\mathbf{b}$ be two distinguishable points in $B(N/4)$. It follows from (6.44) that

$$\|f_\mathbf{a} - f_\mathbf{b}\|_p = C(N) \,\|\sum_{j=0}^{N-1} (a_j - b_j)\psi_j\|_p$$

$$= 2^{1/p}\, C(N)\, N^{-1/p} \{\operatorname{card}\{j : a_j \neq b_j\}\}^{1/p}\|\varphi\|_p \quad (6.46)$$

$$\geq \left(\frac{1}{2}\right)^{1/p} C(N)\|\varphi\|_p.$$

Let us consider the case where $|h| \leq \frac{1}{4N}$. Then, by (6.44), and since the functions $\psi_j^{(r)}(\cdot + h) - \psi_j^{(r)}(\cdot)$ are supported on disjoint intervals, we have

$$\|\psi^{(r)}\|_p = C(N)N^r \|\varphi^{(r)}(\cdot + h) - \varphi^{(r)}(\cdot)\|_p \leq \{C(N)N^r\} C|h|^\alpha. \qquad (6.47)$$

Now , if $|h| \geq \frac{1}{4N}$, we have

$$\begin{aligned}
\|\psi^{(r)}(\cdot + h) - \psi^{(r)}(\cdot)\|_p &\leq 2\|\psi^{(r)}\|_p = 2C(N)N^r \|\varphi^{(r)}\|_p \\
&\leq 2 \{C(N)N^{r+\alpha}\} \|\varphi^{(r)}\|_p |h|^\alpha.
\end{aligned} \qquad (6.48)$$

Let us suppose, for simplicity, that the function φ has been chosen in such a way that $\|\varphi^{(r)}\|_p \leq 1/2$. Then from (6.47) and (6.48), it follows that

$$\|\psi^{(r)}(\cdot + h) - \psi^{(r)}(\cdot)\|_p \leq \frac{1}{2}C|h|^\alpha, \quad \text{if} \quad C(N) \leq \frac{N^{-\beta}}{2}. \qquad (6.49)$$

These results are summarized in the following

Lemma 3. *Let $C(N) = L N^{-\beta}$. If $L \leq 1/2$ and if N is sufficiently large, then*

1. *the functions $f = f_\mathbf{a}$ defined by (6.42) are densities;*
2. *$\mathcal{N} = \operatorname{card} \mathcal{F}_N = \{f : f = f_\mathbf{a}, \mathbf{a} \in B(N/4)\} \geq C_* \exp\{N/8\};$*
3. *elements of the set $\mathcal{F}_N$ are ε-distinguishable when*

$$\varepsilon = L_* N^{-\beta}, \ L_* = \left(\frac{1}{2}\right)^{1/p} L\|\varphi\|_p;$$

4. *$\|f_\mathbf{a} - f_*\|_p = C(N)\|\psi\|_p = L N^{-\beta}\|\varphi\|_p;$*
5. *the functions f in $\mathcal{F}_N$ belong to $\mathcal{F}$.*

The last statement requires some explanation. One must show that the functions f in $\mathcal{F}_N$ satisfy condition (6.41). Indeed,

$$\|f^{(r)}(\cdot + h) - f^{(r)}(\cdot)\|_p \leq \|f_*^{(r)}(\cdot + h) - f_*^{(r)}(\cdot)\|_p + \|\psi^{(r)}(\cdot + h) - \psi^{(r)}(\cdot)\|_p.$$

Since the functions f_* satisfy (6.41) with constant $C/2$, then by using (6.49) we obtain (6.41) for the function f.

6.5.6 The lower bound for the risk in the estimating problem for smooth density

We will now apply remark 6.5.1. In order to do this, we must bound $\tau(\mathcal{F}_N)$ above. We have

$$\tau(\mathcal{F}_N) \leq \sup_{\mathbf{a} \in B(N/4)} K(f_\mathbf{a}, f_*).$$

Then,

$$K(f_{\mathbf{a}}, f_*) \leq K(f_{\mathbf{a}}, f_*) + K(f_*, f_{\mathbf{a}}) = \int\limits_0^1 \ln\left(1 + \frac{\Delta(x)}{f_*}\right) \Delta(x)\, dx,$$

where $\quad \Delta(x) = f_{\mathbf{a}}(x) - f_*(x)$.

Using the inequality $\ln(1 + x) \leq x$ and the condition $f_*(x) \geq 1/2$ (for $x \in [0,1]$), we deduce that if $p \geq 2$,

$$K(f_{\mathbf{a}}, f_*) \leq 2 \int\limits_0^1 \Delta^2(x)\, dx \leq 2\|\Delta(x)\|_p^2 = 2L^2\, N^{-2\beta}\|\varphi\|_p^2.$$

Thus, our aim (see remark 6.5.1) is to modify n and N in such a way that

$$\frac{n\tau(\mathcal{F}_N)}{\ln(\mathcal{N} - 1)} \leq C < 1.$$

For N sufficiently large, $\ln(\mathcal{N} - 1) > N/16$, and therefore we have

$$\frac{n\tau(\mathcal{F}_N)}{\ln(\mathcal{N} - 1)} \leq k(L, \|\varphi\|_p) n N^{-(2\beta+1)}.$$

The obvious solution $N = N_n = C_1 n^{\frac{1}{2\beta+1}}$, with a suitable constant C_1, gives the following asymptotic expression for $\varepsilon = \varepsilon_n$:

$$\varepsilon_n \asymp C_2 n^{-\frac{\beta}{2\beta+1}}.$$

From this expression and using remark 6.5.1, we obtain a lower bound for the risk:

$$R(\hat{f}_n) \geq C_3 n^{-\frac{\beta}{2\beta+1}} \tag{6.50}$$

where $C_3 > 0$ is a suitable constant, and for n sufficiently large.

For $s \in \mathbb{N}$, a function $\mathcal{K} : \mathbb{R} \to \mathbb{R}$ is said to be a *kernel* of order s if it satisfies the following properties:

1. $\int |x|^{s+1} |\mathcal{K}(x)|\, dx < \infty$;
2. $\int x^j \mathcal{K}(x)\, dx = 0$ for $1 \leq j \leq s$;
3. $\int \mathcal{K}(x)\, dx = 1$.

The estimator of the density associated to the kernel $\mathcal{K}$ is defined by

$$\hat{f}_n(x) = \frac{1}{h_n} \int \mathcal{K}\left(\frac{x-t}{h_n}\right) dP_n(t), \tag{6.51}$$

where the $h_n > 0$ are chosen in such a way that $h_n \to 0$ and $nh_n \to \infty$ when $n \to \infty$. For suitable choices of s, the kernel $\mathcal{K}$ of order s and the ("window-width") h_n, the risk $R(\hat{f}_n)$ of the estimator (6.51), under the conditions of our example, tends to zero at the same rate as the lower bound in (6.50). This lower bound is in fact the desired lower bound. More details can be found in the article by J. Bretagnolle and C. Huber (1979).

6.6 Consistency of the estimation

6.6.1 Consistency with a certain rate of convergence

One must choose a suitable approach for measuring the accuracy of the estimate of the unknown law $P \in \mathcal{P}$ by the estimator $\hat{P}_n$. Let $(\mathcal{Y}, d)$ be a metric space and let θ be a function from $\mathcal{P}$ to $\mathcal{Y}$. Suppose that we have to estimate the value $\theta(P)$ at a point $P \in \mathcal{P}$. We shall first measure the rate of convergence, in some sense, of the estimator towards the true probability law.

Let $X_1, \ldots, X_n, \ldots$ be independent random variables with the same probability law $P \in \mathcal{P}$. Let $r_n = r_n(X_1, \ldots, X_n)$ be a observable sequence of strictly positive variables such that

$$r_n \to 0 \quad \text{when } n \to \infty \quad \text{(in probability)},$$

d is the chosen metric for measuring the accuracy of the estimate. We say that the estimator $\hat{\theta}_n$ of the value $\theta(P)$ is consistent with rate r_n (one should rather say: with rate *at least* r_n) if

$$\sup_{P \in \mathcal{P}} P \left\{ \frac{d(\hat{\theta}_n, \theta(P))}{r_n} > t \right\} \to 0, \quad \text{uniformly on } n, \text{ when } t \to +\infty . \quad (6.52)$$

Under this condition it is possible to localized the value of unknown parameter θ on observations since for the observable ball

$$V = V(X_1, \ldots, X_n) = \left\{ \theta : d(\hat{\theta}_n, \theta) \leq t r_n(X_1, \ldots, X_n) \right\}$$

we have

$$P \{V \ni \theta(P)\} \to 1, \text{ as } t \to \infty.$$

We illustrate this on a simple example. Let $X_1, X_2, \ldots, X_n$ be a sample of random variables from (the same) normal distribution: $X_j \sim N(\theta, \sigma^2)$, $\theta \in \mathbb{R}$, $\sigma > 0$. Then $\overline{X} = \frac{1}{n} \sum_{j=1}^{n} X_j$ is an estimator of the parameter θ which is consistent with rate r_n defined by:

$$r_n = \frac{S_n}{\sqrt{n}}, \quad \text{where} \quad S_n^2 = \frac{1}{n} \sum_{j=1}^{n} \left(X_j - \overline{X}\right)^2 .$$

6.6.2 Consistency and empirical process

The following lemma, which belongs to Sara van de Geer (1993), shows us, as the accuracy of estimating can be connected with the rate of convergence of the empirical process. Let $\hat{f}_n$ be the maximum likelihood estimator of unknown density f constructed on observations $X_1, \ldots, X_n$ with common density $f \in \mathcal{F}$

and distribution P. We assume that all functions $f \in \mathcal{F}$ are supported on the same set Λ and positive on Λ. We set

$$\Phi = \left\{ \varphi : \ \varphi = \sqrt{g/f} \text{ for some } g, f \in \mathcal{F} \right\}$$

Lemma 4.

$$h^2(\hat{f}_n, f) \leq \int_\Lambda \left(\sqrt{\frac{\hat{f}_n}{f}} - 1 \right) d(P_n - P) \leq \sup_{\varphi \in \Phi} | \int_\Lambda \varphi \, d(P_n - P)|. \qquad (6.53)$$

Proof.

$$0 \leq \frac{1}{2} \int_{f>0} \ln\left(\frac{\hat{f}_n}{f} \right) dP_n \leq \int_{f>0} \left(\sqrt{\frac{\hat{f}_n}{f}} - 1 \right) dP_n$$

$$= \int_{f>0} \left(\sqrt{\frac{\hat{f}_n}{f}} - 1 \right) d(P_n - P) - h^2(\hat{f}_n, f). \qquad (6.54)$$

□

6.7 The estimator of the minimal distance

6.7.1 Construction of the metric

Let $X_1, ..., X_n, \cdots$ denote independent random elements of measurable space $\{\mathcal{X}, \mathcal{B}\}$, with the same probability law $P \in \mathcal{P}$ with density $f \in \mathcal{F}$ to the measure μ. We wish to construct a reasonable estimator $\hat{P}_n$ for the unknown law $P \in \mathcal{P}$. The method of estimation of course depends on the information which is known *a priori* about the unknown law (in other words the set $\mathcal{P}$) and the way in which the accuracy of the estimation is measured (in other words, the choice of the metric d). The quality of the estimation will therefore be given by the rate of convergence (r_n) of the estimator $\hat{P}_n$ towards P, as defined by (6.52).

We will take as our estimator of the unknown law $P \in \mathcal{P}$ (after a suitable choice of metric d_*: see below) the *estimator of minimal distance* $\hat{P}_n$, in other words an element $\hat{P}_n \in \mathcal{P}$ such that

$$d_*(\hat{P}_n, P_n) \leq d_*(Q, P_n) \quad \text{for all } Q \in \mathcal{P}, \qquad (6.55)$$

without dealing, for the time being, with the question of whether it exists.

First, we must choose a metric d_* which will measure the proximity of the two probability laws $\hat{P}_n$ and P_n. For a subset $\mathcal{A} \subset \mathcal{B}$ of our σ-algebra, we set

$$d_\mathcal{A}(P, Q) = \|P - Q\|_\mathcal{A} = 2 \sup_{A \in \mathcal{A}} |P(A) - Q(A)|. \qquad (6.56)$$

In the case where $\mathcal{A} = \mathcal{B}$, we will simply write d instead of $d_{\mathcal{B}}$. Observe that if P and Q have densities f and q with respect to the measure μ, we have

$$
d(P,Q) = \int |f - q| \, d\mu = \int_{f>q} (f-q) \, d\mu - \int_{f<q} (f-q) \, d\mu
$$

$$
= 2 \int_{f>q} (f-q) \, d\mu = 2(P(A) - Q(A)),
$$

(6.57)

where $A = \{f > q\}$. In other words, $d(P,Q) = d(f,q)$ is the *total variation* of $f - q$. Since the measures P_n and Q ($Q \in \mathcal{P}$) can be orthogonal (this is typically the case), we obtain $d\left(\hat{P}_n, Q\right) = 2$, and therefore this metric (which corresponds to the case $\mathcal{A} = \mathcal{B}$) cannot be used to estimate the proximity of the empirical law to the laws Q in $\mathcal{P}$. Thus, in order to construct a metric d_* which is suitable for our purposes, we should not take a set $\mathcal{A}$ which is too large inside $\mathcal{B}$. Let us see how to choose $\mathcal{A}$.

Let $P_1, P_2, \ldots, P_N$ ($N = N(\varepsilon)$) be the centers of N balls $S_1, S_2, \ldots, S_N$ of radius ε, with respect to the metric d, which cover $\mathcal{P}$. We set

$$
\mathcal{P}_\varepsilon = \{P_1, P_2, \ldots, P_N\} .
$$

Let

$$
f_j = \frac{d\,P_j}{d\,\mu}, \quad A_{ij} = \{f_i > f_j\} \quad (i,j = 1, 2, \ldots, N).
$$

We shall take $\mathcal{A} = \mathcal{A}_\varepsilon = \{A_{ij}, \, 0 < i < j \le N\}$. It is clear that

$$
\mathrm{card}(\mathcal{A}_\varepsilon) = \frac{N(N-1)}{2} \le N^2.
$$

(6.58)

We set

$$
d_*(P,Q) = d_*^\varepsilon(P,Q) = d_{\mathcal{A}_\varepsilon}(P,Q) = 2 \sup_{A \in \mathcal{A}_\varepsilon} |P(A) - Q(A)|.
$$

(6.59)

Suppose that the measures $P, Q \in \mathcal{P}$ and the centers $P_i, P_j \in \mathcal{P}_\varepsilon$ of the balls S_i, S_j of radius ε are chosen in such a way that $P \in S_i, Q \in S_j$. Then

$$
d(P,Q) \le d(P, P_i) + d(P_i, P_j) + d(P_j, Q).
$$

Since (see (6.57))

$$
d(P_i, P_j) = 2(P_i(A) - P_j(A)) \quad \text{with } A = \{f_i > f_j\} = A_{ij},
$$

we therefore have

$$
d(P,Q) \le 2\varepsilon + d_{\mathcal{A}}(P_i, P_j).
$$

(6.60)

Then,

$$
d_{\mathcal{A}}(P_i, P_j) \le d_{\mathcal{A}}(P_i, P) + d_{\mathcal{A}}(P,Q) + d_{\mathcal{A}}(Q, P_j).
$$

Finally, we have

$$
d(P,Q) \le 4\varepsilon + d_{\mathcal{A}}(P,Q) = 4\varepsilon + d_*(P,Q), \quad \text{for all } P, Q \in \mathcal{P}.
$$

(6.61)

6.7.2 Choice of the estimator and rate of convergence

Let $\hat{P}_n$ be the estimator of minimal distance for P, in the following sense : it is an element $\hat{P}_n \in \mathcal{P}$ such that

$$d_*(\hat{P}_n, P_n) \le d_*(Q, P_n) \quad \text{for all } Q \in \mathcal{P}. \tag{6.62}$$

Since

$$d_*(\hat{P}_n, P_n) \le d_*(P, P_n)$$

where P is a true law, then, by (6.61), we have

$$d(\hat{P}_n, P) \le 4\varepsilon + d_*(\hat{P}_n, P)$$
$$\le 4\varepsilon + d_*(\hat{P}_n, P_n) + d_*(P, P_n) \le 4\varepsilon + 2d_*(P, P_n). \tag{6.63}$$

Thus,

$$d(\hat{P}_n, P) \le 4\varepsilon + 2d_*(P, P_n). \tag{6.64}$$

Hoeffding's inequality enables us to estimate the probabilities

$$P\{d_*(P, P_n) > y\}.$$

In fact, since the number of subsets $A \in \mathcal{A}_\varepsilon$ is no greater than N^2, it follows from (6.4) that

$$P\{d_*(P_n, P) > y\} = P\{ \sup_{A \in \mathcal{A}_\varepsilon} |P_n(A) - P(A)| > y/2\}$$
$$\le \sum_{A \in \mathcal{A}_\varepsilon} P\{|P_n(A) - P(A)| > y/2\} \le 2N^2 \exp\{-ny^2/2\}. \tag{6.65}$$

In order to find the optimal rate of convergence r_n for the estimator $\hat{P}_n$, the value of the radius ε should be adapted to the number of observations n. Let us choose a value $\varepsilon = \varepsilon_n$ such that

$$\varepsilon_n \asymp \sqrt{\frac{\ln N(\varepsilon_n)}{n}} \quad \text{when} \quad n \to \infty. \tag{6.66}$$

We will show in this case that the estimator $\hat{P}_n$ is consistent with the rate r_n given by

$$r_n = \sqrt{\frac{\ln N(\varepsilon_n)}{n}}. \tag{6.67}$$

By (6.64) and (6.65), with

$$y_n = \frac{r_n t - 4\varepsilon_n}{2}, \quad N = N(\varepsilon_n),$$

we deduce that

$$P\left\{\frac{d\left(\hat{P}_n,P\right)}{r_n}>t\right\}\leq P\left\{d_*(P,P_n)>y_n\right\}\leq 2N^2\exp\{-ny_n^2/2\}$$

(6.68)

$$=2\exp\left\{-\frac{n}{2}\left(y_n^2-4\frac{\ln N}{n}\right)\right\}.$$

Therefore, in a case, as for some positive C_1,C_2

$$\varepsilon_n\leq C_1\sqrt{\frac{\ln N(\varepsilon_n)}{n}},\ r_n\geq C_2\sqrt{\frac{\ln N(\varepsilon_n)}{n}},$$

we obtain by (6.68) that for true distribution P

$$P\left(\frac{d\left(\hat{P}_n,P\right)}{r_n}>t\right)\leq 2\exp\left\{-Ct^2\ln N\right\}$$

for sufficiently large t and some $C>0$. Hence

$$\sup_{P\in\mathcal{P}}P\left(\frac{d(\hat{P}_n,P)}{r_n}>t\right)\to 0\quad\text{when }t\to\infty,$$

uniformly on n, which completes the proof.

6.8 Using entropy to estimate a density

Let $X_1,\ldots,X_n,\ldots$ be a sequence of independent identically distributed random variables, with probability law $P\in\mathcal{P}$ and density $f\in\mathcal{F}$ with respect to the measure μ.[7] In order to specify the density of a probability law, we will write, for example, $P=P_f$. The set $\mathcal{P}$ is thus parametrised by the elements of $\mathcal{F}$.

Let us consider the problem of estimating the density $f\in\mathcal{F}$. The losses incurred in estimating f by $\hat{f}_n$ will be measured by the quantity $\|\hat{f}_n-f\|_1$. Here, $\|\cdot\|_1$ is the L^1 norm:

$$\|f\|_1=\int|f(x)|\,dx.$$

We write $R(\hat{f}_n)=\sup_{f\in\mathcal{F}}\mathbf{E}_f\|\hat{f}_n-f\|_1$. We will use the same method for constructing an estimator as in the previous section, with a few obvious changes in

---

[7] Recall that

$$\mathcal{F}=\left\{f:\ \exists P\in\mathcal{P}\ \text{s.t.}\ f=\frac{dP}{d\mu}\right\}$$

notation, because we are now interested in estimating functions in $\mathcal{F}$, whereas previously we were concerned with estimating functions in $\mathcal{P}$. In this way, we choose an economic ε-covering $\mathcal{F}_\varepsilon = \{f_1, \ldots, f_N\}$ $(N = N(\varepsilon))$ in $\mathcal{F}$, i.e. satisfying

$$N(\varepsilon) \asymp N(\varepsilon, \mathcal{F}, \|\cdot\|_1) \quad \text{when } \varepsilon \to 0.$$

We take $\mathcal{A} = \mathcal{A}_\varepsilon = \{A_{ij}, \ 0 < i < j \leq N\}$, where, as above,

$$A_{ij} = \{f_i > f_j\} \qquad (i, j = 1, 2, \ldots, N).$$

Instead of the L^1 metric, $\mathcal{F}$ is equipped with the metric introduced in the previous section:

$$d_*(f, g) = d_*(P_f, P_g) = 2 \sup_{A \in \mathcal{A}} |P_f\{A\} - P_g\{A\}|.$$

Let $\hat{f}_n$ denote the point of $\mathcal{F}_\varepsilon$ which minimizes the quantity $d_*(P_f, P_n)$ over $\mathcal{F}_\varepsilon$:

$$\forall f \in \mathcal{F}_\varepsilon, \quad d_*(P_{\hat{f}_n}, P_n) \leq d_*(P_f, P_n).$$

Relation (6.64) can be rewritten (with $P = P_f$):

$$\|\hat{f}_n - f\|_1 \leq 4\varepsilon + 2d_*(P_f, P_n).$$

Here (as above) P_n is an empirical distribution. Hence

$$R(\hat{f}_n) \leq 4\varepsilon + 2\mathbf{E}\, d_*(P_f, P_n). \tag{6.69}$$

Let us now estimate $\mathbf{E}_f\, d_*(P_f, P_n)$. We need the following lemma:

Lemma 5. *Let ξ be a random variable which is non-negative, and*

$$P\{\xi > y\} \leq Cx^2 \exp\left\{-\frac{ny^2}{2}\right\}, \quad y > 0. \tag{6.70}$$

Then there exists a constant $C_1 = C_1(C, x_0)$, which only depends on C and x_0, such that for $x \geq x_0 > 1$,

$$\mathbf{E}\xi \leq C_1 \sqrt{\frac{\ln x}{n}}. \tag{6.71}$$

Proof. If ξ is a non-negative random variable, then it is well known that

$$\mathbf{E}\xi = \int_0^\infty P\{\xi > t\}\, dt.$$

For this reason,

$$
\begin{aligned}
\mathbf{E}\xi &= \sqrt{\frac{\ln x}{n}} \int_0^\infty P\left\{\xi > t\sqrt{\frac{\ln x}{n}}\right\} dt \\
&\leq \sqrt{\frac{\ln x}{n}} \left(y + \int_y^\infty P\left\{\xi > t\sqrt{\frac{\ln x}{n}}\right\} dt\right), \quad y > 0.
\end{aligned}
\tag{6.72}
$$

It follows from this inequality and from (6.70) that

$$
\mathbf{E}\xi \leq \sqrt{\frac{\ln x}{n}} \left(y + C \int_y^\infty \exp\left\{-v\left(\frac{t^2}{2} - 2\right)\right\} dt\right), \quad v = \ln x \geq \ln x_0 = v_0 > 0.
$$

Since $v \geq v_0$, on setting $y = 2$ we obtain (6.71) with the constant

$$
\begin{aligned}
C_1 = C_1(C, x_0) &= \left(2 + C x_0^2 \int_2^\infty \exp\left\{-\frac{v_0 t^2}{2}\right\} dt\right) \\
&\geq \left(2 + C \int_2^\infty \exp\left\{-v\left(\frac{t^2}{2} - 2\right)\right\} dt\right).
\end{aligned}
\tag{6.73}
$$

□

Theorem 1. *Consider the estimator $\hat{f}_n$ defined by the relation*

$$
d_*(P_{\hat{f}_n}, P_n) \leq d_*(P_f, P_n), \quad f \in \mathcal{F}_\varepsilon.
$$

Then, for all $N \geq 2$,

$$
R(\hat{f}_n) \leq 4\varepsilon + 2C_1\sqrt{\frac{\ln N}{n}}.
\tag{6.74}
$$

Proof. By inequality (6.65),

$$
P\left\{d_*(P_f - P_n) > y\right\} \leq CN^2 \exp\left\{-\frac{ny^2}{2}\right\}.
$$

Hence, using Lemma 5, we have for $N \geq 2$:

$$
\mathbf{E}_f d_*(P_f - P_n) \leq C_1\sqrt{\frac{\ln N}{n}} \quad (\text{here } C_1 = C_1(C, 2)),
$$

and so inequality (6.74) follows from (6.69).   □

Conclusion

In summary, the development of mathematical statistics over the last fifty years has stimulated very deep research in the area of non-parametric statistics, and in particular on the problem of the estimation of densities. Very many publications have been devoted to this fascinating subject, and we have tried to present the results of several statisticians in a single, unified, consistent article.

We would like to thank our colleagues for their remarks, suggestions, advice and friendly encouragement, and especially Catherine Huber, Ildar Ibragimov, Valentina Nikouline, Jérome Poix, and Vincent Couallier.

References

[Ass83]   Assouad, P.: Deux remarques sur l'estimation. Comptes Rendus de l'Académie des Sciences de Paris, Sér. I. Math., **296**, 1021–1024 (1983)

[Bir83]   Birgé, L.: Approximation dans les espaces métriques et théorie de l'estimation. Zeitschrift für Wahrscheinlichkeitstheorie und Verwandte Gebiete, **65**, 181–237 (1983)

[BH79]   Bretagnolle, J., Huber, C.: Estimation des densités : risque minimax. Zeitschrift für Warscheinlichkeitstheorie und Verwandte Gebiete, **47**, 119–137 (1979)

[BL87]   Bosq, D., Lecoutre, J.-P.: Théorie de l'estimation fonctionnelle. Economica, Paris (1987)

[BM98]   Birgé, L., Massart, P.: Minimum contrast estimators on sieves: exponential bounds and rates of convergence. Bernoulli, 4:3, 329–375 (1998)

[CM04]   Ceci, C., Mazliak, L.: Optimal design in nonparametric life testing. Statistical Inference for Stochastical Processes, **7**, 395–325 (2004)

[DM92]   Deheuvels, P., Mason, D.M.: Functional laws of the iterated logarithm for the increments of empirical and quantile processes. Annals of Probability, **20**, 1248–1287 (1992)

[Dud76]   Dudley, R.: Probability and metrics — convergences of laws on metric spaces with a view to statistical testing. Lecture Notes, **45**, Matematisk Institut Aarhus Universitet, Denmark (1976)

[Dud97]   Dudley, R.: Empirical processes and p-variation. In: Pollard, D., Torgersen, E., Yang, G.L. (ed): Festschrift for Lucien Le Cam. Research Papers in Probability and Statistics, 219–234. Springer, New York (1997)

[Hal48]   Halmos, P.: The theory of unbiased estimation. Ann. Math. Statist., **17**, 34–43 (1948)

[HHOC94]   Hall, P., Huber, C., Owen, A., Coventry, A.: Asymptotically Optimal Balloon Density Estimates. Journal of Multivariate Analysis, **51**, 352–371 (1994)

[Hoe63]   Hoeffding, W.: Probability inequalities for sums of bounded random variables. Journal Amer. Statist. Assoc., **58**, 13–30 (1963)

[Hub97]   Huber, C.: Lower bound for function estimation. In: Pollard, D., Torgersen, E., Yang, G.L. (ed): Festschrift for Lucien Le Cam. Research Papers in Probability and Statistics, 245–258. Springer, New York (1997)

[IK81]   Ibragimov, I., Khas'minskii, R.: Statistical estimation: asymptotic theory. Springer-Verlag (1981)

[Kha78]   Khas'minskii, R.Z.: A lower bound on the risk of non-parametric estimates of densities in the uniform metric. Theory Probab. Appl., **23**, 794–796 (1978)

[KL51]   Kullback, S., Leibler, R.: On information and sufficiency. Ann. Math. Statist., **22**, 79–86 (1951)

[Kol33]   Kolmogorov, A.N.: Sulla determinazione empirica di una legge di distribuzione. Giornal Instit. Ital. Attuari, **4**, 83–91(1933)

[Kol50]   Kolmogorov, A.N.: Unbiased estimates (in Russian). Izvestia Acad. Nauk SSSR Math., **14**, 303–326 (1950)

[Kol51]   Kolmogorov, A.N.: A statistical acceptance control for an admissible number of defective products equal to zero (in Russian). Leningrad House of Scientific and Technical Propaganda. Leningrad (1951)

[KT59]   Kolmogorov, A.N., Tikhomirov, V.: ε-entropy and ε-capacity of sets in functional spaces. Russian Mathematical Surveys, American Mathematical Society Translation (series 2), **17**, 277–367 (1959)

[LeC73]   Le Cam, L.: Convergence of estimates under dimensionality restrictions. Ann. Statist., **19**, 633–667 (1973)

[LeC86]   Le Cam, L.: Asymptotic Methods in Statistical Decision Theory. Springer-Verlag (1986)

[Leh83]   Lehmann, E.L.: Theory of Point Estimation. Springer-Verlag, New York (1983)

[LY90]   Le Cam, L., Yang, G.L.: Asymptotics in Statistics. Springer, New York (1990)

[NS02]   Nikulin, M., Solev, V.: Testing problem for increasing function in a model with infinite dimensional nuisance parameter. In: Huber-Carol, C., Balakrishnan, N., Nikulin, M., Mesbah, M. (ed): Goodness-of-Fit Tests and Model Validity, 477–493. Birkhauser, Boston (2002)

[She97]   Shen, X.: On Methods of Sieves and Penalization. The Annals of Statistics, **25**:6, 2555–2591 (1997)

[vandG93]   Van de Geer, S.: Hellinger-consistency of certain nonparametric maximum likelihood estimators. The Annals of Statistics, **21**:1, 14–44 (1993)

[vandG95]   Van de Geer, S.: Exponential inequalities for martingales, with applications to maximum likelihood estimation for counting processes. The Annals of Statistics, **23**:5, 1779–1801 (1995)

[vandV89]   Van der Vaart, A.: On the asymptotic information bound. The Annals of Statistics, **17**:4, 1487–1500 (1989)

[vandV00]   Van der Vaart, A.: Asymptotic statistics. Cambridge University Press (2000)

[VN93]   Voinov, V., Nikulin, M.: Unbiased estimators and their applications. V. 1: Univariate Case. Kluwer Academic Publishers (1993)

[VN96]   Voinov, V., Nikulin, M.: Unbiased estimators and their applications. V. 2: Multivariate Case. Kluwer Academic Publishers (1996)

[VW96]   Van der Vaart, A., Wellner, J.: Weak convergence and empirical processes. Springer (1996)

[WS95]   Wong, W.H., Shen, X.: Probability inequalities and convergence rates of sieve MLES. The Annals of Statistics, **23**:2, 339–362 (1995)

[Yat85]   Yatrakos, Y.G.: Rates of convergence of minimum distance estimators and Kolmogorov's entropy. The Annals of Statistics, **13**:2, 768–774 (1985)

[Yu97]   Yu, B.: *"Assouad, Fano and Le Cam"*. In: Pollard, D., Torgersen, E., Yang, G.L. (ed): Festschrift for Lucien Le Cam. Research Papers in Probability and Statistics, 423–435. Springer (1997)

A.N. Kolmogorov in Helsinki, 1965. With friendly permission of Mathematisches
Forschungsinstitut Oberwolfach/photo collection of Prof. Konrad Jacobs

Kolmogorov and topology

Victor M. Buchstaber

Steklov Mathematical Institute of Russian Academy of Sciences, and Moscow
State University, Russia
http://www.mi.ras.ru/~buchstab
buchstab@mi.ras.ru
and School of Mathematics, The University of Manchester, UK
Victor.Buchstaber@manchester.ac.uk

Translated from the French by Emmanuel Kowalski

At the beginning of the nineteen thirties, Andreï Nikolaevich Kolmogorov
published a few papers on topology, totaling roughly thirty pages. These works
immediately made him one of the main creators of modern algebraic topology. In this short survey, I have tried to present Kolmogorov's remarkable
results (not only for a public of topologists), and to explain the source of
some of the ideas leading to those results. This survey is based on a lecture by
the author, entitled *Kolmogorov, cohomology and cobordisms*, which was presented during the conference *Kolmogorov and contemporary mathematics* in
commemoration of the centennial of Kolmogorov, held in Moscow from June
16 to June 21, 2003.

7.1 Prelude

Pavel Samuilovich Urysohn (1898–1924) and Pavel Sergeevich Aleksandrov
(1896–1982) are recognized as the founders of the Soviet school of topology.
In the years 1920–1924, Urysohn's enthusiasm and the remarkable results
he had obtained played a great role in bringing talented young researchers
to this area of mathematical research. In [AK53], Aleksandrov remembers
that Kolmogorov, still a student, first attracted Urysohn's attention when,
during a lecture, he pointed out a mistake in the complicated constructions
Urysohn was using to prove his theorem concerning the topological dimension
of Euclidean space of dimension three[1]:

---

[1] Urysohn is the creator of a famous theory of topological dimension, and it was
natural to check that this notion was consistent with the usual dimension in the
case of the Euclidean spaces $\mathbb{R}^n$. (Editor's note.)

> "P. S. Urysohn corrected this mistake a few days later, but the mathematical acuity of the eighteen year old student Kolmogorov had made a great impression on him."

(Something similar happened during N. N. Lusin's special course. Kolmogorov himself described it in this manner: "Although my contribution was rather childish, it helped me to become known in the lusitanian[2] circle" [Kol72].)

Fifty years later, in his article "A scientific teacher" [Kol72] devoted to Urysohn, Kolmogorov describes his first scientific steps and his conversations with Urysohn. He describes his hesitations concerning the choice of a research subject, due among other things

> "to a vague desire to do mathematics, but with strong applications to physics and natural sciences ... As far as possible, Pavel Samuilovich tried to enlist me in his researches concerning Poincaré's problem about closed geodesics on surfaces ... All these questions appealed to me in themselves, they corresponded to the idea I had of what a mathematician should most occupy himself with ... The internal logic of my personal investigations brought me to topology only much later, after mathematical passions for logic and probability theory."

In November 1925, at Moscow University, the famous *Moscow topological circle* was created, Aleksandrov being its permanent leader. Until the beginning of the nineteen-sixties, this was the center of the Moscow school of topology.

The programme of the first ten years of the circle is described in the paper [Nem36], in which one finds in particular a very impressive *list of lectures given at the topological circle between 1925 and 1935*, in which Kolmogorov's lectures appear:

N° 52. The group of homeomorphisms in metric spaces, April 1929.
N° 53. The group of homeomorphisms in metric spaces, May 1929.
N° 55. The group of homeomorphisms in a topological space, February 1930.
N° 69. The topological axiomatics of projective space, June 1931.
N° 83. On the theory of continuous repartitions, May 1933.

This list indicates that Kolmogorov, who already enjoyed a world-wide reputation, was participating actively in the topology seminar, but had not yet found his own topological subject. The ensuing events are all the more surprising.

---

[2] At the end of the 1910s, Lusin, with Egorov, had created a brilliant research group at Moscow University, which was called "Lusitania" by its members, as a pun on Lusin's name. *Lusitania* is the name of an antique Roman province which corresponds roughly to present-day Portugal; it is also the name of a British ocean liner which had been much in the news in 1915 and the following years: after it was sunk by a German submarine, with the loss of more than a thousand civilian lives, international public opinion was scandalized, so that when American troops landed in Europe two years later, a cartoon showed a picture of Kaiser Wilhelm II asking one of his officers: "How many ships did it take to bring them here?", and the officer answering: "Only one, the *Lusitania*"! (Editor's note.)

From September 4 to September 10, 1935, at the Institute of Mathematics of Moscow University, the first international topology conference was held. It was an extraordinary moment in the history of topology. The results which were presented at this conference delineated the main directions of research for many years.

Here is a selection of the lectures:

- J. W. Alexander (Princeton), *On the rings of complexes and the combinatorial theory of integration*;
- A. N. Kolmogorov (Moscow), *On the homology rings of closed sets*;
- E. Čech (Brno), *On the Betti groups with coefficients in an arbitrary field*;
- H. Freudenthal (Amsterdam), *On topological approximations of spaces*;
- A. W. Tucker (Princeton), *On discrete spaces*;
- M. H. Stone (Cambridge, Mass.), *The theory of maps in general topology*;
- P. S. Aleksandrov (Moscow), *Some solved and unsolved problems in general topology*;
- W. Hurewicz (Amsterdam), *Homology and homotopy*;
- K. Borsuk (Warsaw), *On spherical spaces*;
- S. Lefschetz (Princeton), *On locally-connected manifolds*;
- H. Hopf (Zurich), *New investigations on n-dimensional manifolds*;
- G. Nöbeling (Erlangen), *On triangulation of manifolds and the Main Hypothesis of combinatorial topology*;
- H. Whitney (Cambridge, Mass.), *Topological properties of differentiable manifolds*;
- P. Smith (New York), *On 2-periodic maps*;
- P. Heegaard (Oslo), *On the four-color problem*;
- G. de Rham (Lausanne), (1) *On Reidemeister's new topological invariants*; (2) *Topological aspects of the theory of multiple integrals*;
- A. A. Markov (Leningrad), *On equivalence of closed braids*;
- L. S. Pontriaguin (Moscow), *Topological properties of compact Lie groups*;
- E. R. van Kampen (New Heaven), *The structure of compact groups*;
- J. von Neumann (Princeton), *The theory of integration in continuous groups*;
- A. Weil (Paris), (1) *A topological proof of a theorem of Cartan*; (2) *The systems of curves on a torus*.

A complete list of lectures is found in [Ale36a], and the programme of the conference is in [Shi03, n° 2, pp. 590–593].

Among all those wonderful subjects and all those mathematicians – all internationally-renowned leaders of topology – the lectures of Kolmogorov and Alexander on the construction of dual complexes for quite general spaces, the homology of which has a natural structure of ring, attracted widespread attention. In modern terminology, this was the construction of the cohomology ring of topological spaces.

All ulterior developments of algebraic topology have confirmed the extraordinary importance of the results of those two authors.

7.2 The main topological results of A. N. Kolmogorov

7.2.1 Algebraic topology

During the nineteen-thirties, a branch of topology, called *algebraic topology* by Solomon Lefschetz (1884–1972), emerged. Lefschetz himself created the deep theory of homology of projective complex algebraic varieties, including the fundamental intersection theory of algebraic cycles; he then carried his ideas over to topology. Another approach to homology theory was developed by Élie Cartan (1869–1951), based on Poincaré's ideas and on Riemannian geometry. Cartan constructed the "tensor theory of homology", and conjectured that it leads to ordinary homology theory. Cartan's programme was implemented by de Rham (see, e.g. [Nov04], [Die89], about the history and the development of the main ideas of algebraic topology).

The results obtained both by Kolmogorov and Alexander are considered as a complete solution of this fundamental problem, with a global mathematical reach. In the monography [Lef42] which introduced the name "algebraic topology", the theory of Kolmogorov and Alexander plays a prominent role. Concerning the origins of the ideas of this theory, Lefschetz wrote:

> "Chiefly for purposes of extending the concepts of *differential* and *integral* to general topological spaces Alexander [Ale36b, Ale38], and later Kolmogorov [K4, K5, K6, K7], have developed a type of theory based directly upon chains and cochains."

Kolmogorov himself explained the main idea in [K9]

> "The author's goal is to construct a particular difference calculus which, on the one hand, leads to differential operators acting on anti-symmetric tensors (multivectors) by a limit process, and on the other hand is closely related to the concepts of combinatorial topology.
>
> In particular, it is possible to define new invariants of complexes and closed sets using this difference calculus, and to prove some generalizations of the known duality theorems."

Kolmogorov wrote the following comments concerning his work of 1936–37 on homology theory in the edition of his selected works published in honor of his eightieth birthday [Kol85]:

> "The initial impulse for these works was reading the thesis of Georges de Rham [DeR31] (1931), in which the duality of Betti groups of differentiable manifolds and Betti groups generated by currents was established. After the 1930's, I did not work on those subjects anymore; yet, the idea presented in the four notes in Comptes Rendus de l'Académie des Sciences de Paris, which is to exploit the duality between the groups of antisymmetric functions of n points and of additive antisymmetric functions of n sets, still seems to me to have some pedagogical interest."

Except for those notes and the articles [K2, K3, K9], Kolmogorov produced other works of algebraic topology which he did not publish, despite having a great opinion of this subject:

"... Homologic topology interested me a lot, and during the years 1934–36, I should have worked on more on this ..."

writes Kolmorogov in a letter to N. N. Lusin about his research plans, dated October 7, 1945 [Shi03, 1, p. 227].

The theory of Kolmogorov and Alexander on a space X relates the set of chains with the boundary operator Δ (which is called the chain complex) and its dual, the set of cochains with the coboundary operator ∇ (called the cochain complex).

Fix an abelian group G. Let $\{A\}$ be the family of closed sets of the space X. A p-chain (or chain of dimension p) on G is an antisymmetric function φ^p with values in G, depending on $p+1$ sets $(A_0, \ldots, A_p)$, with the following properties:

(1) it is equal to zero whenever $\bigcap A_i = \varnothing$ or $A_i = A_j$ for some $i \neq j$;
(2) it is multilinear, in the sense that if the interiors of A_i and A_i' are disjoint, we have

$$\varphi^p(\ldots, A_i \cup A_i', \ldots) = \varphi^p(\ldots, A_i, \ldots) + \varphi^p(\ldots, A_i', \ldots).$$

The boundary of the chain φ^p is the $(p-1)$-chain

$$(\Delta\varphi^p)(A_0, \ldots, A_{p-1}) = \varphi^p(X, A_0, \ldots, A_{p-1}).$$

Obviously, we have $\Delta\Delta\varphi^p = 0$. A p-chain φ^p is called a p-boundary if it is of the form $\Delta\varphi^{p+1}$, where φ^{p+1} is a $(p+1)$-chain. And a p-chain φ^p is a p-cycle if $\Delta\varphi^p = 0$. The relation $\Delta\Delta\varphi^{p+1} = 0$ guarantees that p-boundaries are p-cycles. The p-dimensional homology (*in degree p*) of X with values in G is then the set of p-cycles modulo p-boundaries.

The dual complex, formed with cochains, is described in detail in [Lef42], for instance. Here, we only need to know that p-cochains are constructed from antisymmetric functions ψ_p with values in the *Pontryagin dual H* of G (i.e. the character group of G), depending on $p+1$ points $(x_0, \ldots, x_p)$, $x_i \in X$, and that the coboundary operator[3] is given by the formula

$$(\nabla\psi_p)(x_0, \ldots, x_{p+1}) = \sum_q (-1)^q \psi_p(\ldots, x_{q-1}, x_{q+1}, \ldots),$$

and obviously $\nabla\nabla\psi^p = 0$. In the dual complex formed with cochains, a p-cochain ψ^p is called a p-coboundary if it is of the form $\nabla\psi^{p-1}$, where ψ^{p-1} is a $(p-1)$-cochain. A p-cochain ψ^p is called a p-cocycle if $\nabla\varphi^p = 0$. The

---

[3] In [K9], Kolmogorov, who was writing for the public of the *Seminar on vector and tensor analysis and its applications to mechanics and physics*, used the notation rot for this operator

relation $\nabla\nabla\psi^{p-1} = 0$ guarantees that p-coboundaries are p-cocycles. The p-dimensional cohomology (*in degree* p) of X with values in H is the set of p-cocycles modulo p-coboundaries.

The applications of Kolmogorov's approach to the construction of homology and cohomology theory with duality theorem for homology and cohomology are discussed in the comments of G.S. Chogoshvili "Homology Theory" in [Kol85].

In their lectures at the first international topology conference in 1935, Kolmogorov and Alexander gave an explicit multiplication formula, which associates a $(p + q)$-cochain to a p-cochain and a q-cochain. This operator, like Cartan's exterior product of differential forms, is anticommutative: Kolmogorov uses the notation $[\,\cdot\,,\,\cdot\,]$ for this operation, and we have then

$$[\psi_p, \psi_q] = (-1)^{pq}[\psi_q, \psi_p].$$

Kolmogorov and Alexander proved that this operation induces an associative and anticommutative product on cocyles, and in this manner, they introduced a ring structure on the cohomology.

Shortly after the conference, Čech [Čec36], Whitney [Whi37] and Alexander [Ale36b] showed the existence of another operation on cochains, denoted $\smile$, given by a formula which is close to the original formula of Kolmogorov and Alexander. This operation induces also an associative and anticommutative product on cocyles. In cohomology, there is a relation

$$[\psi_p, \psi_q] = \frac{(p+q)!}{p!q!}\, \psi_p \smile \psi_q. \tag{$*$}$$

Hence, in cohomology with coefficients in $\mathbb{Q}$, both products are equivalent, but already in integral cohomology, if there are cocycles of finite order, one may have cocycles, say a and b, such that (cohomologically) we have $[a, b] = 0$ and at the same time $a \smile b \neq 0$. In textbooks of algebraic topology, when the Kolmogorov-Alexander product for cohomology is mentioned, what is meant is the operation $\smile$. In what follows, we will say that $\smile$ is the *standard* operation.

The operation $\smile$ in combinatorial topology leads to a combinatorial analogue of Lefschetz's intersection theory, whereas Cartan's theory of products of differential forms corresponds to Kolmogorov's $[\,\cdot\,,\,\cdot\,]$ operation.

At the level of cochains, in contrast with the cocycles, the formula for the standard operation $\smile$ *depends of the choice of an ordering of the vertices* in a simplicial complex. The operation $[\,\cdot\,,\,\cdot\,]$ is deduced from the operation $\smile$ by averaging

$$[\psi_p, \psi_q](x_0, \ldots, x_{p+q}) = \sum_\sigma \psi_p \smile \psi_q\,(x_{\sigma(0)}, \ldots, x_{\sigma(p+q)}) \tag{$**$}$$

where the sum is over all permutations σ of $(0, \ldots, p + q)$. Thus, the formula $(**)$ expresses the relation between the two products of combinatorial topology, which are philosophically related with the theories of Cartan and Lefschetz.

The important distinction between integral cohomology (with coefficients in $\mathbb{Z}$) and rational cohomology (with coefficients in $\mathbb{Q}$) is expressed in the fundamental rule of combinatorial topology:

It is impossible to obtain the standard multiplication using anticommutative operations on cochains with integral coefficients.

We must emphasize that *at the level of cochains* the operation $[\cdot, \cdot]$ *is anticommutative, but not associative*, whereas the operation $\smile$ *is not anticommutative, but is associative.*

The defect of anticommutativity of the standard operation at the level of cochains was used by Steenrod [Ste62] to construct "cohomological operations", and also to define structures of modules over the algebra of cohomological operations on the cohomology ring.

A question is raised: *what additional structure may be introduced in cohomology, using the fact that Kolmogorov's operation is not associative at the level of cochains?*

The ulterior progress of algebraic topology has led to extraordinary cohomology theories, the best known of which (in terms of their numerous applications) are K-theory and cobordism theory. Almost all those applications use in an essential way the product structure on the cohomology ring and the powerful algebras of cohomological operations (see [Nov04]).

The transformation of the standard product to the Kolmogorov product by means of the formula $(*)$ is the unique non-trival (non-invertible[4]) transformation of the product in the case of classical cohomology. As shown in [BBNY00], there is a great variety of transformations of products in the case of complex cobordism theory. Those transformations are related to important structures in analysis, representation theory, and commutative and non-commutative algebra.

7.2.2 General topology

A number of fundamental notions are also due to Kolmogorov, who introduced them in order to solve some important problems of general topology. These notions are commonly used today, appearing constantly in new applications.

- Kolmogorov's name is attached to the separation Axiom $\mathcal{T}_0$ — the weakest separation axiom in use. A topological space X is called a $\mathcal{T}_0$-space, or a *Kolmogorov space* if, for any pair of distinct points $x, y \in X$, $x \neq y$, there exists an open set of X containing one of the points but not the

---

[4] The formula $(*)$ is not invertible in cohomology with *integral* coefficients because $\frac{p!q!}{(p+q)!}$ is not an integer (except when it is equal to 1)

other[5]. An important non-trivial example of Kolmogorov space is the set of simplexes of a simplicial complex, with the topology in which the closure of a point (i.e. of a simplex) is the set of all its "faces" of arbitrary dimension (including itself, of course): see [AH35] (for instance, p. 132 of the 1974 edition) or [Kur66] (Chap. 1, Sect. 5, IX: T_0-spaces).

- Kolmogorov found a necessary and sufficient condition for a general topological vector space to be normable [K1][6]. In the course of solving this problem, he introduced the notion of *bounded* subset in topological vector spaces. This concept turned out to be of fundamental importance in their duality theory, and in the applications of topological vector spaces to analysis.

- Recall that a map from a topological space X to a topological space Y is *open* if the image of any open set in X is open in Y. Kolmogorov constructed a surprising example of an open map from a topological space of dimension 1 onto a space of dimension 2. This had a great importance in general topology. It shows, in particular, that the notion of "dimension of a topological space" is not trivial. At the root of the construction of this map is an explicit open map from the 2-dimensional torus onto the Möbius strip, such that the equator of the torus is mapped to the boundary of the strip. Later on, Kolmogorov's example found an application in the theory of group actions on topological spaces: it was remarked that it describes an effective action of a 0-dimensional abelian group (namely, the compact group of 2-adic integers) on a 1-dimensional compact space (the Menger's curve), such that the space of orbits is of dimension 2 (the Pontryagin 2-dimensional surface) (see [Wil63]: Kolmogorov's example is one of the "three famous examples in topology" mentioned in the title).
Kolmogorov wrote the following comments concerning the origins of this result in [Kol85] (see p. 476):

> "The possibility of an increase in dimension under open mappings ([K8]) interested P. S. Aleksandrov very much. For some time we together tried to prove that increase in dimension is impossible. In these attempts we gradually understood why we failed. An analysis of the failure led us to a counterexample."

Further investigations in this direction are described in the recently survey "Problems of dimension raising" (see [Che05], §2).
- A. N. Kolmogorov in collaboration with I. M. Gel'fand published the paper [K10]. In introduction to this paper we read:

---

[5] Recall that a topological space is *separated*, or *Hausdorff*, if it satisfies the (stronger) Axiom T_2: *any two distinct points $x, y \in X$ have disjoint neighborhoods*. There is also an Axiom T_1: *any singleton is a closed set*. These axioms express properties of separation of varying strength: it is obvious that T_2 implies T_1, and that T_1 implies T_0. (Editor's note.)

[6] See Theorem 4 of Chap. 8 (by Vladimir Tikhomirov) in this volume. (Editor's note.)

"... we consider the ring of continuous functions on a topological space as a purely algebraic object without defining any topological relations in it. It turns out that in the case of bicompact spaces, considered by M. H. Stone, and also in some much more general cases, even the purely algebraic structure of the ring of continuous functions determines the topological space to within a homeomorphism."

Many monographs and textbooks containing basic general topology and functional analysis include this result. Further investigations in this direction are described in the recently survey [BR04].

7.3 A topological idea of Kolmogorov

In May 1929, having defended his thesis four years before, Kolmogorov had already published 18 works, which brought him worldwide renown. In the paragraph concerning this period in his *remembrances of P. S. Aleksandrov*, we read [Kol86]:

"Our personal contacts with Pavel Sergeevich were very limited at that time, although we often met at the concerts in the Small Room of the conservatory ... In 1929, we met again during a trip on the Volga. I do not remember very well how I had decided to suggest to Pavel Sergeevich that he be the third companion[7]. However, he immediately accepted ... One may consider the day of departure − June 16 − as the starting point of our friendship."

To this friendship, mathematics owes the correspondence between Kolmogorov and Aleksandrov. A selection of this correspondence is now accessible in the commemorative edition of Kolmogorov's works [Shi03]. In Kolmogorov's letters, one finds many remarks on mathematical results, which give insights into his ideas and scientific projects. Some of these ideas and projects were much in advance on their time.

In the letter to Aleksandrov dated September 22, 1932, "from Dnepropetrovsk to Zurich, Switzerland" (see [Shi03], Vol. 2, p. 439), Kolmogorov wrote:

"Questions of topology. It seems to me that it is not difficult to prove that an n-dimensional closed set may be embdedded in a space of sufficiently large dimension, and in only one way. I know how to prove this for polyhedra, which embed in (Euclidean) space of dimension $4n + 2$. It seems that the properties that can be expressed in terms of the complement of the set in this space should be called homologic?

---

[7] It is a rowing-boat trip, which Kolmogorov and his *gymnasium* friend Nikolaï Njuberg had first thought of doing together only

In other words, F and F_1 are homologically equivalent if, being embedded in a space of sufficiently large dimension, their complements are homeomorphic. Then, the question of determining whether homologic invariants are complete will have a well-defined meaning — namely, to characterize completely, or not, the complement in a space of sufficiently large dimension."

Almost fourty years later, Borsuk developed a new direction in homotopic topology — shape theory [Bor71]. Using the results of this theory [Cha72, GS73], one can find the answers to the questions raised in Kolmogorov's letter.

For instance, let F and F_1 be compact polyhedra (or even compact absolute neighborhood retracts[8]) of dimension n. Then, under some restrictions on their position in Euclidean space $\mathbb{R}^N$ with sufficiently large dimension N, they are homotopy-equivalent if and only if their complements $\mathbb{R}^N \setminus F$ and $\mathbb{R}^N \setminus F_1$ are homeomorphic.

References

[AH35]   Alexandroff, P. (Aleksandrov, P.S.), Hopf, H.: Topologie, I (in German), vol. 45 of Grundlehren der Math. Wiss. Springer, Berlin (1935) Corrected printing: Springer, Berlin, New York (1974)

[AK53]   Aleksandrov, P.S., Khinchin, A.J.: Andreï Nikolaevich Kolmogorov: in honor of his 50th birthday (Russian). Uspekhi Mat. Nauk, **8**, 178–200 (1953)

[Ale36a]   Aleksandrov, P.S.: First international congress on topology, Moscow (Russian). Uspekhi Mat. Nauk, **1**, 260–262 (1936)

[Ale36b]   Alexander, J.W.: On the connectivity ring of an abstract space. Annals of Math., **37**, 698–708 (1936)

[Ale38]   Alexander, J.W.: A theory of connectivity in terms of gratings. Annals of Math., **39**, 883–912 (1938)

[BBNY00]   Botvinnik, B.I., Buchstaber, V.M., Novikov, S.P., Yuzvinsky, S.A.: Algebraic aspects of the theory of multiplications in complex cobordism theory. Uspekhi Mat. Nauk, **55**:4, 5–24 (2000). Russian Math. Surv., **55**:4, 613–633 (2000)

[Bor71]   Borsuk, K.: Theory of shape. Aarhus, Lecture Notes Series vol. 28 (1971)

[BR04]   Buchstaber, V.M., Rees, E.G.: Rings of continuous functions, symmetric products, and Frobenius algebras. Uspekhi Mat. Nauk, **59**:1, 125–144 (2004). Russian Math. Surv. **59**:1, 125–146 (2004)

[Čec36]   Čech, E.: Multiplication on a complex. Annals of Math., **37**, 681–697 (1936)

[Cha72]   Chapman, T.A.: On some applications of infinite-dimensional manifolds to the theory of shape. Fund. Math., **76**, 181–193 (1972)

---

[8] Absolute neighborhood retract: an important notion invented by Borsuk in 1931. In [Bor71], he explains his motivation as follows: the use of absolute neighborhood retracts helps to bring topology closer to geometric intuition. See [Mar99] for the history of this notion, and of shape theory (and their relations)

[Che05]   Chernavskii, A.V.: On the work of L.V. Keldysh and her seminar. Uspekhi Mat. Nauk, **60**:4, 11–36 (2005). Russian Math. Surv., **60**:4, 589–614 (2005)

[DeR31]   de Rham, G.: Sur l'Analysis situs des variétés à n-dimensions. J. Math pures et appl., **10**, 115–200 (1931)

[Die89]   Dieudonné, J.: *A history of algebraic and differential topology: 1900–1960.* Birkhäuser (1989)

[GS73]   Geoghegan, R., Summerhill, R.R.: Concerning the shapes of finite-dimensional compacta. Trans. Amer. Math. Soc., **179**, 281–292 (1973)

[Kol85]   Kolmogorov, A.N.: Selected works, Volume I. Mathematics and Mechanics, Tikhomirov, V.M. (ed). Kluwer, Boston, Mathematics and its Applications, vol. 25 (1991). Original Russian edition: Nauka, Moscow (1985)

[Kol86]   Kolmogorov, A.N.: Remembering P.S. Aleksandrov (Russian). Uspekhi Mat. Nauk, **41**, 187–203 (1986). Russian Math. Surveys, **41**, 225–246 (1986)

[Kol72]   Kolmogorov, A.N.: A scientific tutor (remembrances of P.S. Urysohn) (Russian). In: Neiman, L.: The joy of discovery. Literature for children, Moscow, 160–164 (1972). Nauka, collection "Quantum", vol. 64, 10–14 (1988)

[Kur66]   Kuratowski, K.: Topology, vol. I. Academic Press, New York, London (1966)

[Lef42]   Lefschetz, S.: Algebraic Topology. New York (1942)

[Mar99]   Mardešić, S.: Absolute neighborhood retracts and shape theory. In: James, I.M. (editor): *History of topology*, 241–269. North-Holland, Amsterdam (1999)

[Nem36]   Nemytskij, V.V.: The Moscow topological circle after 10 years (Russian). Uspekhi Mat. Nauk, **2**, 279–285 (1936)

[Nov04]   Novikov, S.P.: Algebraic topology. Steklov Mathematical Institute, Russian Academy of Sciences, ser. Contemporary problems in Mathematics, No. 4 (2004). Completed version: Topology in the 20th century: a view from the inside. Uspekhi Mat. Nauk (Russian). Russian Math. Surveys, vol. **59**:5, 803–829 (2004)

[Shi03]   Shiryaev, A.N. (editor): A.N. Kolmogorov, Commemorative edition (Russian), vol. 1–3, Moscow, Physmatlit (2003)

[Ste62]   Steenrod, N.E.: Cohomology operations (lectures written and revised by D.B.A. Epstein). Princeton Univ. Press, Princeton, New Jersey (1962)

[Whi37]   Whitney, H.: On products in a complex. Proc. Nat. Acad. Sci. USA, **23**, 285–291 (1937)

[Wil63]   Williams, R.F.: A useful functor and three famous examples in topology. Trans. Amer. Math. Soc., **106**, 319–329 (1963)

Topological articles by A.N. Kolmogorov

[K1]   Kolmogorov, A.N.: Zur normierbarkeit eines allgemeinen topologischen linearen Raumes. Stud. math., **5**, 29–33 (1934). (Art. n° 23 in [Kol85].)

[K2]   Kolmogorov, A.N.: Über die dualität im Aufbau der kombinatorischen Topologie. Math. Sb., **1**, 97–102 (1936). (Art. n° 29 in [Kol85].)

[K3]   Kolmogorov, A.N.: Homologierung des Komplexes und des lokalbikom-
        pakten Raumes. Math. Sb., **1**, 701–706 (1936). (Art. n° 30 in [Kol85].)

[K4]   Kolmogorov, A.N.: Les groupes de Betti des espaces localement bicom-
        pacts. C.R. Acad. Sci. Paris, **202**, 1144–1147 (1936). (Art. n° 32 in
        [Kol85].)

[K5]   Kolmogorov, A.N.: Propriétés des groupes de Betti des espaces locale-
        ment bicompacts. C.R. Acad. Sci. Paris, **202**, 1325–1327 (1936). (Art.
        n° 33 in [Kol85].)

[K6]   Kolmogorov, A.N.: Les groupes de Betti des espaces métriques. C.R.
        Acad. Sci. Paris, **202**, 1558–1560 (1936). (Art. n° 34 in [Kol85].)

[K7]   Kolmogorov, A.N.: Cycles relatifs. Théorème de dualité de M.
        Alexander. C.R. Acad. Sci. Paris, **202**, 1641–1643 (1936). (Art. n° 35
        in [Kol85].)

[K8]   Kolmogorov, A.N.: Über offene Abblidungen. Ann. Math., **38**, 36–38
        (1937). (Art. n° 36 in [Kol85].)

[K9]   Kolmogorov, A.N.: Skew-symmetric forms and topological invariants
        (Russian). Proc. Seminar on Vector and Tensor Analysis and Applica-
        tions in Geometry, Mechanics and Physics, Moscow, Leningrad: GONTI,
        vol. 1, 345–347 (1937). (Art. n° 37 in [Kol85].)

[K10]   Kolmogorov, A.N., Gel'fand, I.M.: On rings of continuous functions on
        topological spaces (in Russian). Dokl. Akad. Nauk SSSR **22**:1, 11–15
        (1939). (Art. n° 41 in [Kol85].)

Geometry and approximation theory in A. N. Kolmogorov's works

Vladimir M. Tikhomirov

Moscow State University and Independent University of Moscow, Russia

Translated from the Russian by Dmitry Chibisov

8.1 Geometric motives in Kolmogorov's works

8.1.1 Introduction

One day I happened to organize a discussion on the topic: "Development of geometry in the 20th century". Naturally, the following question was raised: What is the geometry in general and what is it nowadays?

The 19th century was the age of geometry. What names! Carl Friedrich Gauss, Nikolay Lobachevsky, János Bolyai, Bernhard Riemann, the discoverers of new geometries; Eugenio Beltrami and Arthur Cayley, their successors; Felix Klein who put together different geometries on the basis of a unified conception; Sophus Lie and Elie Cartan, the creators of geometry of homogeneous manifolds; Henri Poincaré, a universal genius comprising the ideas of all the famous geometers mentioned above; and moreover: Augustin Cauchy and Hermann Minkowski, the founders of convex geometry; Karl Jacobi and Hermann Grassmann, the creators of multidimensional geometry; Ferdinand Minding and Karl Peterson who continued Gauss' works in differential geometry; Julius Plücker and August Möbius who developed analytic-algebraic approaches to geometry; the outstanding synthetic geometer Jacob Steiner, and, finally, David Hilbert with his "Foundations of Geometry" (which appeared in 1899) — all this is an incomplete list of great geometers of the 19th century.

And, among mathematicians of the 20th century, who can continue such a list? Who of our contemporaries may be called a great geometer? What happened to geometry? Is there a single geometry nowadays or are there many of them? All these questions became the subject of a spirited discussion.

There were different opinions regarding what geometry is. Of course, the discipline originating from antiquity whose subject matter is the *mathematical description of figures, which we see by our own eyes* (and which may be subject to all conceivable transformations, say, affine, projective, conformal, isometric, continuous, etc.), this discipline continues to exist. But we don't have the gift of seeing the multidimensional world, which is explored by the methods of algebra, calculus, combinatorics, — is it geometry as well? On the other hand, does the study of two- or three-dimensional geometric figures lie within the scope of the modern fundamental research?

There was also an opinion that geometry is a *way of thinking*. This opinion was supported by recent research in physiology, which distinguished two components in the structure of our brain. It is regarded now as a well established fact that a half of our brain is responsible, so to speak, for "harmony", i.e. it handles intuition, imagination, perception of shape and color, whereas the other one is responsible for "algebra". This "algebraic" part of the brain is in charge of logic and formal analysis, it governs all the algorithmic components of our behavior and thought. So, a widely spread opinion is that geometry is the part of mathematics where the "imaginative" part of our brain plays the most important role.

Here we stop this discussion and proceed to our first subject: Kolmogorov as a geometer. As we will see, it is closely related to the second subject, which is Kolmogorov and approximation theory.

During all his life, Andrey Nikolaevich continued to ponder over the problems of talent. He distinguished three (rather than two) components of mathematical talent: *algorithmic* ("in the sense of skilful transformation of complicated expressions, finding successful ways of solving equations which did not fit to usual rules, etc."), *logical* (meaning the art of consistent, correctly structured logical reasoning), and, finally, *geometric*. He wrote that "geometric imagination, or, as they say, 'geometric intuition' plays a great role when working in almost all areas of mathematics, even the most abstract ones". Thinking of himself, speaking of his abilities, Kolmogorov always emphasized that he had this "geometric intuition". In one of his draft notebooks Andrey Nikolaevich assessed his algebraic-analytic abilities as "very moderate", logic and intuition were marked by a simple plus sign, whereas "geometric constructions" were assessed by a plus with an exclamation sign. It should be added that Andrey Nikolaevich liked to draw pictures. In his letters and diaries there are a lot of pen and ink drawings. P. S. Aleksandrov entrusted drawing figures for his books to Andrey Nikolaevich. So, there were essential prerequisites for application of geometric ideas and constructions in Kolmogorov's work. However we will also encounter Kolmogorov's works where the logical or analytical components prevail.

8.1.2 Two geometric papers by Kolmogorov

We begin with discussion of Kolmogorov's geometric papers. It is generally agreed that there are only two of them: *Zur topologish-gruppenteoretischen Begründung der Geometrie* (On the topological group-theoretic foundation of geometry) [Kol30] (1930), and *Zur Begründung der projektiven Geometrie* (On the foundation of projective geometry) [Kol32] (1932).

These are indeed geometric papers in the spirit of the 19th century, since they treat the two most important geometric objects studied at that time, namely, the *spaces of constant curvature* (which appeared in conjunction of works by Gauss, Lobachevsky, Bolyai, Riemann, Beltrami, Minding, and others) and the *projective spaces*, which played an extremely important role in geometry, so that Cayley wrote: "Projective geometry is all of geometry".

But both these Kolmogorov's papers in their ideas and methodology belong to the first half of the 20th century, namely, they relate to axiomatic theories. In each of them a system of axioms is introduced, which describes the objects mentioned above from some standpoint novel for that time. These papers required logical constructions rather than geometric ones, i.e. they are closer related to logic rather than to geometry[1].

In the first of the papers mentioned above, Kolmogorov characterizes "spaces of constant curvature are the only type of topological spaces with a sufficient amount of freedom of motion". Kolmogorov provides an elegant and very economical axiomatics of such spaces. It is as follows.

Let X be a *metrizable, locally compact and connected topological space on which a group Γ is given of single-valued continuous mappings of X on itself, possessing the property of uniform continuity.* The first interesting Kolmogorov's observation is that whenever X has the above properties, one can introduce a metric on X such that all the mappings in Γ become isometric. This enables one to define *spheres $S(x)$ centered at x,* and then Kolmogorov introduces one more axiom: *for any two different spheres with common center, one of them always separates the other one from the center.* Using this axiom,

---

[1] In what follows we will repeatedly use the terms involving the words "topology", "topological", "compact", "continuous", and so on. Let us recall what they mean.

Topological spaces are characterized by the property that in these spaces the notions of continuity and open and closed sets are defined. Nowadays topological spaces are defined usually in terms of open sets. A set X is called a *topological space* is there is a class τ of its subsets with the following properties: (a) the empty set and the set X itself belong to τ, and (b) the union of any family of sets in τ and the intersection of finitely many sets in τ belong to τ. The elements of τ are called *open sets*, their complements are *closed sets*, and for a given set, the least closed set containing it is called its *closure*. A subset of X is *compact* if any its open covering contains a finite subcovering. A space is said to be *locally compact* if every its point has a neighborhood such that its closure is compact. A mapping of a topological space to another one is said to be *continuous* if the inverse image of any open set is open.

A special case of topological spaces are *metric spaces* in which a distance $d(x_1, x_2)$ between any two points is defined with the properties: (a) $d(x_1, x_2) \geq 0$ and $d(x_1, x_2) = 0$ if and only if $x_1 = x_2$; (b) $d(x_1, x_2) = d(x_2, x_1)$, and (c) $d(x_1, x_3) \leq d(x_1, x_2) + d(x_2, x_3)$. A set A in a metric space (X, d) is said to be open if for any point $a \in A$ there is a number $\varepsilon > 0$ such that the ball $U(a, \varepsilon) = \{x \in X \mid d(x, a) < \varepsilon\}$ belongs to A.

it is proved that the invariant distance has the following property: *for any two points x and y of X there is a point z whose distances from x and y are equal to one half of the distance between x and y.* Then, if X is one-dimensional, this implies that this space is homeomorphic either to a straight line or to a circle, and in the general case X can be mapped on a line or a sphere so that Γ becomes the isometry group of a line or a circle respectively. Kolmogorov conjectures that these algebraic-topological axioms combined with the separability axiom already provide the spaces of constant curvature. But then the group of isometries may be thinner than for classical constant curvature manifolds (hyperbolic, Euclidean, elliptic, or spherical). He points out that the permutation group of quaternions $x' = ax + b$, $|a| = 1$, depending only on seven (rather then ten) parameters, satisfies all the axioms. Then, in order to ensure completeness of the group of motions, the separability axiom is modified. Kolmogorov defines spheres of different ranks. A sphere $S(x)$ is called a sphere of the first rank. For a point belonging to $S(x_1)$ a sphere $S(x_1, x_2)$ of the second rank is defined in a similar way, and so on up to rank n, and it is required that out of two spheres of the same rank n with parameters $(x_1, x_2, \ldots, x_k)$ one of them separates the other one from the center x_k. Then the following theorem is formulated.

Theorem 1. *If the axioms listed above hold, then X is homeomorphic to a finite-dimensional space of constant curvature, and it can be mapped on this space in such a way that Γ will go into the complete group of motions and reflections.*

All this looked so transparent that no proof followed: the author apparently believed that the interested reader would be able to reconstruct all the details himself. The famous German mathematician Heinz Hopf asked Heinrich Titz to restore Kolmogorov's arguments. The latter fulfilled the task, though, having somewhat extended Kolmogorov's axiomatics.

That was the first Kolmogorov's geometric paper.

Whereas in this paper Kolmogorov originated the use of the methods related to topological groups in geometry, his second paper initiated the development in our country of both topological algebra and topological geometry (because a projective space is simultaneously a geometric and algebraic object; this concerns any classical geometry, but "especially" projective).

In the second paper Kolmogorov introduces three groups of axioms, which describe properties of three systems of elements, viz., points, lines, and planes. The first is the *compactness axiom* saying that *each of these systems of elements is a connected compact topological space.* Next goes the group of *incidence* axioms, where points, lines, and planes satisfy the usual incidence axioms of projective geometry. (Namely: *for any two different points there is a unique line passing through them and each line contains at least three points; for any three non-collinear points there is a unique plane passing through them; any line and plane have at least one common point; any two planes have at least one common line, and there exist four non-coplanar points such that any three of them are not collinear.*) The third group are *continuity* axioms saying that the relations described by the incidence axioms are continuous (for instance, a line is a continuous function of two different points, etc.)

Now, if we select three points on one of the lines to take them for zero, unity, and infinity, then we can uniquely define (based on incidence axioms, as is usually done in projective geometry) the addition and multiplication operations on the set of points of the line so that this set becomes a skew-field with the geometry of the plane over this skew-field being isomorphic to the initial one.

But when the compactness and continuity axioms hold, the skew-field so constructed is connected and locally compact. Having realized this, Andrey Nikolaevich posed to L. S. Pontryagin the problem *to describe all connected locally compact skew-fields.* And Pontryagin proved that there are only three of them: the fields of real and complex numbers and the skew-field of quaternions. This was one of the first results of topological algebra. As a consequence of Kolmogorov's and Pontryagin's theorems the following theorem obtains:

Theorem 2. *There exist only three different types of projective spaces which possess the connectedness, compactness, and continuity properties, namely, the real, complex, and quaternion projective (three-dimensional) spaces.*

That was the initial result of topological geometry.

Let us briefly comment on these two papers. Among his teachers, Kolmogorov always reckoned A. K. Vlasov who gave lectures on projective geometry which Kolmogorov attended when being a second course student. The paper on manifolds of constant curvature is dated June 18, 1930. This was the time of Kolmogorov's first visit to Germany. Possibly he produced this paper in anticipation of meeting Hilbert, who exerted a profound influence on all the initial period of Kolmogorov's work. The second paper is dated May 26, 1931, so that it was completed right after his return from this trip, which might have influenced writing it. In both these papers Kolmogorov combines geometric and algebraic ideas with topological concepts. This will happen repeatedly during 30-s. The reason for that was, most likely, the influence of his friend Pavel Sergeevich Aleksandrov. Thus in these two papers Kolmogorov's interests in axiomatic methods, projective geometry, and topology were put together.

Both papers were met with approval and continued. The reader can find an account of the subsequent developments related to the axiomatics of the projective space in the comment by A. V. Mikhalev in Kolmogorov's *Selected Works* [Kol], Vol. 1, pp. 479–481.

8.1.3 Definition of measure of a set and normability of a topological vector space

Now we will speak of geometric motives in non-geometric papers by A. N. Kolmogorov.

In his old age, Andrey Nikolaevich remembered about his first outstanding result, viz., construction of an example of an integrable function with almost everywhere divergent Fourier series[2], as follows: "For a rather long time I worked in both ways, trying by turns to construct an example or to prove its nonexistence. The last stage was a week of persistent thinking, which resulted in a suddenly appeared construction. Somewhat later, without much effort there appeared an analytic version of the idea, which was initially purely geometric."

The paper presenting the construction of the example contains only the analytic proof, and it is hard to understand now what Andrey Nikolaevich meant when speaking of the "purely geometric idea". But he told me more than once that the cornerstone of his solution was the intuitive geometric image that suddenly appeared to him.

The next two papers we will speak of are devoted to comprehension of the notion of measure (in particular, the area of a surface) and topological vector space.

The first of this papers is largely related to geometry. This is: *Beiträge zur Masstheorie* (On measure theory) [Kol33] (1933). This paper deals with the logical structure of the notion of the surface area, more precisely, of the measure of a k-dimensional set lying in an n-dimensional space.

Let us discuss Kolmogorov's approach in the intuitively clearer case when $k = 2$ and $n = 3$, in other words, let us try to see how he treats the area of a fairly arbitrary two-dimensional set in the three-dimensional space.

Kolmogorov defines two measures, the upper and lower ones. Their construction is of geometric nature and very clear intuitively. Suppose you have a surface (the boundary of a three-dimensional body) and you want to measure its area. Take a piece of cloth which can crease, but cannot stretch, and apply it to the surface, possibly with wrinkles, but without breaks. The least area of a piece of cloth sufficient to cover the entire surface (more precisely, the infimum of the areas of such pieces) is referred to as the *upper area* of this surface.

And the lower area is defined as follows. Suppose we are allowed to cut the cloth into pieces and apply these pieces to the surface so that they are disjoint. Kolmogorov calls the supremum of the sums of the areas of such pieces, which cover the entire surface, the *lower area*.

The paper contains the proof of the following main result.

Theorem 3. *Any measure, which possesses the most natural properties for this notion, lies between the upper and lower measures so constructed.*

Of course, in the general setup the piece of cloth applied to the surface is replaced by something very abstract (*the infimum of the Lebesgue measure of the inverse image of the surface under a nonexpanding mapping* defined on the widest and favorite at that time class of *Suslin's sets*), but the essential part

---

[2] See Chap. 1 (by J.-P. Kahane) in this book. (Editor's note.)

of the matter was explained above. It has a transparent and purely geometric meaning, and the paper on the measure theory may be regarded as geometrical in its "way of thought".

Kolmogorov was preparing the paper on measure theory for discussion with Carathéodory who was the first to consider the very problem of defining the measure of a k-dimensional set in an n-dimensional space, which led, according to Kolmogorov, "to a number of definitions of measure". As Kolmogorov wrote afterwards, Carathéodory "liked my paper on measure theory and insisted on its fastest publication".

For the subsequent developments of this paper see the comment by V. A. Skvortsov in Kolmogorov's *Selected Works* [Kol], Vol. 1, pp. 428–435.

The next paper has played a considerable role in building of functional analysis. It is mentioned, in particular, in N. Bourbaki's survey of the history of mathematics. We mean the paper: *Zur Normierbarkeit eines allgemeinen topologishen linearen Raumes* (On normability of a general topological linear space) [Kol34] (1934). In this paper an algebraic (namely, vector) structure was again combined with topology and for the first time the definition of bounded sets in linear topological spaces (or "topological vector spaces" as they are termed nowadays) was given[3].

In Bourbaki's treatise quoted above one of the two mentions of Kolmogorov's name is in the section dealing with topological vector spaces. Bourbaki points out therein the importance of "the notion of a bounded set introduced by Kolmogorov and von Neumann in 1935" (as is seen from the year of publication, Kolmogorov introduced this notion in 1934).

Functional analysis had arisen shortly before Kolmogorov's paper: the famous book by S. Banach "Theory of Linear Operators", which in fact gave birth to the new direction in mathematics, appeared in 1931 (French translation was published in 1932). One copy of Banach's book in French was sent to Kolmogorov and it gave an impact to development of functional analysis in our country. Kolmogorov's paper was one of the first responses to Banach's book[4].

Kolmogorov writes that "it seems quite natural to develop the general theory of linear functionals and operators in linear topological spaces. However the considerable part of this theory has been developed by now for *normed*

---

[3] A vector space is called a *topological vector space* if it is a topological space and the basic operations, viz., addition of vectors and multiplication of a vector by a number, are continuous

[4] The basic notion in Banach's book is that of a normed space. Recall that a vector space X is said to be *normed* if on X a function $\| \cdot \| \colon X \to \mathbb{R}$ is defined with the following properties:
  (a) $\|x\| \geq 0$ and $\|x\| = 0$ if and only if $x = 0$;
  (b) $\|\alpha x\| = |\alpha| \, \|x\|$ $\forall x \in X, \ \alpha \in \mathbb{R}$;
  (c) $\|x + \xi\| \leq \|x\| + \|\xi\|$ $\forall x, \xi \in X$.
  A normed space becomes a metric space if we set $d(x_1, x_2) = \|x_1 - x_2\|$

spaces [...] Hence the question arises: which linear topological spaces can be normed?" The paper answers this question by the following theorem.

Theorem 4. *A linear topological space can be normed if and only if there exists at least one convex bounded neighborhood of zero.*

The necessity part is clear: the open unit ball in a normed space is a convex bounded neighborhood of zero. On the other hand, the Minkowski function[5] of the intersection of a bounded convex set with a symmetric set is a norm, which, as can be easily shown, specifies the same topology as the one given initially in the linear topological space.

The Minkowski function is one of the basic functions of convex geometry, and the proof that the Minkowski function of a convex set containing zero is a convex function homogeneous of first order (which is the key point of the normability theorem) is a fundamental fact of convex geometry.

8.1.4 Widths of ellipsoids and octahedrons

The next Kolmogorov's paper with geometric content has initiated a new direction in the approximation theory. This is the paper: *Über die beste Annäherung von Funktionen einer gegebenen Funktionenklasse* (On the best approximation of functions of a given functional class) [Kol36] (1936). We will comment here on the geometric aspect of this paper and then discuss its role in the approximation theory.

In this paper the notion was introduced, which was subsequently called the *Kolmogorov n-width* (of a centrally-symmetric set C in a normed space X). It is defined as follows:

$$d_n(C, X) = \inf_{L_n} \sup_{x \in C} \inf_{\xi \in L_n} \|x - \xi\|,$$

where the leftmost infimum is taken over all subspaces L_n of dimension n. The Kolmogorov n-width characterizes the accuracy of approximation of a set by n-dimensional subspaces.

In the paper at hand the following geometric problem was solved: *to describe the n-dimensional subspace approximating with the best possible accuracy a compact ellipsoid C lying in a Hilbert space.*

An *ellipsoid* in the Hilbert space is the image of a ball under a linear mapping. Hilbert's theorem on reduction to principal axes of a quadratic form generated by a completely continuous operator in a Hilbert space implies that a compact ellipsoid is isometric to the following ellipsoid E_a in the space l_2:

$$E_a = \left\{ x \mid \sum_{i \in \mathbb{N}} \frac{x_i^2}{a_i^2} \leq 1 \right\} \quad (a = (a_1, a_2, \ldots),\ a_i \downarrow 0)$$

(where $\mathbb{N} = \{1, 2, \ldots\}$).

---

[5] The *Minkowski function* of a convex neighborhood V of 0 in a topological vector space E is the function $\rho_V : E \to [0, +\infty[$ defined by $\rho_V(x) = \inf\{\lambda > 0,\ x \in \lambda V\}$. If E is a normed space, it is easily seen that the norm is the Minkowski function of the unit ball. (Editor's note.)

Theorem 5. *The following equalities hold:*

$$d_n(E_a, l_2) = a_{n+1}, \quad n = 0, 1, \ldots$$

A geometric proof of this theorem in the three-dimensional case can be easily illustrated by a picture. Now we will give an analytic proof.

Upper bound. Let us approximate a vector $x \in E_a$, $x = (x_1, x_2, \ldots)$, by the vector $S_n x = (x_1, \ldots, x_n, 0, \ldots)$. We obtain:

$$d_n^2 = |x - S_n x|^2 \overset{\text{def}}{=} \sum_{i \geq n+1} x_i^2 \overset{\text{Id}}{=} \sum_{i \geq n+1} \frac{x_i^2}{a_i^2} a_i^2 \overset{a_i \downarrow}{\leq} a_{n+1}^2 \sum_{i \in \mathbb{N}} \frac{x_i^2}{a_i^2} \overset{x \in E_a}{\leq} a_{n+1}^2.$$

This gives an upper bound for the approximation accuracy.

Lower bound. The lower bound is obtained by the method of "imbedded ball". Consider the set of vectors

$$B^{n+1}(0, a_{n+1}) = \left\{ x = (x_1, \ldots, x_{n+1}, 0 \ldots) \mid \sum_{i=1}^{n+1} x_i^2 \leq a_{n+1}^2 \right\}. \tag{i}$$

We see that

$$\sum_{i=1}^{n+1} \frac{x_i^2}{a_i^2} \overset{a_i^{-1} \uparrow}{\leq} a_{n+1}^{-1} \sum_{i=1}^{n+1} x_i^2 \overset{(i)}{\leq} 1.$$

Hence the $n + 1$-dimensional ball $B^{n+1}(0, a_{n+1})$ lies in the ellipsoid E_a. But for any n-dimensional subspace L_n there is a vector of length a_{n+1} which lies in the orthogonal complement to L_n and belongs to $B^{n+1}(0, a_{n+1})$, i.e. the deviation of our ellipsoid from L_n is no less than a_{n+1}, hence $d_n \geq a_{n+1}$. Thus the upper and lower bounds coincide. Hence the theorem.

Eleven years after his 1936 paper Andrey Nikolaevich became interested in Gauss' paper on the method of least squares, which led him to some problems of finite-dimensional Euclidean geometry. Kolmogorov engaged in his studies young mathematicians A. A. Petrov and Yu. M. Smirnov. Their cooperation resulted in the joint paper *"A Gauss' formula in the theory of least squares method "* [KPS47] (1947). The authors obtained an upper bound for some geometric quantity, but did not succeed in getting a lower bound. Then Kolmogorov proposed this problem to his pupil A. I. Mal'tsev (one of the leading algebraists of the last century) who proved the required bound by an elegant algebraic construction. Mal'tsev's proof took 2 pages and was published as a short note right after the paper by Kolmogorov–Petrov–Smirnov. Kolmogorov did not focus attention on the fact that the resulting quantity was a certain width. After another seven years S. B. Stechkin, who worked in the approximation theory, restated the result by Kolmogorov–Petrov–Smirnov–Mal'tsev in terms of widths to obtain the following theorem:

Theorem 6. *Let $\mathcal{O}^n$ be a regular octahedron (the convex hull of a system of n pairwise orthogonal unit vectors in $\mathbb{R}^n$ along with those opposite to them). Then*

$$d_k(\mathcal{O}^n) = \sqrt{\frac{n - k}{n}}.$$

In the three-dimensional case this is a beautiful stereometric problem. It is natural to conjecture that a line and a plane providing the best approximation to the octahedron (or, equivalently, to its vertices) must be equidistant from all the vertices. It is not hard to prove this in the three-dimensional case. But this fact holds in the general setting as well. We will give its analytic proof in the n-dimensional case.

The *lower bound* is obtained by the *averaging method*, which was repeatedly used by Kolmogorov and was actually applied by the three authors mentioned above. Let $\{e_k\}_{k=1}^n$ be the standard basis in $\mathbb{R}^n$ ($e_1 = (1, 0, \ldots, 0), \ldots, e_n = (0, \ldots, 0, 1)$), let L_k be some k-dimensional subspace of $\mathbb{R}^n$, and let $\{f_j\}_{j=1}^k$ be an orthonormal basis in L_k. Then the squared distance from some point $x \in \mathbb{R}^n$ to L_k is given by the formula:

$$d^2(x, L_n) = \|x\|^2 - \sum_{j=1}^n (x, f_j)^2$$

where (x, y) is the scalar product of vectors x and y, i.e. $\sum_{i=1}^n x_i y_i$. Denote (f_i, e_j) by f_{ij}, then we obtain

$$\sum_{j=1}^k d^2(e_i, L_k) = n - \sum_{j=1}^k \sum_{i=1}^n f_{ij}^2 = n - k$$

(since $\sum_{i=1}^n f_{ij}^2 = \|f_j\|^2 = 1$). But since the largest distance between a point of the octahedron and a subspace is attained at one of its vertices, this formula implies

$$\max_{1 \le j \le n} d^2(e_j, L_k) \ge n^{-1} \sum_{j=1}^k d^2(e_j, L_k) = \frac{n-k}{n}.$$

Since L_k is an arbitrary k-dimensional subspace, we obtain the lower bound: $d_k(\mathcal{O}^n) \ge \sqrt{\frac{n-k}{n}}$.

It is seen from the above derivation that if $\widehat{L}_k$ is equidistant from all the vertices, then the deviation of the octahedron from $\widehat{L}_k$ is equal to $\sqrt{\frac{n-k}{n}}$.

Upper bound. Denote by T the transformation acting by the following rule: $Te_i = e_{i+1}$, $i = 1, \ldots, n-1$, $Te_n = e_1$. It is clear that the one-dimensional subspace $L_1 = \text{span}\{e_1 + \cdots + e_n\}$ is invariant relative to T ($TL_1 = L_1$). Consider now all the two-dimensional invariant subspaces into which the transformation T decomposes. It is easily seen that one can compose a k-dimensional subspace $\bar{L}_k$ out of them, which, like L_1, will be invariant relative to T. Hence we obtain (denoting by Px the orthogonal projection of x on $\bar{L}_k$, $Pe_i = \xi_i$, and $T^{-1}\xi_i = \xi_i'$):

$$d(e_i, \bar{L}_k) = \|e_i - Pe_i\| = \|Te_{i-1} - T(T^{-1}\xi_i)\|$$
$$= \|e_{i-1} - \xi_i'\| \ge \inf_{\xi \in \bar{L}_k} \|e_i - \xi\| = d(e_{i-1}, \bar{L}_k).$$

The reverse inequality is obtained in a similar way, hence the required upper bound holds. The proof is completed.

8.1.5 More about geometric and visual motives in Kolmogorov works

Topological results. Topology can naturally be regarded as a chapter of geometry, but it has been developed and extended to such an extent that now it overshadows its ancestor. In this book there is a separate chapter devoted to topology, so I will only briefly comment on Kolmogorov's works.

1. *Can an open mapping increase topological dimension?* In thirties this question interested many topologists. Kolmogorov constructed an example of such a mapping in his paper: *Über offene Abbildungen* (On open mappings) [Kol37] (1937). Kolmogorov's construction is based on the following geometric fact: there exists an open mapping of a torus on the Möbius strip which carries the equator of the torus into the edge of the strip. Iterating such mappings and taking the projective limit, Kolmogorov constructs an open mapping of the one-dimensional continuum on the so-called "Pontryagin's surface", which is the compact set constructed shortly before that by Pontryagin and giving the negative answer to the question whether the formula $\dim(X \times Y) = \dim X + \dim Y$ always holds true. Pontryagin's surface is a two-dimensional continuum, whose square is of dimension 3.

2. Kolmogorov introduced one of the most important topological notions, namely, that of *cohomology*. This is discussed in detail in the chapter on topology[6], and I will not tackle this matter.

I will only mention the "visual" images, which helped this discovery. Andrey Nikolaevich told V. I. Arnol'd that he "invented his topological homology theory having in mind fluid flows in hydrodynamics or magnetic fields rather than combinatorial or algebraic objects". He wanted to model these physical phenomena in the combinatorial setup of an abstract complex and succeeded in it.

Spirals. Working on the theory of stochastic processes in 30-s, Kolmogorov came to necessity of *describing the curves in the Hilbert space invariant relative to a one-parameter group of motions.*

Not doubt he imagined initially a spiral sliding along itself. But the proof does not show any geometric ideas: the description of such spirals reduced to an elegant analytic problem, which was solved by Kolmogorov.

As one of essential components of his creative endowments, Kolmogorov regarded also the "processes intuition", which undoubtedly has very much in common with geometric imagination.

I will quote Kolmogorov's comment about one of his most remarkable papers related to an applied problem as a manifestation of such "processes intuition".

---

[6] See Chap. 7 (by V. M. Buchstaber). (Editor's note.)

In his paper *"A study of the diffusion equation combined with increase in the amount of substance and its application to a biological problem"* (jointly with I. G. Petrovsky and N. C. Piskunov) [KPP37] (1937), the problem setting was due to Kolmogorov. As he wrote, "this problem appeared owing to my long-time contacts with A. S. Serebrovsky and a group of his collaborators N. P. Dubinin, A. A. Malinovsky, and D. D. Romashov". A characteristic feature of this paper is that it contains an invariant solution of traveling wave type. It was striking for me to hear from him how he guessed the solution. He said: "you know, I had seen a Bickford's fuse burning". Is not it wonderful that Kolmogorov perceived the nature of evolution of a biological system having seen a burning Bickford's fuse!

8.1.6 The 13th Hilbert problem

The 13th Hilbert problem[7] is devoted to one of the central questions of analysis: *are there functions of many variables?* Functions studied in high school are mostly functions of a single variable: quadratic trinomial and other polynomials, trigonometric functions, exponential and logarithmic functions, etc. But, of course, we often encounter functions of two and many variables. For example, the distance from a point to the origin on a plane and in the space is given by the functions $\sqrt{x^2 + y^2}$ and $\sqrt{x^2 + y^2 + z^2}$ of two and three variables. The simplest function of two variables is the sum, which associates with a pair of numbers (x, y) the number $x + y$.

The experience of classical analysis shows that functions of two variables have much more complicated structure than those of a single variable, functions of three variables are much more complex than those of two variables, and so on. How can it be expressed? One possible way is as follows. Some functions of three variables can be specified as a *superposition* of functions of two variables, as, say, $f(x, y, z) = \varphi(x, \psi(y, z))$. (For example, the function $\sqrt{x^2 + y^2 + z^2}$ is the superposition of the functions $u \to u^2$ and $v \to \sqrt{v}$ of a single variable and the addition function of two variables.) Hilbert was sure that functions of three variables cannot be reduced to functions of two variables, so in the 13th problem he put the question in the most radical form. He has chosen a specific *algebraic* function of three variables (namely, the solution to the polynomial equation $(x, y, z) \mapsto w^5 + xw^2 + yw + z = 0$) and asked *whether it can be expressed as a superposition of continuous functions of two variables.* (Supposedly, he expected a negative answer and assumed that the proof of this fact would rely on the methods of algebraic geometry.) But the answer turned out be positive and the solution based on geometry.

The story of how Hilbert's 13th problem was solved is very amusing. In the spring of 1956 A. N. Kolmogorov arranged a seminar for students of the second course of the Faculty of Mechanics and Mathematics of the Moscow State University, where he discussed some problems bearing in mind to come in the

---

[7] See Chap. 13 (by V. Brattka) in this book. (Editor's note.)

long run to the solution of Hilbert's 13th problem. This seminar was attended by the second course student Dima Arnol'd (as Dmitry Igorevich was called by fellow students). Arnol'd's participation in this seminar resulted in his first scientific paper. The seminar lasted only for one semester, during this time some interesting results were obtained, but Hilbert's problem looked still a far away matter. However, when the seminar was over, Kolmogorov somewhat unexpectedly even for himself managed to concentrate an immense impulse of energy on solution of this very problem. As a result of about two weeks of intensive thought he proved that *any continuous function of four variables is a superposition of functions of three variables.* Kolmogorov presented this result at the 3rd All-Union Mathematical Congress in the summer of 1956. Then he gave up these studies letting his disciples to continue and complete them.

After about half a year, once in spring of 1957, I visited Andrey Nikolaevich in his *dacha.* Andrey Nikolaevich showed me a copy-book with inscription on the cover: "Term paper by the 3rd course student Arnol'd". Kolmogorov said: "I am checking this paper now, but it is not improbable that it contains the solution of Hilbert's 13th problem". And this was the case indeed.

But this was not the end of the story. In summer of 1957 Kolmogorov succeeded in strengthening Arnol'd's result to prove the following theorem.

Theorem 7. *Any continuous function of n variables (defined on the n-dimensional unit cube) is representable as a superposition of functions of one variable and the only function — addition — of two variables.*

Let us state a result by Kolmogorov, which implies Theorem 7 in the two-dimensional case:

Let f be a continuous function defined on the unit square $Q = \{(x, y) \mid 0 \leq x, y \leq 1\}$. Then it is representable in the form

$$f(x, y) = \sum_{i=1}^{5} \chi_i(\varphi_i(x) + \psi_i(y)),$$

where the φ_i, ψ_i, and χ_i are continuous functions of one variable.

The proof of the general theorem dealing with functions of n variables is illustrated by this two-dimensional case quite well.

Here we will outline the proof of the theorem stated above. It consists of three stages, the first two of which are purely geometric.

1. CONSTRUCTION OF SYSTEMS OF SQUARES. Consider the system S_1 of intervals on the real line $\{\Delta_i\}_{i \in \mathbb{Z}} = [i, i + \frac{4}{5}]$ separated from each other by intervals of length $1/5$. Next, consider on the plane the direct product of S_1 by itself (i.e. the set of pairs (x, y), where $x \in \Delta_i$, $y \in \Delta_j$, $i, j \in \mathbb{Z}$). Thus we obtain something like a map of a city with avenues and streets of equal width. Denote the coordinate origin by O_0 and consider also the points O_k, $1 \leq k \leq 4$, with coordinates $(k/4, k/4)$. Now we make

four translations of the initial "map" which shift O_0 to the points O_k, $1 \leq k \leq 4$. And, finally, we make l homothetic transformations of the whole picture with homothety coefficient γ. As a result, we obtain the system of squares $Q_{ij}^{k,l}$, $i, j \in \mathbb{Z}$, $0 \leq k \leq 4$, $l \in \mathbb{Z}_+$. The construction of the systems of squares is completed.

2. CONSTRUCTION OF FUNCTIONS φ_k AND ψ_k. These functions do not depend on the function f. The main requirement on these functions is that the functions $\Phi_k(x, y) = \varphi_k(x) + \psi_k(y)$ *separate any two squares of the l-th system of squares*, i.e. that the segments $\Phi(Q_{ij}^{k,l})$ and $\Phi(Q_{i'j'}^{k,l})$ for $(i, j) \neq (i', j')$ are disjoint.

We will demonstrate only the first step in construction of these functions to be defined on the entire plane. We have the zero-system of squares Q_{ij}^{00} consisting of the squares with side $4/5$ and containing the "initial" square with a vertex at the origin. Construct a continuous function $\Phi_i^0(x, y)$ representable as a sum of two functions of one variable $\varphi^0(x)$ and $\psi^0(y)$, which separates the squares Q_{ij}^{00}. The square $\varphi^0(x)$ is the product of two intervals, Δ_i on the Ox-axis and Δ_j on the Oy-axis. Define the functions $\varphi^0(x)$ and $\psi^0(y)$ so that the values of $\varphi^0(x)$ on the square Q_{ij}^{00} be close to the integer i and the values of $\psi^0(x)$ on the same square be close to $\sqrt{2}j$. In other words, we enclose the points i into segments $[i - \varepsilon_i, i + \varepsilon_i]$ and the points j into segments $[\sqrt{2}j - \eta_j, \sqrt{2}j + \eta_j]$ so that the intervals $\delta_{ij}^{00} = [i - \varepsilon_i + \sqrt{2}j - \eta_j, i + \varepsilon_i + \sqrt{2}j + \eta_j]$ are disjoint. Then the functions $\varphi^0(x)$ and $\psi^0(x)$ are completed by linearity. This is the first step, after which the construction of our functions continues in a similar manner successively, by induction.

3. END OF THE PROOF (CONSTRUCTION OF FUNCTIONS χ_i). We will again demonstrate only one step of the inductive construction. Suppose we are given a function $f(x, y)$ on the unit square Q and let $M = \max_{(x,y) \in Q} |f(x, y)|$. Construct the functions χ_i^1 of one variable so that the function $f_1(x, y) = f(x, y) - \sum_{k=1}^{5} \chi_k^1(\Phi_k(x, y))$ (where the $\Phi_k(x, y)$ are the functions constructed at step 2) fulfills the relation

$$\max_{(x,y) \in Q} |f_1(x, y)| \leq \frac{5}{6} M$$

and, moreover,

$$\max_{(x,y) \in Q} |\chi_k^1(\Phi_k(x, y))| \leq \frac{1}{3} M.$$

To this end, we choose the rank l so that the difference between the largest and smallest values of f on any square Q_{ij}^{kl} would be no greater than $\frac{M}{6}$. Now we set the function χ_k^1 on the interval δ_{ij}^{kl} equal to $1/3$ of the value of f at some (arbitrary) point of the square Q_{ij}^{kl}. Then we continue the function χ_k^1 by linearity. It is not hard to show that the function f_1 satisfies the required condition. Then we have to iterate our constructions and pass to the limit.

In this way the proof of Kolmogorov's theorem on superpositions in the two-dimensional case is completed.

It is natural to stop here the description of geometric ideas and constructions in the non-geometric works of Andrey Nikolaevich Kolmogorov. But we have to tackle one more subject.

8.1.7 Kolmogorov's geometry course for high school

Andrey Nikolaevich Kolmogorov devoted the last years of his life to high school mathematical education (see Fig. 8.1). In particular, he intended to reorganize the school course in geometry. Andrey Nikolaevich wanted to describe the Euclidean plane for schoolchildren in precise and intuitively clear manner. Kolmogorov's visual description involved the ancient conception of the plane originating from Euclid (when the plane is modeled by the blackboard on which one can make ruler and compass constructions, i.e. to draw straight lines through two points or circles of given radius, in other words, to do everything what is usually done in school) combined with modern views related to Felix Klein's Erlangen program. According to the Erlangen program the Euclidean plane is characterized by the *group of motions* (i.e. *isometric transformations*). Isometry allows for measuring distances, which is a very common notion: the distances between points are measured with a scale.

And the motion can be easily conceived with the aid of a sheet of glass, which is commonly used to cover a desk. It may play the role of a *tangent plane*, which may be moved along the surface of the blackboard. If there is a figure drawn on the blackboard (say, a triangle), then applying the sheet of glass to it, one can copy the figure to the glass, then to move the glass to another place and copy the figure again to the blackboard. Thus we obtain an isometric

Fig. 8.1. Kolmogorov in 1963 at the high school founded by him (Moscow). Photo Alexander Zvonkin

(preserving distances) transformation of the figure. We can "turn over" the glass, then the transformed fugure will be symmetric to the initial one.

A motion of the plane is an abstraction, since it is a transformation which moves the entire plane rather than a part of it. In the mathematical language it is an isometry transformation of the plane (in the first paper discussed above Kolmogorov also considered the group of isometry transformations).

Motions of the plane include, as particular cases, parallel translations, rotations about some point of the plane, symmetries about some straight line.

The notions of distance and motion enable us to provide new proofs to many theorems and to obtain intuitively and visually clear solutions of various geometric problems.

For example, one of the first theorems of geometry saying that the angles at the base of an isosceles triangle are equal (attributed to Thales who according to a legend was the first to realize what a proof was) can be proved using the idea of motion ("with the aid of a glass") as follows. Let us copy the triangle to the glass, then turn it over and apply it to our triangle on the blackboard "with the opposite side". Now we copy the triangle from the glass again to the blackboard. The two triangles (initial and copied from the glass) will coincide with each other, hence the angles at the base are equal (since they have just traded places). This proof can be found in writings of Lewis Carrol, who was perplexed: is it a proof? Certainly it is a proof if we can use the concept of motion. But for this purpose a special axiomatics is needed. Such axiomatics was devised by A. N. Kolmogorov. It is designed to give an exact description of the Euclidean plane.

Such a description can be achieved by introducing coordinates with subsequent use of algebraic methods (this way was outlined by Hermann Weyl, Issai Schur and others; Dieudonné, remembering Euclid's words which he said to Ptolemy, called this treatment of planar geometry "Royal").

But this is not the only way. It is possible to give a precise axiomatic description of the plane without recourse to the vector model. Such an axiomatic description of the plane goes back to Euclid, and it was completed by David Hilbert in his "Foundations of Geometry".

The axiomatics of the plane devised by Kolmogorov has much in common with the Euclid–Hilbert axiomatics, but it contains also a "Klein's component". This axiomatics is rather simple, natural, and "geometric". It involves four undefined notions: *a point, a line, a set, and an element of a set*. The axioms are divided into five groups: *incidence, distance, order, mobility, and parallelizm.*

... According to some words in Kolmogorov's publications he envisioned that the teachers who love mathematics would be able starting from his course to extend for interested schoolchildren the limits of Euclidean geometry and to open for them the worlds of other geometries: the Euclidean world of finite and infinite dimensions, the world of Lobachevsky's geometry, the convex Minkowski's world, affine world, and Andrey Nikolaevich's favorite projective world.

8.2 Kolmogorov's works on the approximation theory

8.2.1 Introduction

A. N. Kolmogorov published in the first volume of his *Selected Works* (Mathematics and Mechanics) only six papers which may be regarded as dealing with approximation theory. (Sometimes his paper on superpositions is also viewed as related to this field.) Indeed, approximation theory did not belong to directions of his primary interest. However he exerted a profound influence on formation and development of this field. This influence gave rise to several directions, which formed new stages in the development of approximation theory, and to several scientific schools. The topics originated by Kolmogorov were further elaborated by hundreds of researchers, there were certainly more than a thousand papers dealing with this subject, and dozens of Doctor of Science theses were defended.

Our aim here is to review these six Kolmogorov's papers and to give a brief survey of subsequent research. For a more detailed exposition of this matter the reader is referred to my paper *"Kolmogorov and the approximation theory"* [Tik89].

8.2.2 Widths of functional classes and sets in finite-dimensional spaces

Kolmogorov's 1936 paper on widths discussed above begins as follows: "We assume that for the functions under consideration a distance is introduced. If we consider the problem of approximating f by linear forms $\phi = c_1\phi_1 + \cdots + c_n\phi_n$ with fixed $\phi_1, \ldots, \phi_n$, then we obtain the (Chebyshev's) problem: minimize the distance $\rho(f, \phi)$ by a suitable choice of the coefficients $c_1, \ldots, c_n$. Denote by $E_n(f)$ the infimum of these distances $\rho(f, \phi)$. Further, for a class F of functions f, denote by $E_n(F)$ the supremum of $E_n(f)$ over all f in F. The quantity $E_n(F)$ is determined by the class F and the functions $\phi_1, \ldots, \phi_n$. Now we set another problem: for given F and n minimize $E_n(F)$ over the choice of the functions $\phi_1, \ldots, \phi_n \ldots$ The infimum $D_n(F)$ of $E_n(F)$ may be referred to as the *n-width*".

This passage outlines briefly the main stages of the approximation theory. At the first stage *approximation of an individual element by a fixed approximation tool* was studied (such problems were dealt with within Chebyshev's school). The next stage was to study the relationships between smoothness and the rate of approximation, where approximations of *classes of elements by a fixed approximation tool* became of importance (this problem for many years was discussed by the schools of S. N. Bernstein, S. N. Nikol'skii, S. B. Stechkin, B. K. Dzyadyk, N. P. Korneichuk and others, and this direction was influenced, in particular, by Kolmogorov's 1935 paper). And finally, the problem of *optimal choice of the approximation tool* was posed. This new direction was stimulated by Kolmogorov's 1936 paper. In this paper the optimal n-dimensional subspace was sought for. The deviation of this optimal subspace from the approximated class became known as the *Kolmogorov n-width*, for which the notation d_n was adopted.

Let us repeat the definition of the n-width of a set C in a normed space X, which is denoted by $d_n(C, X)$. According to the above, the Kolmogorov n-width is defined as

$$d_n(C, X) = \inf_{L_n} \sup_{x \in C} \inf_{\xi \in L_n} \|x - \xi\|_X$$

(where the L_n are all possible n-dimensional subspaces; the quantity

$$\sup_{x \in C} \inf_{\xi \in L_n} \|x - \xi\|_X$$

is called the *deviation* of C from L_n). It is seen that this is the same definition as was given in Part I.

The studies on estimation and evaluation of n-widths formed a fruitful field of research in the approximation theory, which attracted activity of many mathematicians.

It began with Kolmogorov's results obtained in his 1936 paper which we discuss. In that paper Kolmogorov considered the classes of functions, which were subsequently denoted by $W_2^r(\mathbb{T})$ and $W_2^r([0, 1])$ and were called *Sobolev's classes*. The class $W_2^r(\mathbb{T})$ $(W_2^r([0, 1]))$ consists of the functions on the circle $\mathbb{T}$ (on the interval $[0, 1]$) which have $r - 1$ continuous derivatives such that the $(r-1)$th derivative is absolutely continuous and the rth derivative satisfies the inequality $\|x^{(r)}(\cdot)\|_{L_2(\mathbb{T})} \le 1$ $(\|x^{(r)}(\cdot)\|_{L_2([0,1])} \le 1)$. Kolmogorov obtained the following result.

Theorem 8. *The following formulas hold:*

a) $d_{2n-1}(W_2^r(\mathbb{T}), L_2(\mathbb{T})) = d_{2n}(W_2^r(\mathbb{T}), L_2(\mathbb{T})) = 1/n^r$, $d_0 = \infty$,
b) $d_n(W_2^r([0, 1]), L_2([0, 1])) = \lambda_{n-r}^{-1}$, $n \ge r$, $d_n = \infty$, $0 \le n \le r - 1$,

where $\lambda_0 < \lambda_1 < \ldots < \lambda_k \ldots$ are the eigenvalues of the Sturm–Liouville problem $(-1)^r x^{(2r)} - \lambda x = 0$, $x^{(k)}(0) = x^{(k)}(1) = 0$, $0 \le k \le r - 1$.

This result was actually proved in Theorem 5. Namely, on taking the Fourier expansions of the functions contained in the classes under consideration in some orthonormal systems, these classes become ellipsoids.

Indeed, if we decompose a function $x(\cdot) \in W_2^r(\mathbb{T})$ in the Fourier series

$$x(t) = \frac{a_0}{2} + \sum_{k \in \mathbb{N}} (a_k \cos kt + b_k \sin kt),$$

then

$$\|x(\cdot)\|_{L_2(\mathbb{T})}^2 = \frac{a_0^2}{4} + \frac{1}{2} \sum_{k \in \mathbb{N}} (a_k^2 + b_k^2), \quad \|x^{(r)}(\cdot)\|_{L_2(\mathbb{T})} = \frac{1}{2} \sum_{k \in \mathbb{N}} k^{2r}(a_k^2 + b_k^2).$$

Hence the class $W_2^r(\mathbb{T})$ is isometrically imbedded into l_2 as the product of the space of constants by the ellipsoid with axes $1, 1, 2^{-r}, 2^{-r}, \ldots, n^{-r}, n^{-r}, \ldots$ orthogonal to this space. Applying Theorem 5 we arrive at the statement a) of Theorem 8.

For the proof of part b) of the theorem A. N. Kolmogorov employed the results by M. G. Krein who at that time studied a class of equations of Sturm–Liouville type containing the equations of the form $x^{(2r)} + (-1)^r x = 0$ with boundary conditions $x^{(k)}(0) = x^{(k)}(1) = 0$, $r \le k \le 2r - 1$. Their eigenvalues $\{x_k(\cdot)\}_{k \in \mathbb{N}}$ completed by orthogonal polynomials of degrees $0, \ldots, r - 1$ form a complete orthogonal system in $L_2([0,1])$, whereas the functions $\{x^{(r)}(\cdot)\}_{k \ge r}$ form an orthogonal basis in $L_2([0,1])$. The eigenvalues are positive and of multiplicity 1, i.e. $0 < \lambda_r < \lambda_{r+1} < \ldots$ This implies that $W_2^r([0,1])$ can be isometrically imbedded in l_2 as the product of the space of polynomials of degree $= r - 1$ by an ellipsoid with axes $\lambda_r, \lambda_{r+1}, \ldots$ orthogonal to this space. Applying again Theorem 5 we obtain Theorem 8.

The papers by Kolmogorov, Petrov, Smirnov, and Mal'tsev of 1947, where the n-widths of octahedrons were evaluated, were reviewed in Part I.

8.2.3 Estimation of the accuracy of the Fourier method on a class of functions

The first Kolmogorov's paper on the approximation theory was the one written a year before the paper on n-widths. This was the paper: *Zur Grössenordnung des restgliedes Fourierschen Reihen differenzierbarer Funktionen* (On the order of the remainder terms in the Fourier series of differentiable functions) [Kol35] (1935). In this paper Kolmogorov considers the class $W_\infty^r(\mathbb{T})$ of functions on the circle $\mathbb{T}$ which are continuous along with their derivatives up to order $r - 1$ and such that $x^{(r-1)}(\cdot)$ satisfies the Lipschitz condition with constant 1 (i.e. $\|x^{(r)}\|_{L_\infty(\mathbb{T})} \le 1$).

Kolmogorov seeks to evaluate the quantity

$$C_{nr} = \sup\{\|x(\cdot) - S_n x(\cdot)\|_{C(\mathbb{T})} \mid x(\cdot) \in W_\infty^r(\mathbb{T})\},$$

where $S_n x(\cdot)$ is the nth Fourier sum of the function $x(\cdot)$.

The main Kolmogorov's result is as follows:

Theorem 9. *The following exact asymptotics holds:*

$$C_{nr} = \frac{4}{\pi^2} \frac{\log n}{n^r} + O\left(\frac{1}{n^r}\right).$$

We will give the proof for an even $r = 2m$, in which case it is simpler. We employ the well-known integral representation for the difference between a function and its Fourier sum:

$$x(t) - S_n x(t) = \pi^{-1} \int_{\mathbb{T}} D_{n,2m}(t - \tau) x^{(r)}(\tau) d\tau,$$

where
$$D_{n,2m}(t) = (-1)^m \sum_{k \geq n} k^{-2m} \cos kt.$$

Then $C_{nr} = \pi^{-1} \int_{\mathbb{T}} |D_{n,2m}(t)| dt$. To evaluate this integral Kolmogorov applies twice the Abel transform to obtain $D_{n,2m}(t) = (-1)^{m+1} n^{-2m} D_n(t) + R(t)$, where $D_n(t) = \frac{1}{2} + \sum_{k=1}^{n} \cos kt$ is the Dirichlet kernel and $|R(t)| = |n\Delta(n)F_n(t) + \sum_{k \geq n}(k+1)\Delta^2(k)F_k(t)|$. Here

$$\Delta(k) = \frac{1}{k^m} - \frac{1}{k^{m+1}}, \qquad \Delta^2(k) = \frac{1}{k^m} - 2\frac{1}{k^{m+1}} + \frac{1}{k^{m+2}}$$

and $F_n(t) = n^{-1} \sum_{k=0}^{n-1} D_k(t)$ is the Fejér kernel.

Now we use the well-known asymptotics of the L_1-norm of the Dirichlet kernel

$$\frac{1}{\pi} \int_{\mathbb{T}} |D_n(t)| dt = \frac{4}{\pi^2} \log n + O(1),$$

the asymptotic relations[8] $\Delta(k) \asymp k^{-(2m+1)}$, $\Delta^2(k) \asymp k^{-(2m+2)}$, and the properties of the Fejér kernel, namely, its nonnegativity and the equality $\frac{1}{\pi} \int_{\mathbb{T}} F_n(t) dt = 1$, to obtain

$$C_{n,2m} = \frac{1}{n^{2m}} \int_{\mathbb{T}} |D_n(t)| dt + \int |R(t)| dt = \frac{4}{\pi^2} \frac{\log n}{n^{2m}} + O\left(\frac{1}{n^{2m}}\right).$$

8.2.4 Kolmogorov's inequality for an intermediate derivative

In 1939 Kolmogorov publishes his famous paper *"On inequalities between suprema of successive derivatives of an arbitrary function on an infinite interval"* [Kol39].

In 1914 Hadamard proved that a function $x(\cdot)$ with derivative satisfying the Lipschitz condition on the real line $\mathbb{R}$ (thus $\ddot{x}$ exists almost everywhere and belongs to $L_\infty(\mathbb{R})$) fulfills the inequality

$$\|\dot{x}(\cdot)\|_{C^b(\mathbb{R})} \leq \sqrt{2}\|x(\cdot)\|_{C^b(\mathbb{R})}^{1/2}\|\ddot{x}(\cdot)\|_{L_\infty(\mathbb{R})}^{1/2}$$

(where $C^b(\mathbb{R})$ is the space of continuous bounded functions on $\mathbb{R}$ with supnorm). Kolmogorov regarded this Hadamard's result as the one of fundamental importance for the subject[9].

In late thirties Kolmogorov posed to his student G. Shilov[10] the problem to extend Hadamard's inequality to the case of n times differentiable

---

[8] The notation $u_n \asymp v_n$ means that there exists constants $c > 0$, $c' > 0$ and an integer n_0 such that $0 < cv_n \leq u_n \leq c'v_n$ when $n \geq n_0$. (Editor's note.)

[9] In fact, the inequality above is due to Edmund Landau (Proc. London Math. Soc. ser. 2, Vol. 13, 1914: p. 44), as noticed by Hadamard himself. (Editor's note.)

[10] Whose name at that time was Yuri Bosse after the name of his stepfather

functions. Shilov, having proved a number of partial results, formulated a general conjecture that the extremal functions in this problem should be the ones that appeared shortly before in papers by G. Favard and a subsequent paper by N. I. Akhiezer and M. G. Krein on approximation of functional classes by trigonometric polynomials. Now these functions are known as *Euler's splines*[11]. But Shilov did not succeed in fully justifying his conjecture. Then Kolmogorov became interested in this problem himself and solved it by proving the following result.

Theorem 10. *If a function $x(\cdot)$ is $n-1$ times continuously differentiable on the real line and the $(n-1)$th derivative satisfies the Lipschitz condition, then for $0 < k < n$ the following inequality holds:*

$$\|x^{(k)}(\cdot)\|_{C^b(\mathbb{R})} \le C_{kn}\|x(\cdot)\|_{C^b(\mathbb{R})}^{\frac{n-k}{n}}\|x^{(n)}(\cdot)\|_{L_\infty(\mathbb{R})}^{\frac{k}{n}}, \tag{8.1}$$

where C_{kn} are some definite constants.

This theorem does not look as a result of the approximation theory. It became regarded as part of this theory after S. B. Stechkin used the whole series of results on inequalities for derivatives on the real line and half-line to solve problems on approximation of unbounded operators by bounded ones.

Kolmogorov's proof is very complicated. It contains an important fragment which is referred to as the "comparison theorem". This result is a remote predecessor of the reasoning which became common after formulation of Pontryagin's maximum principle. Kolmogorov's proof could not be simplified for a long time. Only in 70-s the American mathematician Cavaretta provided a simplified proof based on reduction to the periodic case (for details, see my 1989 paper mentioned above).

8.2.5 Criterion and uniqueness of the elements of the best approximation

In 1948 A. N. Kolmogorov published a short note *"Remark on Chebyshev's polynomials of least deviation from a given function"* [Kol48]. This paper is related to the first, Chebyshev's, stage of the approximation theory, which deals with approximation of individual functions by a fixed approximation tool. In this note Kolmogorov establishes a "complex analog of the characteristic property due to Chebyshev of the best approximation polynomials"

---

[11] They are the functions $\phi_n(t) = af_n(bt + c)$, with $a > 0$, $b > 0$, c arbitrary, and

$$f_n(t) = \frac{4}{\pi}\sum_{m\ge0}\frac{\sin\left((2m+1)t - \frac{\pi}{2}n\right)}{(2m+1)^{n+1}}.$$

Shilov conjectured (and Kolmogorov proved) that they are exactly the functions such that equality holds in (8.1) of theorem 10. (Editor's note.)

and shows that "Haar's theorem [on uniqueness of the best approximation element] can be easily extended to complex functions".

In this note the following theorem was proved.

Theorem 11. *Let $L_n \in C(\mathbb{T}, \mathbb{C})$ (i.e. L_n is an n-dimensional subspace of the space of continuous functions on a compact set $\mathbb{T}$ with sup-norm) and let $x(\cdot) \in C(\mathbb{T}, \mathbb{C}) \setminus L_n$. Then an element $\widehat{y}(\cdot) \in L_n$ is least deviating from $x(\cdot)$ if and only if for any $y(\cdot) \in L_n$ the following inequality holds: $\min_{t \in \mathbb{T}_0} \mathrm{Re}\,(\bar{y}(t)(\widehat{y}(t) - x(t))) \le 0$, where $\mathbb{T}_0 = \{t \in \mathbb{T} \mid |x(t) - \widehat{y}(t)| = \|x(\cdot) - \widehat{y}(\cdot)\|_{C(\mathbb{T}, \mathbb{C})}$. Moreover, if any non-zero polynomial in L_n vanishes on $\mathbb{T}$ at most at $n+1$ points, then the best approximation element is unique.*

In 1947 the remarkable monograph "Lectures on the approximation theory" [Akh56] by N. I. Akhiezer appeared. The author presented to Kolmogorov a copy with dedicatory inscription. Andrey Nikolaevich expressed a very positive opinion about this book, time and again he invoked it in his research and repeatedly quoted it. In the paper we discuss now he wrote: "I present complete proofs of all results, though some of them are only minor modifications of Sects. 44–45 of Akhiezer's book. My interest to this field of research was caused by the appearance of this book".

Unfortunately both N. I. Akhiezer and A. N. Kolmogorov did not pay attention to a paper by L. G. Shnirel'man in which the approximation problem of a continuous function on a compact set was automatically reduced to the finite-dimensional case, where everything becomes obvious (for more details, see my paper of 1989).

Finally, we will review Kolmogorov's papers on ε-entropy of functional classes[12]. After the short note "On some asymptotic characteristics of completely bounded metric spaces" [Kol56] (1956) there appeared a survey paper "ε-entropy and ε-capacity of sets in functional spaces" (jointly with the author of the present review) [KT59] (1959). Let us set out the origins and development of this field.

8.2.6 ε-Entropy

During 50-s there appeared new reasons for Kolmogorov to turn once more to the approximation theory. One of the incentives was his interest to problems of the information theory. This interest arose apparently in 1953 or 1954. At the same time Andrey Nikolaevich started to think over the paper by Vitushkin [Vit54] (published in Doklady Akad. Nauk SSSR, 1954, by Kolmogorov's presentation), which dealt with the problems of superposition of functions. Kolmogorov found another way to prove the main Vitushkin's theorem based on bounds for ε-nets in classes of smooth functions of many variables. In this connection, he remembered that the very "idea of possibility to describe the

---

[12] The Chap. 6 (by M. Nikouline and V. Solev) in this book explains the use of ε-entropy in statistics. (Editor's note.)

"massiveness" of sets in metric spaces by means of the rate of growth of the cardinality of their most economical ε-coverings as $\varepsilon \to 0$ was elaborated in a paper by Pontryagin and Shnirel'man" in 30-s (where this method was used to define the very *dimension* of compact sets in metric spaces). Combining these conceptions with Shennon's ideas, Kolmogorov formulated an extensive programme of investigating the ε-entropy and ε-capacity of stochastic processes as well as of "compact sets in functional spaces which are of interest for the theory of functions".

A. N. Kolmogorov immediately appreciated the significance of Shannon's ideas. He also highly appreciated Shannon himself as a creative personality (he wrote that "his [Shannon's] mathematical intuition is amazingly precise"). Kolmogorov continued and elaborated Shannon's conceptions.

An element of a finite set C of cardinality $N(C)$ may be specified by $[\log_2 N(C)] + 1$ binary digits. "Therefore, Kolmogorov says, the quantity $H(C) = \log_2 N(C)$ may be viewed as a measure of "amount of information" needed to single out a given element of C. In case of infinite sets it is natural to consider some ways of approximate specification of these elements".

This idea gave rise to the concept of ε-*entropy*, which is the logarithm of the least number of points in an ε-net of a compact set C in a metric space X which approximate the elements of C within ε. This quantity is now commonly denoted by $\mathcal{H}_\varepsilon(C, X)$.

Vitushkin's results implied a lower bound for the ε-entropy of the class $W_\infty^r(I^n)$ of functions of n variables defined on the unit cube I^n in $\mathbb{R}^n$ and having uniformly bounded partial derivatives up to order r. Studying Vitushkin's paper Kolmogorov realized that its main result (obtained with the aid of the brilliant but rather complicated theory of multivariate variations constructed by Vitushkin) can be much simpler derived from the upper bound for the ε-entropy of this class.

In his paper [Kol56] in Doklady Akad. Nauk of 1956 Kolmogorov announced the following result.

Theorem 12. *There exist constants c and C such that*

$$c\left(\frac{1}{\varepsilon}\right)^{n/r} \leq \mathcal{H}_\varepsilon(W_\infty^r(I^d), C(I^d)) \leq C\left(\frac{1}{\varepsilon}\right)^{n/r}.$$

A slightly worse upper bound can be obtained very simply. Let us show this for r-smooth functions (i.e. functions having bounded derivatives up to rth order) of a single variable (defined on the interval $I = [0, 1]$). How many elements do we need in order to construct a table which would enable us to recover any function of this class with accuracy ε? (The logarithm of this number will be an upper bound for the ε-entropy of this class.) It can roughly be counted as follows. Divide the interval into subintervals of equal length $\varepsilon^{1/r}$ and calculate at their end-points the values of the function and its derivatives up to rth order within accuracy ε. Then, if you want to calculate the value of some function of this class at some point with accuracy ε times a readily obtainable constant, you can find the subinterval containing the given point and apply the Taylor expansion about its left end-point using the approximate values of the derivatives. The remainder in the Taylor formula will be of order ε, hence the total accuracy will also be of order ε. To specify a derivative at a given point we need in our table of order $1/\varepsilon$ numbers and of order $(1/\varepsilon)^{r+1}$ numbers for all derivatives, then totally we need of order $(1/\varepsilon)^{(r+1)\varepsilon^{-1/r}}$ numbers. Taking logarithm yields the upper bound $C(1/\varepsilon)^{1/r}\log\frac{1}{\varepsilon}$ for the ε-entropy, where C is some constant. The correct upper bound obtains by a more accurate calculation (one has to proceed as above only at the initial point, then the number of elements sufficient to recover the function on the next interval does not depend on ε since the derivatives at the end-point of the second interval are already approximately known, and the same is repeated all over the entire interval). The upper bound thus obtained together with Vitushkin's lower bound yield the quantity of order $(1/\varepsilon)^{1/r}$ (and of order $(1/\varepsilon)^{n/r}$ for r-smooth functions of n variables).

Thus the "massiveness" of r-smooth functions of n variables is determined by the exponent $q = n/r$. And it is not hard to derive from Baire's theorem that among the functions of n variables with exponent q there is a function which cannot be represented by superpositions of functions of smaller number of variables with smaller exponent than q. Now this is exactly Vitushkin's theorem.

At the same time Kolmogorov estimated the rate of approximation accuracy for analytic functions of n variables. The order of magnitude of ε-entropy for analytic functions turned out to be essentially different, namely, it was $\left(\log\frac{1}{\varepsilon}\right)^{n+1}$. The information capacity of analytic functions is much less than that of functions of finite smoothness.

Kolmogorov presented the results about the bounds for the ε-entropy of smooth and analytic functions and their proofs in a special course of lectures in fall of 1956. The "extensive programm of investigation of the ε-entropy and ε-capacity [...] of compact sets in functional spaces which are of interest for the theory of functions" began to be accomplished by Babenko, Vitushkin, Erokhin, the present author, and others. Andrey Nikolaevich discussed their results with interest, but did not work in this direction himself any more.

Kolmogorov developed his result on the entropy of analytic functions in the same direction as for smooth functions when proving Vitushkin's theorem. He proved that in some respect analytic functions are the more "massive" the larger is their domain of definition. Namely, he introduced the concept of linear dimension, somewhat specializing Banach's definition (which shows once more that Kolmogorov had studied very deeply Banach's treatise sent to him in 30-s just as the book had appeared). Kolmogorov conceived the idea of entropy invariant for the linear dimension when traveling by train to Leningrad; it was published with supplements in 1958. Somewhat earlier similar ideas occurred to the young (world-renowned nowadays) Polish mathematician A. Pelczinski. At that time they were actively discussed in Warsaw. In 1958 Kolmogorov makes one more striking discovery: he constructs the entropy invariant in the theory of dynamical systems. This is discussed in Chap. 12.

8.2.7 Concluding remarks

Here we list some names of mathematicians whose research was strongly influenced and motivated by Kolmogorov's works on the approximation theory.

The studies on estimation of the remainder of the Fourier series on a class of functions were continued primarily by S. M. Nikol'ski and S. B. Stechkin. Their progeny formed several scientific schools in Ukraine and Sverdlovsk (nowadays Ekaterinburg).

The subject "bounds for accuracy of a fixed approximation tool on a class of functions" was continued by S. P. Baiborodov, V. A. Baskakov, and others.

The reader can learn of the developments in this field from the monographs by Dzyadyk, Korneychuk, Stepanec.

The research on n-widths was continued by many dozens of mathematicians. Among them: Rudin, Stechkin, Tikhomirov, Babenko.

See also monographs by Tikhomirov and Pinkus.

The work on inequalities for derivatives on the real line and half-line, where Kolmogorov's result remains the most brilliant, were continued by Arestov, Berdyshev, and others.

See reviews by Arestov, Magaril-Ilyaev & Tikhomirov.

Nagy, Stein, Arestov.

The subject of Kolmogorov's paper on approximation of individual functions by subspaces was explored further by Zukhovitski, Krein, Stechkin, and others.

Kolmogorov's progeny in the theory of entropy is now countless.

References

[Akh56] Akhiezer, N.I.: Lectures on the approximation theory (in Russian). OGIZ, Moscow (1947). Engl. Transl.: Theory of approximation. F. Ungar, New York (1956)

[Kol] Shiryayev, A.N., Tikhomirov, V.M. (ed): Selected works of A.N. Kolmogorov. Kluwer Academic Publishers (1991–1993)

[Kol30] Kolmogorov, A.N.: Zur topologisch-gruppenteoretischen Begründung der Geometrie. Nachr. Ges. Wiss. Göttingen, Fachgr. I (Mathematik), H.2, 208–210 (1930)

[Kol32] Kolmogorov, A.N.: Zur Begründung der projektiven Geometrie. Ann. Math., **33**, 175–176 (1932)

[Kol33] Kolmogorov, A.N.: Beiträge zur Masstheorie. Math. Ann., **107**, 351–366 (1933)

[Kol34] Kolmogorov, A.N.: Zur Normierbarkeit eines allgemeinen topologischen linearen Raumes. Stud. Math. **5**, 29–33 (1934)

[Kol35]   Kolmogorov, A.N.: Zur Grössenordnung des restgliedes Fourierschen Reihen differenzierbarrer Funktionen. Ann. Math., **36**, 521–526 (1935)

[Kol36]   Kolmogorov, A.N.: Über die beste Annäherung von Funktionen einer gegebenen Funktionenklasse. Ann. of Math. **37**, 107–110 (1936)

[Kol37]   Kolmogorov, A.N.: Über offene Abbildungen. Ann. Math., **38**, 36–38 (1937)

[Kol39]   Kolmogorov, A.N.: On inequalities between suprema of successive derivatives of an arbitrary function on an infinite interval (in Russian). Uchen. Zap. MGU Mat., **30**, 3–16 (1939)

[Kol48]   Kolmogorov, A.N.: Remark on Chebyshev's polynomials of least deviation from a given function (in Russian). Uspekhi Mat. Nauk, **3**, 216–231 (1948)

[Kol56]   Kolmogorov, A.N.: On some asymptotic characteristics of completely bounded metric spaces (in Russian). Doklady Akad. Nauk SSSR, **108**, 385–388 (1956)

[KPP37]   Kolmogorov, A.N., Petrovski, I.G., Piskunov, N.S.: Étude de l'équation de la diffusion avec croissance de la quantité de matière et son application à un problème biologique. Bull. MGU, Mat. Mekh., **1**, 1–26 (1937)

[KPS47]   Kolmogorov, A.N., Petrov, A.A., Smirnov, Yu.M.: A formula of Gauss in the theory of the method of least squares (in Russian). Izv. Akad. Nauk SSSR, **11**, 561–566 (1947)

[KT59]   Kolmogorov, A.N., Tikhomirov, V.M.: ε-Entropy and ε-capacity of sets in functional spaces (in Russian). Uspekhi Matem. Nauk, **14**, 3–86 (1959)

[Tik89]   Tikhomirov, V.M.: Kolmogorov and the approximation theory (in Russian). Math. USSR Uspekhi, **44**, 83–122 (1989). Engl. Transl.: Russian Math. Surveys, **44**, 101–152 (1989)

[Vit54]   Vitushkin, A.G.: On Hilbert's Thirteenth Problem (in Russian). Doklady Akad. Nauk SSSR, **95**, 701–704 (1954)

Kolmogorov and population dynamics

Karl Sigmund

Faculty for Mathematics, University of Vienna, and International Institute for
Applied Systems Analysis, Austria
http://homepage.univie.ac.at/Karl.Sigmund
karl.sigmund@univie.ac.at

Translated from the French by Elizabeth Strouse

9.1 Introduction

During his childhood, Andreï Nikolaievitch Kolmogorov found biology very
interesting. In fact, in the book *Kolmogorov in Perspective*, one can read that
he made the following comment about himself as a schoolboy:

"*I was one of the best in my class in mathematics, but my real scientific
passions were, first of all biology, and then russian history*" ([Kol00], p. 5)

He kept these centers of interest for the rest of his life. Thus, in 1940,
Kolmogorov dared to confront the feared Lysenko by defending Mendel's laws
— a very dangerous move to make in the middle of Stalin's regime. And,
according to V. I. Arnold, the last research done by Kolmogorov [KB67],
published in 1967, was motivated by biological ideas about the structure of
the brain ([Kol00], p. 94).

In fact, Kolmogorov made only a few isolated contributions to biomathematics; but they all demonstrate, as one would expect, a remarkable origi-
nality. In particular, the short note [Kol36] about the predator-prey equation
is a model of perspicacity and has had great influence on the deterministic
theory of population dynamics. It is one of the rare articles that Kolmogorov
published in Italian, doubtlessly in honor of the mathematician Vito Volterra
who inaugurated what would later be called *The Golden Age of biomathematics* [SZ78]. Kolmogorov's note represents a qualitative jump in the theory of
predator-prey systems.

9.2 From Volterra equations to Gause equations

The beginning point of the study of Kolmogorov discussed here is the famous model that Volterra[1] used, as early as 1925, to explain a surprising discovery of his son in law, the ecologist Umberto D'Ancona [Kin85]. Because of his research in marine biology, based on statistics from fish markets, D'Ancona noticed that during World War 1, the number of predators among Adriatic fauna had increased while the number of prey had diminished. This seemed to be an effect of the reduction of fishing due to the Austro-Italian hostilities: but why did it work in this manner and not in another?

Volterra based his argument on an ordinary differential equation: if $x(t)$ and $y(t)$ are the densities of prey and of predators, respectively, then the rate of increase $\dot{x}/x$ of the prey should be a decreasing function of y, positive for $y = 0$, and the rate of increase $\dot{y}/y$ an increasing function of x, negative for $x = 0$. If we suppose that these functions are linear, we see that

$$\dot{x} = x(a - by) \tag{9.1}$$

$$\dot{y} = y(-c + dx) \tag{9.2}$$

where the constants a, b, c, d are positive. In the positive quadrant, the phase portrait consists of periodic orbits around the equilibrium position $(\bar{x}, \bar{y}) = (c/d, a/b)$. Volterra showed that the temporal averages of $x(t)$ and $y(t)$ along periodic orbits coincide with the values $\bar{x}$ and $\bar{y}$, which gave him a way to explain D'Ancona's observation: in fact, the supplementary contribution due to the fishermen's work diminishes the quantity a (the rate of increase of the prey in the absence of predators) and increases c (the rate of decrease of predators in the absence of prey), without affecting the values of the coefficients b and d, which measure the effects of the interaction between the predators and their prey. The corresponding effect on the temporal averages of the densities of the two populations is just that which D'Ancona observed.

The elegance of Volterra's reasoning stands in clear contrast with the plausibility of his equation. In fact, (9.1)–(9.2) is unstable from many points of view. In particular, the model implies that a prey population, in the absence of predators, would grow exponentially towards infinity. This evident flaw in the (9.1) can be easily corrected, one way is to introduce a self-limiting term for the growth of the prey, reducing the equation, for $y = 0$, to a logistic model $\dot{x} = ax(1 - x/K)$. Georgii Frantsevitch Gause proposed another system of much more general equations ([Gau34], [GSW36]), which, using modern notation, take the following form:

$$\dot{x} = xg(x) - yp(x) \tag{9.3}$$

$$\dot{y} = yq(x) \tag{9.4}$$

---

[1] We note that Alfred Lotka introduced the same model, at approximately the same time, independently of Volterra, and in a different context. (Editor's note.)

Here, g describes the rate of increase of prey when the predators are absent: it is a function which is positive on an interval $[0, K]$ and negative for $x > K$ (because, for example, the food resources being limited, the prey are in competition with each other when there are too many of them), and the density of prey thus converges, in the absence of predators, towards the limit K. The functions p and q, which are called the response functions, describe the predator-prey interaction: we suppose that p is a positive function with $p(0) = 0$, while q is strictly increasing for $x > 0$, has a negative limit when x decreases to 0, and a positive limit when x increases to $+\infty$ (an abundance of prey). These equations are much more reasonable and more flexible than (9.1)–(9.2).

9.3 The Kolmogorov equations

Kolmogorov did not mention these equations in his note [Kol36], even though he must have known about the work of Gause — who also lived in Moscow in the thirties and, at the age of twenty-two, revolutionized mathematical biology with his book *The Struggle for Existence* [Gau34]. After noticing that, in Volterra's work, there was an arbitrary postulate of linear rates of increase that could not be justified as anything but a first approximation of real rates of increase, Kolmogorov considered the most general case possible

$$\dot{x} = xS(x, y) \tag{9.5}$$
$$\dot{y} = yW(x, y) \tag{9.6}$$

and was led to postulate (assuming that the rates of increase S and W were continuously differentiable) some minimal conditions which are satisfied in any realistic predator-prey interaction.

The first group of conditions requires that, if the number of predators increases, then the rates of increase of the two populations decrease:

$$\frac{\partial S}{\partial y} < 0, \qquad \frac{\partial W}{\partial y} < 0. \tag{9.7}$$

These conditions (the second of which was not satisfied in Volterra's and Gause's models) are, in general, accepted without objection by ecologists (even though, for example, one can imagine predators who only manage to attack their prey when there are enough of them to surround it, which would imply that the second condition is not valid for small values of y). The immediate consequence of the postulate (9.7) is that the two isoclines in the interior of the positive quadrant, $\{(x, y) : \dot{x} = 0\}$ and $\{(x, y) : \dot{y} = 0\}$, can be viewed as the graphs of two functions of x. Their intersections are, evidently, the fixed points of (9.5)–(9.6), which correspond to equilibria of the system with coexistence of the two populations.

The other conditions of [Kol36] describe the behavior of (9.5)–(9.6) on the boundary of the positive quadrant, that is, in the absence of one population or of the other. They imply, more precisely, that the unique equilibrium $(K, 0)$ which is composed of prey but not of predators, can be invaded by predators, so that

$$W(K, 0) > 0. \tag{9.8}$$

This implies that there is at least one equilibrium with coexistence of the two species.

What is still missing is a condition to guarantee that there is only one such equilibrium, that is, a unique point in the intersection of the two isoclines in the interior of the first quadrant. This would be a simple consequence of a condition analogous to (9.7):

$$\frac{\partial S}{\partial x} < 0, \qquad \frac{\partial W}{\partial x} > 0. \tag{9.9}$$

Kolmogorov noticed that the validity of this condition is not clear if x, the number of prey, is small. Today, all ecologists are familiar with the Allee effect, which is the fact that the rate of increase (of a prey population for example) can decrease and even become negative if the density of the population is sufficiently small. Kolmogorov, who could not yet know of this effect, seems to have suspected that there was such a mechanism, even though the argument he gave, invoking the presence of a large number of predators, does not seem very clear. According to a note on the bottom of a page, as long as the density x of the prey is small, the probability of survival of the prey would be an increasing function of x (maybe because the predators have eaten their fill, while the competition effect within the prey species can not yet be felt). Having understood that (9.9) is not necessarily valid, Kolmogorov introduced another condition, that S decrease and W increase along rays starting at the origin. This is therefore a condition concerning directional derivatives. Today this condition is often written in the form

$$x \frac{\partial S}{\partial x} + y \frac{\partial S}{\partial y} < 0, \qquad x \frac{\partial W}{\partial x} + y \frac{\partial W}{\partial y} > 0. \tag{9.10}$$

Imagine, indeed, that a large habitat become smaller because of some external force. In this case the densities x and y increase while the ratio between them stays constant. The individuals would be obliged to get closer to each other, the predators would have less distance to travel to find their meal, and each prey would be tracked by more predators. Thus, the predators would have the benefit of the new circumstances, while the prey would suffer from them.

The condition (9.10) implies that there is only one equilibrium $\mathbf{Z}$ in the interior of the positive quadrant, and that it is divided by the isoclines (which intersect in $\mathbf{Z}$) into four regions depending on the signs of $\dot{x}$ and of $\dot{y}$.

The rest of the reasoning is surprisingly simple. The equilibrium with only the prey population present, which is given by $\mathbf{B} = (K, 0)$, is necessarily made up of a saddle point, and the two orbits which converge there are situated on the x-axis. Thus the positive quadrant contains a unique orbit having $\mathbf{B}$ as α-limit, that is, such that $\lim x(t) = K$ and $\lim y(t) = 0$ for $t \to -\infty$. This orbit stays in a compact domain, and thus must have a non-empty ω-limit[2]. If this ω-limit contains a fixed point, it must be $\mathbf{Z}$. In this case, it is easy to see that all the orbits, in the interior of the positive quadrant, converge to $\mathbf{Z}$. If this is not the case, the theorem of Poincaré-Bendixson implies that the ω-limit is a limit cycle around $\mathbf{Z}$. In this case, all the orbits in the positive quadrant which are in the exterior of this cycle converge to it.

9.4 Technical aspects

Kolmogorov, who didn't like to dwell upon technical details, gave only a rough sketch of his idea. Actually, what he wrote was not totally correct. In fact, there is an obvious contradiction between the two parts of conditions (9.7) and (9.10), as can be easily seen by setting $x = 0$ and considering $y > 0$. This error is not, however, very serious and is easy to correct. For example, in [AGW73] and [AGHW74], it is shown that it suffices to assume that (9.7) and (9.10) are valid in the interior of the positive quadrant and to specify the behavior of S and W along the axes. The same type of proof, with all possible details, can be found in [Fre75] and [Fre80]. Kolmogorov didn't feel the need to give the details of his argument as his milieu was extremely well informed about the methods for studying ordinary non-linear differential equations in two dimensions, and in particular conversant with results that had just been obtained by mathematicians and engineers in the Soviet Union such as Pontryagin, Andronov, Krylov, Bogoliubov, Moiseev and Bautin (see for example [ALGM73]). We remark that in 1939 Moiseev showed that, if the functions S and W are affine then an equation of type (9.5)–(9.6) never admits a limit cycle.

The equations of Volterra (9.1)–(9.2) and of Gause (9.3)–(9.4) are not particular cases of Kolmogorov's equation because neither (9.2) nor (9.4) satisfy the second condition of (9.7): the isocline in both cases is vertical. But the conclusion stays the same. This is also the case for the so-called equation of Holling-Tanner

$$\dot{x} = xr(1 - \frac{x}{K}) - yp(x) \tag{9.11}$$

$$\dot{y} = ys(1 - \frac{hy}{x}) \tag{9.12}$$

(with the constants r, s, h and K positive and the response function p as in (9.3)). Here the growth of prey (without predators) is logistic with a constant

---

[2] The ω-limit of the orbit $(x(t), y(t))$ is the set of all points $(\hat{x}, \hat{y})$ such that $\hat{x} = \lim x(t_k)$ and $\hat{y} = \lim y(t_k)$ for a subsequence $t_k \to +\infty$

capacity, and the growth of predators is logistic with a capacity proportional with x. This model seems to be designed to describe real cases of the predator-prey systems, when one has made a suitable choice of the response function p.

Kolmogorov makes clear that his reasoning implies nothing about what happens in the interior of the cycle, but notes that, in the simplest cases, the limit cycle is unique and attracts all the orbits coming from the interior (excepting, of course, equilibrium). Afterwards, a great deal of research was done to find conditions for the global stability of $\mathbf{Z}$, and for the unicity of the limit cycle [Che81], [KF88], [Kua90], [HH95], [GKT97], [Has00]. One often hears the assertion that asymptotic stability of $\mathbf{Z}$ implies its global stability: but this is false, in general (see e.g. [Bul76]).

9.5 The impact

At first the reaction to Kolmogorov's note was rather lukewarm. It was only during the sixties that these results began to be appreciated, mostly because of the appearance at this time of the articles of Rosenzweig and MacArthur [RM63], and of Rescigno and Richardson [RR67]. The detailed study of the explicit form of the response function, in particular that done by Holling [Hol65], also played an important role. It was no longer necessary to convince the ecologists of the reality of non-transitory oscillations in certain systems of predator-prey, or of the robustness of their period and their amplitude. It became evident that models with limit cycles were necessary. Furthermore, by using an elegant criterion proposed by Rosenzweig and MacArthur, one could determine whether the equilibrium $\mathbf{Z}$ of the system (9.3)–(9.4) was locally stable or not. Thus, the emergence of Hopf bifurcation became easy to verify. Sometime around 1972 there was a real stampede toward limit cycles due to three articles in *Science* written by Rosenzweig, Gilpin and May, respectively ([May72], [Gil72], [Ros72]). In particular, the book of May [May73] spread Kolmogorov's message among ecologists, and showed that the linear analysis of an equilibrium does not always allow one to make conclusions about the global behavior of the system.

For his own reasons, Kolmogorov himself came back to his model in a short note written in 1972 ([Kol72]). In particular, he applied his method to the Gause equations (again without citing Gause) and gave a classification of possible phase portraits. These arguments were extended in [Baz74] and [SL78].

We conclude this section with several remarks. From an ecologist's perspective it is less important to know whether a certain equilibrium is stable than to know if the system is permanent, that is, if the species under consideration can survive indefinitely; it is of secondary interest to know if their densities converge or oscillate. The role of this notion of permanence in theoretical analyses of the ecology of populations is increasing [HS98]. The associated notion of stability is more like that of Lagrange than that of Lyapunov: it is formulated as a condition that the border of the phase space (including points at infinity) be repulsive.

It is interesting to notice, in this context, that Kolmogorov explains in his note [Kol36] that the modelization of the ecological system by a deterministic model is not valid if the populations are very small (that is, if one is close to the border of the phase space). No one was better than Kolmogorov at deducing a differential equation from a stochastic model. The fact that he did not try to do this in the case of a predator-prey interaction suggests that he was conscious of the difficulty of doing so: and it is perhaps because he considered it too difficult to deduce the analytic expression corresponding to the vector field given by (9.5)–(9.6) from a stochastic process modelling the encounters between predators and prey, that he decided to do without, and to instead use the general properties (9.7),(9.8),(9.9) and (9.10).

But, what is of primary importance is the general approach used by Kolmogorov: in particular, today, to give a model of biological communities made up of three or more species one frequently uses equations of the type

$$\dot{x}_i = x_i F_i(x_1, ..., x_n) \tag{9.13}$$

(which are called ecological equations, or, more and more often, Kolmogorov equations) and to specify the system, not by giving precise analytic expressions for the rates of growth F_i, but by setting conditions for the signs of their partial derivatives: for example, competitive communities are described by conditions like

$$\frac{\partial F_i}{\partial x_j} \leq 0 \tag{9.14}$$

for $i \neq j$, etc. Thanks to work by Morris Hirsch, Hal Smith and their collaborators (see [Hir88], [Smi95]), this approach now gives some of the results which are most useful for ecological applications and most interesting from a mathematical point of view. More generally, an ordinary differential equation $\dot{x}_i = f_i(x_1, ..., x_n)$ defines a cooperative system if

$$\frac{\partial f_i}{\partial x_j} \geq 0 \tag{9.15}$$

for $i \neq j$, and a competitive system if the inequalities are reversed. One of the principal results concerning such systems is that the flow, restricted to a compact limit set, is topologically equivalent to a flow defined by a system of lipschitzian differential equations on an invariant compact set of $(n - 1)$-dimensional space. In particular, we can use this result to obtain a theorem of Poincaré-Bendixson in three dimensions: a compact limit set for a cooperative or competitive system in $\mathbb{R}^3$ which contains no fixed point is actually a periodic orbit. Zeeman [Zee93] used these results in his attack on the problem of classifying the competitive systems of Lotka-Volterra in three dimensions, thus identifying 33 stable equivalence classes, and Hofbauer, Mallet-Paret and

Smith [HMS91] established the existence of stable periodic orbits for "hypercyclic" systems.

A particular case of Kolmogorov equations is the set of ecological equations describing the three-species food chains. Hastings and Powell [HP91] showed that these equations often present a chaotic behavior (see also [KH94] and [MY94]). Muratori and Rinaldi [MR89] as well as Kuznetsov and Rinaldi [KR96] studied Hopf bifurcations in the prey-predator-superpredator systems.

Our last remark is that recent work by Hofbauer and Schreiber [HS04] (see also Schreiber and Mielcynski [MS02]) show that there are open sets of Kolmogorov equations containing a dense subset of permanent equations and a dense subset of equations having an attractor on the boundary of $\mathbb{R}_+^n$. For these equations, it is impossible to predict whether or not all species will survive.

Such results are well within the tradition inaugurated by Kolmogorov: in general the study of ecological systems cannot be reduced to a local study of stable equilibria. Only a dynamic global study can account for the complexity of ecological feedback.

References

[AGHW74] Albrecht, F., Gatzke, H., Haddad, A., Wax, N.: The dynamics of two interacting populations. J. Math. Anal. Appl., **46**, 658–670 (1974)

[AGW73] Albrecht, F., Gatzke, H., Wax, N.: Stable limit cycles in predator-prey populations. Science, **181**, 1073–1074 (1973)

[ALGM73] Andronov, A., Leontovich, E., Gordon, I., Maier, A.: Qualitative theory of second-order dynamic systems. Halsted Press, New York (1973)

[Baz74] Bazykin, A.D.: The Volterra system and the Michaelis-Menten equation (in Russian). In "Problems in mathematical genetics", Novosibirsk, Siberian Branch of the Acad. Sci. USSR, 103–143 (1974)

[Bul76] Bulmer, M.G.: The theory of predator-prey oscillations. Theor. Pop. Biol., **9**, 137–150 (1976)

[Che81] Cheng, K.S.: Uniqueness of a limit cycle for a predator-prey system. SIAM J. Math. Anal., **12**, 541–48 (1981)

[Fre75] Freedman, H.: A perturbed Kolmogorov-type model for the growth problem. Math. Biosci., **23**, 127–149 (1975)

[Fre80] Freedman, H.: Deterministic mathematical models in population ecology. Dekker, New York (1980)

[Gau34] Gause, G.F.: The struggle for existence. Williams and Wilkins, Baltimore (1934)

[Gil72] Gilpin, M.E.: Science, **177**, 902–904 (1972)

[GKT97] Gasull, A., Kooij, R.E., Torregrosa, J.: Limit cycles in the Holling-Tanner model. Publicaciones Matemàtiques, **41**, 149–167 (1997)

[GSW36] Gause, G.F., Smaragdova, N.P., Witt, A.A.: Further studies of interaction between predator and prey. J. Anim. Ecol., **5**, 1–18 (1936)

[Has00] Hasik, K.: Uniqueness of limit cycle in the predator-prey system with symmetric prey isocline. Math. Biosci., **164**, 203–215 (2000)

[HH95]   Hsu, S.B., Huang, T.W.: Global stability for a class of predator-prey systems. SIAM J. Appl. Math. **55**, 763–783 (1995)

[Hir88]   Hirsch, M.W.: Systems of differential equations which are cooperative or competitive, III: Competing species. Nonlinearity, **1**, 51–71 (1988)

[HMS91]   Hofbauer, J., Mallet-Paret, J., Smith, H.L.: Stable periodic solutions for the hypercycle system. J. Dynamics and Diff. Equs., **3**, 423–436 (1991)

[Hol65]   Holling, C.S.: The functional response of predators to prey density and its role in mimicry and population regulation. Mem. Ent. Soc. Can., **6**, 1–60 (1965)

[HP91]   Hasings, A., Powell, T.: Chaos in a three-species food chain. Ecology, **72**, 896–903 (1991)

[HS04]   Hofbauer, J., Schreiber, S.J.: To persist or not to persist? Nonlinearity, **17**, 1393–1406 (2004)

[HS98]   Hofbauer, J., Sigmund, K.: Evolutionary Games and Population Dynamics. Cambridge Univ. Press (1998)

[KB67]   Kolmogorov, A.N., Barzdin, Ya.M.: On the realization of nets in three-dimensional space (in Russian). Probl. Kibernetiki, **19**, 261–269 (1967)

[KF88]   Kuang, Y., Freedman, H.I.: Uniqueness of limit cycles in Gause-type models of predator-prey systems. Math. Biosci., **88**, 67–84 (1988)

[KH94]   Klebanoff, A., Hastings, A.: Chaos in three-species food chains. J.Math. Biol., **32**, 427–451 (1994)

[Kin85]   Kingsland, S.: Modeling Nature: Episodes in the History of Population Ecology. Univ. of Chicago Press (1985)

[Kol00]   Kolmogorov in Perspective (translated from Russian by Harold H. McFaden). History of Mathematics, **20**. American Math. Soc. (2000)

[Kol36]   Kolmogorov, A.N.: Sulla teoria di Volterra della lotta per l'esistenza. Giornale Istituto Ital. Attuari, **7**, 74–80 (1936)

[Kol72]   Kolmogorov, A.N.: The quantitative measurement of mathematical models in the dynamics of populations (in Russian). Problems of Cybernetics, **25**, 100–106 (1972)

[KR96]   Kuznetsov, Y.A., Rinaldi, S.: Remarks on food chains. Mathematical Biosciences, **134**, 1–33 (1996)

[Kua90]   Kuang, Y. Global stability of Gause-type predator-prey systems. J. Math. Biol., **28**, 463–474 (1990)

[May72]   May, R.M.: Limit cycles in predator-prey communities. Science, **177**, 900–902 (1972)

[May73]   May, R.M.: Stability and Complexity in Model Ecostystems. Princeton Univ. Press (1973)

[MR89]   Muratori, S., Rinaldi, S.: A dynamical system with Hopf bifurcations and catastrophes. Appl. Math. and Comp., **29**, 1–15 (1989)

[MS02]   Mierczynski, J., Schreiber, S.J.: Kolmogorov vector fields with robustly permanent subsystems. Journal of Math. Anal. and Appl., **267**, 329–337 (2002)

[MY94]   McCann, K., Yodzis, P.: Biological conditions for chaos in a model food chain. Ecology, **75**, 561–564 (1994)

[RM63]   Rosenzweig, M.L., MacArthur, R.H.: Graphical representation and stability conditions of predator-prey interactions. Amer. Naturalist, **97**, 209–223 (1963)

[Ros72]   Rosenzweig, M.L.: Science, **177**, 904 (1972)

[RR67]   Rescigno, A., Richardson, I.W.: The struggle for life, I: two species. Bull. Math. Biophys., **29**, 377–388 (1967)

[SL78]   Svirezhev, Y.M., Logofet, D.O.: Stability of biological communities. Nauka (1978). English translation in 1983

[Smi95]   Smith, H.: Monotone Dynamical Systems, an Introduction to the Theory of Competitive and Cooperative Systems. Mathematical Surveys and Monographs. American Mathematical Society (1995)

[SZ78]   Scudo, F., Ziegler, J.: The Golden Age of Theoretical Ecology. Springer Lecture Notes in Biomathematics, **22** (1978)

[Zee93]   Zeeman, M.: Hopf bifurcations in competitve three-dimensional Lotka-Volterra systems. Dynamics and Stability of Systems, **8**, 189–216 (1993)

Resonances and small divisors

Étienne Ghys

Unité de mathématiques pures et appliquées (UMPA), CNRS and École normale
supérieure de Lyon, France
http://www.umpa.ens-lyon.fr/~ghys
etienne.ghys@umpa.ens-lyon.fr

Translated from the French by Kathleen Qechar

During the International Congress of Mathematicians held in Amsterdam in
1954, A.N. Kolmogorov announced an important theorem which was made
precise (and proven!) a few years later by V.I. Arnold and J. Moser [Kol54,
Arn63a, Mos62]. I would like to present a *very elementary introduction to
this Kolmogorov-Arnold-Moser (KAM) theorem* according to which "the solar
system is probably almost periodic". My (modest) aim is to show the role of
resonances and small divisors in celestial mechanics by focusing on a very
simplified example, inspired by the real KAM problem: it is in some sense
a "toy model" of the solar system, much easier to understand. Facing a too
difficult question, the mathematician has the right to simplify the statement
to its maximum, in order to locate the difficulties. I will try to treat this
example in detail with the help of Fourier series. The "real" KAM theory
is much more difficult: the reader may find more information, along with
indications about the proof of the theorem, in J. H. Hubbard's chapter in this
volume (Chap. 11).

10.1 A periodic world

We live in a world full of a great number of periodic phenomena. The Sun
rises about every 24 hours, the new moon comes back every 29.5 days, the
summer about every year... Of course, such examples could be multiplied ad
infinitum. This observation is old and the first scientists tried very early to
measure these cycles. Sometimes the period is not easy to determine and it
is very often only an approximation. Let us think e.g. about the cycle called
saros: every 6 585 days and 8 hours, the Moon, the Sun and the Earth find
themselves in about identical relative positions and there is such a periodicity
in the appearance of eclipses. As a matter of fact, due to the 8 hours, the

periodicity of eclipses in a given place of the Earth is in fact triple (one day =
3 times 8 hours) so that the period is of 19 756 days (54 years and 32 or 33
days depending on leap years). We can only be fascinated by the precision
of the astronomers' observations made during Ancient times which led to the
exact determination of this astronomical cycle. Maybe the existence of these
cycles in our universe is a preliminary condition for the appearance of life and
civilization? Can we imagine the difficulties of living on a planet which would
be the satellite of a double star: the rising and setting of the two suns would
become entangled in a more or less random way.

Mathematicians have always been fascinated by cycles and one did not
have to wait for Fourier to decompose a cyclic phenomenon into a sum of
elementary cyclic phenomena. What is more elementary than a point which
rotates on a circle with a constant angular velocity? It is of course the model
the first observers of the Sun (which rotates "evidently" around the Earth)
were thinking about. The situation is a little more complicated in the case
of planets, as the paths they follow in the sky seem sometimes complex (see
Fig. 10.1).

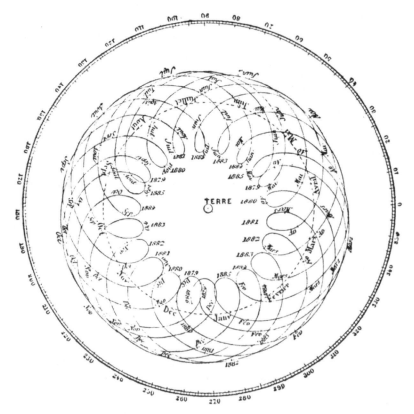

Fig. 10.1. Mercury's orbit seen from the Earth ("Terre"). (From Flammarion's
Astronomie Populaire.)

In Ancient times, astronomers progressively elaborated a remarkably efficient model giving a precise description, extremely close to the measurements
they could perform with their basic instruments. This is the theory of epicycles
(see Fig. 10.2) and equants, dating back at least to the time of Hipparchus,
which I will not describe in detail and which culminates with the marvelous
system of Ptolemy (the *Almageste*, II-nd century). The Earth is in the center and the Sun and the planets turn around the Earth while following finite
combinations of uniform circular motions. The reader interested in detailed
information concerning Hipparchus and Ptolemy's theories could refer to the
article [Gal01].

Ptolemy, one of the greatest geniuses of his time, is only known by contemporary students for his "false" geocentric system theory. And yet! What
is a "correct" theory in the fields of physics or astronomy? Isn't the main aim
to develop a model which explains experiments of a given era? Isn't any question relative to the "correct" nature of space and time only a metaphysical
question which the physicist can ignore?

Copernicus' heliocentric theory superseded Hipparchus/Ptolemy's geocentric theory. Is this new theory more correct than the previous one? One point
is clear: Copernicus' theory is nicer and everything seems to fit in a quite
harmonious and simple way. This suffices to prefer heliocentrism. But if we
take a closer look, Copernicus' theory is not as elementary as it seems. It also
uses cycles and epicycles. Ptolemy used 40 cycles and Copernicus still uses
34 of them... The tables established by Copernicus are not more precise than
those of Ptolemy. Besides, Copernicus does not present his theory as being
"true": he puts at the beginning of his *De Revolutionibus Orbium Coelestium*
(1543) a preface, written by Osiander, about which a lot has been written. Did

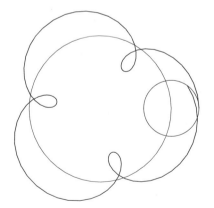 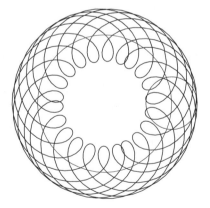

Fig. 10.2. A (simplified) epicycle model. Mercury moves along a small circle ("epicycle") of radius 0.38 (Mercury-Sun distance in Astronomical Units), with period 88
days (Mercurian year), while this epicycle moves along a larger circle ("deferent")
of radius 1 AU, with period 1 Earth's year. Left hand side: after one (Earth's) year;
right hand side: after seven years

Osiander want to protect Copernicus from the pope's ire? Or on the contrary does this preface reflect Copernicus' opinion? Here is an extract from this preface (see [Cop92]):

> ... it is the duty of an astronomer to compose the history of the celestial motions through careful and expert study. Then he must conceive and devise the causes of these motions or hypotheses about them. Since he cannot in any way attain to the true causes, he will adopt whatever suppositions enable the motions to be computed correctly [...] these hypotheses need not be true nor even probable. On the contrary, if they provide a calculus consistent with the observations, that alone is enough.

Let us return to our cycles. If a phenomenon is periodic with period T, all multiples of T can also be considered as a period. Consequently, if two phenomena have respectively a period T_1 and T_2, the combination of these phenomena will be periodic as soon as a multiple of T_1 coincides with a multiple of T_2, in other words as soon as the ratio T_1/T_2 is a rational number. Since we are talking about astronomy and these periods can only be known approximately, we can consider that these ratios are always (almost) rational. The combinations of the cycles that we observe in our universe define therefore a globally periodic phenomenon. A reader could quite rightly notice that this type of argument may easily lead to gigantic periods and that the physical meaning of a period of e.g. one hundred billion years would be questionable. This reader may be reassured: this question is somehow at the heart of this article and our (pre-pythagorician) "physical hypothesis" that all numbers are rational will be discussed and modified all along this article. Let us therefore start by imagining that all physical functions are periodic...

The idea of combining circles to approach a periodic function may not be due to Hipparchus and Ptolemy but in respect for these geniuses, I would like to attribute them the joint property of the following theorem:

Theorem. [Hipparchus-Ptolemy-Fourier] Let $f : \mathbb{R} \to \mathbb{C}$ denote a continuous periodic curve of period T with values in the complex plane. Then f may be arbitrarily closely approximated by a finite combination of uniform circular motions. In other words, for any $\varepsilon > 0$, there exists a function of the form $f_\varepsilon(t) = \sum_{n=-N}^{N} a_n \exp(2i\pi nt/T)$ (with $a_n \in \mathbb{C}$) such that $|f(t) - f_\varepsilon(t)| < \varepsilon$ for all t.

Clearly, Hipparchus and Ptolemy did not prove this theorem in the modern sense of the term but neither did Fourier[1]. For a "modern proof", the reader can refer to e.g. [Kör89].

---

[1] A "theorem" attributed to V.I. Arnold asserts that on one hand no theorem is due to the mathematician which it is named after and on the other hand that this theorem applies to itself.

1. An adelic fantasy.

I would like to allow myself a mathematician's fantasy which is totally useless for the rest of this article and which the reader may skip. The time of contemporary science is modeled by the set $\mathbb{R}$ of real numbers (even if it has been subject to several avatars with the relativity theories). This set does not suggest the idea of successive cycles which we have just mentioned: it flows inexorably from the past to the future. Let us try to formalize time the same way astronomers such as Ptolemy used to think about it, formed by cycles "piled up one on top of the other", in which recurrences are omnipresent.

For any integer $n > 0$, the quotient $\mathbb{R}/n\mathbb{Z}$ formed by real numbers modulo n represents the "cyclic time of period n". If m and n are two integers such that m divides n, there is an obvious projection $\pi_{m,n}$ from the cycle $\mathbb{R}/n\mathbb{Z}$ to the cycle $\mathbb{R}/m\mathbb{Z}$: if we know a real number modulo n, we know it in particular modulo m. Let us define the cyclic time $\mathcal{T}$ as follows: an element t in $\mathcal{T}$ is a map which associates to any integer n an element t_n of $\mathbb{R}/n\mathbb{Z}$ in a way which is compatible with these natural projections, i.e. in such a way that if m divides n, then we have $\pi_{n,m}(t_n) = t_m$. In other words, an element of $\mathcal{T}$ is a way to place oneself in all cycles while respecting the evident compatibilities. Obviously, the "cyclic time" $\mathcal{T}$ contains the "ordinary time" $\mathbb{R}$: to a given real number t, we can associate for every n the point t modulo n in $\mathbb{R}/n\mathbb{Z}$ and these various points are compatible with each other. But $\mathcal{T}$ is much bigger than $\mathbb{R}$ (exercise). We can equip $\mathcal{T}$ with a topological structure which turns it into a compact topological group (exercise). Time as a compact set... a mathematician's (or oriental philosopher's?) dream which illustrates the idea of recurrence. The usual group of real numbers $\mathbb{R}$ is contained as a dense subgroup of $\mathcal{T}$ (exercise). Can one consider $\mathcal{T}$ as a reasonable psychological model for the time we are actually living in? Is this a futile mathematician's exercise? Maybe not. The group $\mathcal{T}$ we have just introduced is the "adelic torus", the study of which is essential in contemporary number theory.

10.2 Kepler, Newton...

I will not describe in detail the marvelous astronomical works of Kepler which are often summarized as Kepler's *three laws*. The first one states that a planet orbits as a conic with the Sun at one focus. The second (law of areas) describes the speed at which this conic is traversed. The third law expresses the period (in the case of an elliptic motion) in terms of the major axis of the ellipse. All of this is far too well-known and can easily be found in many books dealing with rational mechanics. At this point, I would like to insist on two less well-known aspects of Kepler's work.

Kepler is often "blamed" to have only offered a descriptive and nonexplanatory model: what causes the motion of the planets? Newton's law $f = m\gamma$ and the gravitational attraction in $1/r^2$ are wonders but do they explain more than Kepler why objects attract each other? This is similar to the comparison Ptolemy/Copernicus: Newton's laws prevail over those of Kepler by their aesthetic aspect and because they allowed a revolution in physics (and in mathematics). However, they do not explain the cause of the phenomenon (and of course, I could make the same kind of comments on the explanatory character of general relativity).

Kepler's zeroth law : *if the orbit of a planet is bounded, it is periodic, i.e. it is a closed curve.*

If one thinks about this, it is incredible.

Nowadays, one can show the following result (Bertrand's theorem, already known to Newton?). Let us suppose that a material point moves in the plane while being attracted towards the origin of the plane (the Sun) by a force whose modulus $F(r)$ only depends on the distance r to the origin. Let us suppose that all the orbits which are bounded are in fact closed curves. Then, the force $F(r)$ can only be the Newtonian attraction $F(r) = k/r^2$ or the elastic attraction $F(r) = Kr$ (not very reasonable in astronomy!). Why did "mother Nature" "choose" THE law that ensures the periodicity of motion? This is a mystery physics will not explain soon!

How is the motion of a planet if the force of attraction towards the central point is another function $F(r)$? This is a classical question of mechanics and Newton himself studied a great number of cases in his *Principia* (1687). A bounded orbit consists of arcs which join the successive apogees and perigees (see Fig. 10.3). These arcs are obtained from one of them using a symmetry and rotations, the angle of which depends on the considered orbit. Somehow, we can consider that the motion is the result of two periodic phenomena: one relates to the periodic variation of the distance to the Sun and the other

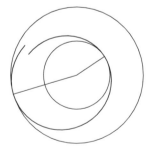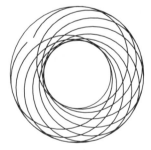

Fig. 10.3. An almost periodic orbit (between the apoapsis and periapsis circles). LHS: an apoapsis and the subsequent periapsis; RHS: after several turns

relates to the periodic variation of the direction of the straight line joining the Sun to the planet. The orbit is periodic if the two periods have a rational ratio and it is almost periodic otherwise. Only the forces in r and $1/r^2$ are such that this ratio is always rational and it happens that it is then equal to 1, so that in these two cases the orbits close themselves in fact after one complete turn. The law $1/r^2$ is a *resonance* of nature since it corresponds to the equality of the radial and angular frequencies.

Kepler must have been filled with wonder when he realized that the orbit of Mars is periodic. This statement is not very obvious when we observe it from the Earth and that does not follow in any way from the epicycle models of Hipparchus-Ptolemy-Copernicus.

I should also mention Kepler's "fourth" law which is rarely cited because it is false, but which Kepler considered as his main discovery. This law was meant to explain the numerical values of the major axes of the orbits of the six planets (which were known at that time). The construction is marvellous, almost philosophical: it is a question of successively encasing the five regular (Platonic) polyhedrons in inscribed and circumscribed spheres (see the beautiful Fig. 10.4 extracted from *Harmonices Mundi* (1619)): the radiuses of the spheres give the radiuses of the orbits (up to similarity of course). Should we make fun of this? Of course not, because it seems that the obtained result is very close to reality and especially because it is an attempt of geometrization of space and motion. Other attempts were very successful later in history. In [Ste69], Sternberg encourages those who make fun of Kepler to also make fun of contemporary theoretical physicists who relate the elementary particles to linear representations of simple Lie groups. The search for groups of symmetries is at the heart of science no matter what the subject is: the icosahedron group, gauge groups, or approximate symmetries in an almost periodic motion, or in a quasicrystal.

Fig. 10.4. Harmonices Mundi

10.3 An almost periodic world

Thus, the world we inherited from Hipparchus, Ptolemy, Kepler and Newton is a *periodic world*. More precisely, each planet is periodic but the solar system is "almost periodic" in its totality since there is of course no reason that the ratios of the periods of the different planets are rational numbers.

Irrational numbers do exist. The sum of two periodic functions whose periods have an irrational ratio is not periodic. But it almost is... The formalization of this idea is recent. Let us begin with two "reasonable" definitions:

Definition. Let f denote a continuous function from $\mathbb{R}$ to $\mathbb{C}$ and $\varepsilon > 0$ a (small) positive real number. A real number T is an ε-period if for every t in $\mathbb{R}$, one has : $|f(t+T) - f(t)| < \varepsilon$.

Definition. Let f denote a continuous function from $\mathbb{R}$ to $\mathbb{C}$. We say that f is almost periodic if for every $\varepsilon > 0$, there exists a number $M > 0$ such that every interval in $\mathbb{R}$ with length greater than M contains at least one ε-period.

The theory of almost periodic functions is rich. The interested reader may read the book [Ste69], in particular for its link with the history of the celestial mechanics. Here are two theorems. The first one is rather an exercise which is left to the reader:

Theorem. Let $a_1, \ldots, a_k$ be complex numbers and $\omega_1, \ldots, \omega_k$ real numbers. The function f from $\mathbb{R}$ to $\mathbb{C}$ defined by $f(t) = \sum_{n=1}^{k} a_n \exp(i\omega_n t)$ is almost periodic.

The second theorem is much more complicated. Formally, it is due to Bohr but for the same subjective reasons as those exposed earlier, I also attribute it to Hipparchus and Ptolemy.

Theorem. [Hipparchus-Ptolemy-Bohr] Every almost periodic function may be arbitrarily closely approximated by functions of the preceding type.

Now that these definitions and theorems are presented, I can start to make the content of this article more precise. *Is the universe in which we live almost periodic?*

10.4 Lagrange and Laplace: the almost periodic world

The proof of Kepler's laws using from those of Newton supposes a "simplified" solar system in which a single planet is attracted by a fixed center. One learns in elementary courses of mechanics that the problem is not much more difficult in the case of two masses which attract each other mutually: each one of them describes a conic. But of course, there is not only one planet in the solar system. Even disregarding many "small" objects, we can consider that nine[2]

---

[2] This paper was written before Pluto was "expelled" from the official list of planets!

2. A remark about the recent history of Physics.

The turbulence of fluids is a quite complex phenomenon which has been puzzling physicists for a long time, at least starting from Leonardo da Vinci, and whose practical applications are more than obvious in aeronautics. How can we understand these eddies of all sizes in turbulent fluids, and the flow of energy from larger eddies towards smaller ones, up to the dissipative scales (Kolmogorov's theory [Kol41])? It is astonishing to note that physicists as eminent and imaginative as Landau and Lifschitz presented for a long time turbulence as an almost periodic phenomenon, of which the number of frequencies depends on the Reynolds number (related in particular to the viscosity of the fluid). It is only with the second edition (1971) of their famous treatise on fluid mechanics that they became aware that the almost periodic functions are too "well behaved" to represent this phenomenon and that it is necessary to call upon much more "chaotic" functions: it is the beginning of the theory of strange attractors, a beautiful example of collaboration between mathematicians and physicists. Old habits are difficult to loose: the epicycles are still present in our scientific subconscious and it is difficult to get rid of them. Should we forget the epicycles and almost periodic functions in the description of our solar system? Are the conservative systems, such as the solar system, also subject to some kind of chaos (and in which sense?), as in the case of the dissipative systems (turbulence)? Somehow, the theorem of Kolmogorov-Arnold-Moser is reassuring: it asserts that under good conditions (explained further on), the almost periodic functions are sufficient to describe the motion of planets.

planets orbit the Sun and attract each other mutually. This N-body problem is mathematically far more complicated and in a sense which I cannot describe precisely here, it has been known since the beginning of the twentieth century that it is impossible to "integrate" it.

For lack of finding "workable" exact solutions for the motion, we are reduced to finding approximate solutions. Lagrange and Laplace are prominent among those who developed best the theory of perturbations. Of course, as a first approximation, the dominant forces in the solar system are the forces of attraction towards the Sun because the mass of the Sun is much bigger than those of the other planets (in a ratio of approximately 10^3). We can thus think that the planets will more or less follow the (periodic) elliptical Keplerian orbits and that those ellipses will change slowly because of the perturbing influence of other planets. How important are these small perturbations? Are they likely to significantly modify the harmony of the Keplerian system? These are difficult questions. We could fear the worst: perhaps a perturbing force of the order of 1/1000 times the principal force could significantly modify the radius of an orbit after a time of about a thousand times the characteristic time of the problem (the year). In other words, we could fear that within a thousand years, the radius of the terrestrial orbit may be divided (or multiplied) by two. This would have important consequences on the history of our civilization! Since we did not notice any catastrophe of this kind in our past, what is the phenomenon that explains why the perturbations perturb less than what we could fear?

The theory of perturbations is complicated and requires many calculations but the basic geometrical idea, such as Gauss explained it, is very simple (like many great ideas). Let us consider a particularly simple case: the Sun, of very large mass, is (almost) fixed; a planet P_1 revolves uniformly on a circular orbit, and another planet P_2 of very small mass compared to P_1 is launched on an orbit around the Sun which is more or less circular and external in comparison to the one of P_1, in the same plane (see Fig. 10.5). Let us imagine that the radius of the orbit of P_2 is really bigger than the one of P_1 so that the angular velocity of P_1 is really bigger than the one of P_2 (according to Kepler's third law). Since the mass of P_2 is very small, one can think that it does not perturb very much P_1 which will therefore stick very closely to its circular trajectory. As for the planet P_2, it is subject to two forces: the main one towards the Sun and a perturbing one towards the planet P_1. The perturbing force is weak but not negligible; its direction oscillates unceasingly because P_1 revolves very quickly. The idea consists of supposing that these oscillations of the direction of the perturbing force can be averaged: in practice, this means that one replaces the revolving planet P_1, by its orbit where one uniformly distributes the mass of P_1. In other words, the planet P_2 is not attracted by a moving planet P_1 but by a circular ring at rest. Is this approximation valid? This is what we will be discussing hereafter. The end of the argument is easy. One knows since Newton that outside the orbit of P_1, the forces of attraction of the Sun and the fixed circular object can be reduced to the force of attraction of a single punctual mass placed in the center. To summarize, everything occurs as if the planet P_2 was subject to the Newtonian force produced by a point whose mass is that of the total mass of the Sun and P_1. Thus, the planet P_2 will almost follow a periodic orbit. In other words, the perturbing forces did not perturb the periodic character of the planet P_2 and this fits with our historical observation: during a few thousand years, the radiuses and the main characteristics of planets did not evolve much.

Many questions are raised by this idea. Is it legitimate to replace a force, whose size and direction vary, by a constant force which is the average of the

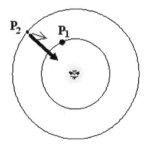

Fig. 10.5. Perturbation of the motion of a small planet P_2 by a planet P_1

varying force? Clearly, there is a situation where this idea cannot work. Let us suppose that the circular orbits which the planets P_1 and P_2 would follow if their masses were infinitely small (and thus unperturbed) are such that the ratio of their periods is rational, 10 for example. This would mean that if the initial positions of P_1 and P_2 are in conjunction e.g. every 10 revolutions of P_1, the two planets are again in exact conjunction. Obviously, to take the average of the perturbation along the orbit of P_1 would not mean much since the angular coordinates of P_1 and P_2 are strongly correlated and the conjunctions are much too regular. On the other hand, if the ratio between the periods is irrational, it seems reasonable to replace the perturbation by its average (see Fig. 10.6). Here is a statement which goes in this direction: it is a particularly simple ergodic theorem (which Lagrange and Laplace did not know, at least explicitly).

Theorem. Let $F(x, y)$ denote a continuous function with real or complex values which depends on two angles x, y considered as elements of $\mathbb{R}/\mathbb{Z}$ (the angle unit is a full turn). Let α and β denote two frequencies whose ratio is irrational. Then, when the time T tends to infinity, the integral $\frac{1}{T} \int_0^T F(x_0 + \alpha t, y_0 + \beta t)\, dt$ converges uniformly to the mean value of F, i.e. to the double integral $\iint F(x, y)\, dx dy$.

Proof. The set of functions F for which the theorem is true is obviously a vector subspace of the space $C^0(\mathbb{R}/\mathbb{Z} \times \mathbb{R}/\mathbb{Z}, \mathbb{C})$ of complex continuous functions on $\mathbb{R}/\mathbb{Z} \times \mathbb{R}/\mathbb{Z}$. This subspace is closed in the uniform topology: a uniform limit of functions which verify the theorem also verifies it. According to Fourier (with two variables), the subspace generated by the functions of the type $\exp(2i\pi(nx+my))$ is dense in $C^0(\mathbb{R}/\mathbb{Z} \times \mathbb{R}/\mathbb{Z}, \mathbb{C})$. It suffices then to verify that each one of these functions $\exp(2i\pi(nx + my))$ satisfies the theorem but this is an explicit and simple calculation which I leave to the reader.   QED

Let us come back to Lagrange and Laplace. Given a real number, it is probably irrational and we may think that the method of Lagrange and Laplace is justified. Still, we should be aware that we took a particularly simple case of only one perturbing planet orbiting on an almost circular path. In its principle, the method applies to the other situations. Let us consider an almost Keplerian solar system, with small perturbations and let us average the perturbations

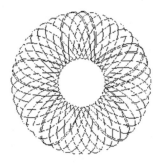

Fig. 10.6. An almost periodic motion

on their configuration spaces. We hope that there are no resonances, i.e. no rational linear relations between the periods which appear. This leads to the *stability theorem* of Laplace which asserts that in the averaged system, the major axes of the orbits remain constant in time, ensuring a certain stability to the system. Finally, this "justifies" the fact that the effects of perturbations are smaller than the ones we could fear *a priori*.

What kind of mathematical credit can we give to this type of "proof"? If we seek "true stability theorems" which are valid for infinitely long times, we will find nothing in Laplace's works which resembles a proof, and the assertions which we sometimes meet according to which "Laplace showed the stability of the solar system" are largely exaggerated. On the other hand, if we seek mathematical statements which are valid for long but finite times, we can hope to transform these methods into theorems, at least in certain particular cases. No matter what, this kind of method lets us think that if the perturbations are of the order of ε (10^{-3} in our system), these perturbations have no global effect at a time $1/\varepsilon$ as we might expect a priori but rather after a time $1/\varepsilon^2$ ("the next term in an asymptotic expansion"). We should have a quiet life for about 10^6 years, which is more reasonable than 10^3. The reader who would like to know more about these perturbation methods may consult some treatises on celestial mechanics if he is brave enough or [Arn89, Arn83, AA68] for a conceptual presentation.

Thus, we inherit from Lagrange and Laplace an almost periodic world, at least for a million years! But they also leave us many questions: what is the role of these resonances between the periods of the planets which put in danger the averaging arguments? Is the stability of the motion perpetual or does it get destroyed after a million years? How can we make this "stability theorem of Laplace" rigorous? It took almost two centuries and the works of mathematicians as powerful as Poincaré, Siegel, Kolmogorov, Arnold and Moser to get to partial answers which themselves raised other questions.

10.5 Poincaré and chaos

At the end of the nineteenth century, Poincaré invented rigorous geometric methods in order to approach a global understanding of the N-body problem. As a matter of fact, he focused on the *restricted three-body problem*: two punctual bodies orbit in a Keplerian way in a plane, around their center of mass, and a third punctual body, with infinitely small mass, is subject to the attraction of the two other masses. Here are some questions studied by Poincaré in his famous article *Sur le problème des trois corps et les équations de la dynamique* (1890) [On the three-body problem and the equations of dynamics]. Is the trajectory of the small mass confined in a bounded domain of the plane if its total energy is sufficiently small? For an initial "generic" condition, is there a risk of collision between the bodies? Is the dynamical behavior of the small body almost periodic? Unfortunately I will not describe this historical article of Poincaré. I will only point out that Poincaré proves the existence of a great number of periodic orbits and that he attempts to understand the dynamics in the vicinity of these periodic orbits. At that point, he makes an error and sins by optimism in a proof (he is used to doing so): his great memoir awarded by king Oscar of Sweden is false. In haste, he has to correct it and this correction will prove to be of considerable scientific richness: Poincaré creates on this occasion the theory of chaos. He highlights trajectories whose behaviors are very far from being almost periodic:

> "Let us try to represent the figure formed by these two curves and their infinite number of intersections each one of which corresponds to a doubly asymptotic solution, these intersections form a kind of web, of fabric, of network with infinitely tight meshes; each one of these curves should never intersect itself, but it must fold up itself in a very complex way to come to cut infinitely often all the meshes of the network. One will be struck by the complexity of this figure, which I do not even attempt to draw. There isn't anything more proper to give us an idea of the complication of the three-body problem and, in general, of all the problems of dynamics where there is no uniform integral and where the Bohlin series are divergent."
> (Poincaré [Poi90])

The history of this error and the way in which Poincaré transforms it into success is fascinating. I recommend the book [Bar97] which is entirely devoted to this question, and the article [Yoc06].

Thus, even if the initial conditions which lead to these examples of chaotic trajectories are not very close to the physical conditions of our solar system, we know thanks to Poincaré that the orbits of the celestial bodies are not necessarily almost periodic. Will we find such orbits in our solar system? In any case, it is necessary for us to be more modest in our search of stability. Previously, we sought to know whether the orbits of planets are almost periodic and we are now much less ambitious since the question becomes the following one. If we launch the planets of a solar system on almost circular orbits around a Sun with very great mass, will the planets remain forever confined in a bounded domain of space? Might it be possible that a planet be ejected from the system for example?

10.6 A "toy model" of the theory of perturbations

We are going to build up a very simple (and even naive) model. On the cylinder $\mathbb{R}/\mathbb{Z} \times \mathbb{R}$, let us consider the transformation f which associates to the point (x, y) the point $(x + \alpha, y)$ where α is an irrational angle. We are going to iterate this transformation and study its dynamics. This is a first simplification: instead of studying dynamics in continuous time (in $\mathbb{R}$), we are going to use a discrete time (in $\mathbb{Z}$). After n iterations, the point (x, y) is sent to the point $(x+n\alpha, y)$ so that the orbits of f spread on the circles $y = const$. We can thus think about f as the dynamics of an almost periodic system. Now, let us try to perturb the motion by imposing to our point of $\mathbb{R}/\mathbb{Z} \times \mathbb{R}$ a "push" towards the top or the bottom which only depends on the first coordinate. In other words, we are now studying a transformation g which associates to the point (x, y) the point $(x + \alpha, y + u(x))$ where u is a certain very regular function defined on $\mathbb{R}/\mathbb{Z}$ (i.e. a periodic function of period 1) which we can think of as being a small perturbation. What is the new dynamics? The n-th iteration of g maps the point (x, y) to the point $(x + n\alpha, y + u(x) + u(x + \alpha) + \cdots + u(x + n\alpha))$.

Lagrange's averaging principle suggests to replace the impulse u by its average on the circle. Of course, if this average is different from 0, we can easily understand that the successive iterations of g will have a tendency to make the second coordinate tend to infinity so that the perturbed system is not stable. Thus, let us study the situation when the average of u on the circle is equal to 0: on average the second coordinate is not modified. Can we deduce that g is stable, in the sense that its orbits stay bounded? This is the simplified problem we are going to study. In symbols, the question is the following:

Let u denote a periodic function of period 1, which is infinitely differentiable, and whose integral on a period is equal to 0. Let α denote an irrational number and x a real number. Are the (absolute values of the) sums $u(x) + u(x + \alpha) + \cdots + u(x + n\alpha)$ bounded when the "time" n tends to infinity?

Let us begin with a lemma which is a special case of a lemma of Gottschalk and Hedlund:

Lemma. Let us fix x_0 in $\mathbb{R}/\mathbb{Z}$. The absolute values of the sums $u(x_0) + u(x_0 + \alpha) + \cdots + u(x_0 + n\alpha)$ are bounded if and only if there exists a continuous function v on $\mathbb{R}/\mathbb{Z}$ such that for all x one has $u(x) = v(x + \alpha) - v(x)$.

Proof. If $u(x)$ is of the form $v(x + \alpha) - v(x)$, the above sum "telescopes" to: $u(x_0) + u(x_0 + \alpha) + \cdots + u(x_0 + n\alpha) = v(x_0 + (n + 1)\alpha) - v(x_0)$. Thus its modulus is bounded by twice the maximum of $|v|$ (which is finite because v is periodic and continuous).

Conversely, let us assume that $|u(x_0) + u(x_0 + \alpha) + \cdots + u(x_0 + n\alpha)|$ is bounded by $M > 0$. This means that the orbit of the point $(x_0, 0)$ in the cylinder $\mathbb{R}/\mathbb{Z} \times \mathbb{R}$ stays confined in the compact cylinder $\mathbb{R}/\mathbb{Z} \times [-M, M]$. Let K denote the closure of this orbit. This is a compact set which is invariant under the transformation g. Among all the non-empty compact sets contained in K and invariant under g, let us choose one which is minimal for inclusion (use the property that the intersection of a family of non-empty compact sets, which is totally ordered for inclusion, is not empty) and let us denote it by $\mathcal{M}$. I claim that $\mathcal{M}$ *is the graph of a continuous function v from $\mathbb{R}/\mathbb{Z}$ to $\mathbb{R}$.*

To justify this assertion, I first observe that the projection of $\mathcal{M}$ on the first coordinate is a non-empty compact set in the circle, which is invariant under the rotation with irrational angle α. All the orbits of such a rotation are dense in the circle. Consequently, the projection of $\mathcal{M}$ on the first coordinate is necessarily the full circle $\mathbb{R}/\mathbb{Z}$.

Now, let me prove that for each x in $\mathbb{R}/\mathbb{Z}$, the "vertical" line $\{x\} \times \mathbb{R}$ only meets the minimal set $\mathcal{M}$ in one point. In order to prove this statement, I consider the vertical translations $\tau_t(x, y) = (x, y + t)$. Obviously, these translations commute with g so that the image by τ_t of an invariant set under g is also an invariant set under g. Consequently, $\tau_t(\mathcal{M})$ is invariant under g and so are the intersections $\tau_t(\mathcal{M}) \cap \mathcal{M}$. We have chosen $\mathcal{M}$ as a minimal non-empty compact invariant set. It follows that for all t, the intersection $\tau_t(\mathcal{M}) \cap \mathcal{M}$ is either empty or equal to $\mathcal{M}$. But if $\tau_t(\mathcal{M})$ would coincide with $\mathcal{M}$ for a t different from 0, then $\mathcal{M}$ would be equal to $\tau_{kt}(\mathcal{M})$ for every integer k and would not be bounded (let k tend to infinity). Therefore $\tau_t(\mathcal{M})$ and $\mathcal{M}$ are disjoint when t is different from 0 and this means that $\mathcal{M}$ meets each vertical $\{x\} \times \mathbb{R}$ at a unique point $(x, v(x))$. Thus, $\mathcal{M}$ is the graph of a function v of $\mathbb{R}/\mathbb{Z}$ to $\mathbb{R}$. As this graph is compact, the function v is continuous (a traditional exercise). The assertion is proven.

We still have to express analytically that the graph of the function v is invariant under the transformation g. The image of $(x, v(x))$ is $(x + \alpha, v(x) + u(x))$ and has to be equal to $(x + \alpha, v(x + \alpha))$. We obtain as expected $u(x) = v(x + \alpha) - v(x)$ and the lemma is proven.   QED

Before continuing, let me restate the lemma in a geometric way. As soon as an orbit of the transformation g is bounded, it remains confined in an invariant circle which is the graph of a continuous function. All the other orbits are then bounded. In other words, in this case, the family of circles $y = \mathit{const}$ which is invariant under the non-perturbed transformation f is replaced by the family of perturbed circles $y - v(x) = \mathit{const}$ which is invariant under the perturbed transformation g.

This leads to a question of harmonic analysis. *Given an infinitely differentiable function u whose integral on the circle is equal to 0, and given also an irrational number α, does there exist a continuous function v on the circle such that $u(x) = v(x + \alpha) - v(x)$ identically?*

The Fourier series are particularly well adapted to study this problem. As the function u is infinitely differentiable, it can be expanded as a Fourier series:

$$u(x) = \sum_{-\infty}^{+\infty} u_n \exp(2i\pi nx).$$

Let us also seek the function v through its Fourier series expansion (we will discuss the convergence of this series afterwards):

$$v(x) = \sum_{-\infty}^{+\infty} v_n \exp(2i\pi nx).$$

(I use the complex notation for convenience: as the function v is real, the complex numbers v_n and v_{-n} are conjugate). Then we have:

$$v(x + \alpha) - v(x) = \sum_{-\infty}^{+\infty} (\exp(2i\pi n\alpha) - 1)v_n \exp(2i\pi nx).$$

Identifying the Fourier coefficients of $u(x)$ and of $v(x + \alpha) - v(x)$, we thus obtain:

$$v_n = \frac{u_n}{(\exp(2i\pi n\alpha) - 1)}.$$

The assumption according to which α is irrational means that $(\exp(2i\pi n\alpha) - 1)$ is different from 0 for n different from 0. Therefore the v_n's are well defined for n different from 0. For $n = 0$, our hypothesis on the average of u means precisely that $u_0 = 0$ so that we can choose any value for v_0 (which of course corresponds to the fact that if v is a solution to our problem, $v + const$ is also a solution).

To summarize, Lagrange's principle seems to work. We have certainly found a function v which is a solution to our functional equation, or at least its Fourier series expansion. But does this series converge and does it define a continuous function as we expect? This is our new problem.

---

3. How can we "see" on a Fourier series that it defines a regular function?

Let us consider a periodic function h of period 1 and let us expand it as a Fourier series:

$$h(x) = \sum_{-\infty}^{+\infty} h_n \exp(2i\pi nx).$$

How can we "see" on the sequence of coefficients h_n that the function h is infinitely differentiable for example? If the function h is supposed to be continuous and not more, is the sequence h_n subject to some constraints? These are delicate questions (which Fourier did not seem to have considered) about which we nowadays know a lot. In this interlude, I will simply give some very elementary observations which will suffice for my discussion. The n-th coefficient h_n is given by Fourier's formula:

$$h_n = \int_{\mathbb{R}/\mathbb{Z}} h(x) \exp(-2i\pi nx)\, dx.$$

If h is continuous, then the sequence h_n must be bounded. Caution: the converse is very far from being valid and my bound is rather crude. One can prove e.g. that the sequence h_n tends in fact to 0 and that the series $(nh_1 + (n-1)h_2 + \cdots + h_n)/n$ is convergent.

If h is continuously differentiable, we can calculate the Fourier coefficients h'_n of its derivative by the well-known formula $h'_n = 2i\pi n h_n$. The continuity of the derivative and the previous observation show that there is an estimate for the decay at infinity of h_n of the form $|h_n| < Cst/|n|$. If h is infinitely differentiable, we can repeat this argument for all derivatives. Thus, the Fourier coefficients of an infinitely differentiable function have *a rapid decay*. This means that for every integer k, there exists a constant $C_k > 0$ such that $|h_n| < C_k |n|^{-k}$.

Conversely, let us consider a rapidly decreasing sequence h_n and let us form its associated Fourier series. It is easy to prove that this series is indeed convergent and defines an infinitely differentiable function.

These simple remarks will suffice but it is a pity to have to leave such a topic without having really gotten into it. The book [Kör89] is magnificent (but requires more mathematical technique).

4. Numbers which are more or less irrationals?

Every irrational number may be arbitrarily approximated by rational numbers. Let us try to make this assertion quantitative. Let α denote an irrational real number. Let us fix a (small) real number $\varepsilon > 0$ and let us seek a rational number p/q (where $q > 0$) such that $|\alpha - p/q| < \varepsilon$. Such a p/q always exists but if ε is very small, a rational p/q which verifies this inequality has necessarily a very large numerator and denominator. What is the minimal value of q as a function of ε? At what speed does this function tend to infinity when ε tends to 0? All depends on the irrational number being considered. In this interlude, we present the basics of the theory of *diophantine approximation*, which is important in our problem.

Some numbers are exceptionally well approximated by rational numbers. The most famous example is the number defined by Liouville:

$$\lambda = \sum_{n=1}^{+\infty} 10^{-n!} = 0.11000100000000000000000010000000000000000000000000 \ldots$$

If we truncate the series at order n, we find a rational number whose denominator is $10^{n!}$ and which approximates λ with a difference smaller than $2.10^{-(n+1)!}$, which is extraordinarily small in comparison to the inverse of the denominator $10^{n!}$. For any physicist, this number is rational since it is different from $0.110001000000000000000001$ by less than 10^{-120} which is a lot smaller than any physically observable number. Nevertheless, not only does the mathematician know that λ is irrational (its decimal expansion is not periodic) but also that Liouville has proven that λ is in fact a transcendental number. If the reader is not impressed by the approximation speed of λ, he may replace the factorials $n!$ by double factorials $n!!$ or even by any increasing function from $\mathbb{N}$ to $\mathbb{N}$, which may even be non-recursive. Thus, given any function $\varepsilon(q)$ from positive integers to positive numbers, tending to zero when q tends to infinity, we can always find irrational numbers α which are approximated by rationals "better than $\varepsilon(q)$", i.e. for which there exists infinitely many rationals p/q such that $|\alpha - p/q| < \varepsilon(q)$. Some irrational numbers resist to the approximation as much as they possibly can. A lemma of Dirichlet shows that every irrational number may be approximated by rationals "up to $1/q^2$":

Lemma. *For any irrational number α, there exists infinitely many rationals p/q ($q > 0$) such that $|\alpha - p/q| < 1/q^2$.*

Proof. Let us project the first $N + 1$ multiples $0, \alpha, \ldots, N\alpha$ in the circle $\mathbb{R}/\mathbb{Z}$. At least two of these projections are at a distance smaller than $1/(N + 1)$ in the circle. This means that we can find $0 \leq k_1 < k_2 \leq N$ such that $(k_2 - k_1)\alpha$ is at a distance less than $1/(N + 1)$ of an integer p. Writing $q = k_2 - k_1 \leq N$, we obtain $|q\alpha - p| < 1/(N + 1) < 1/q$. We observe that $|q\alpha - p| < 1/(N + 1)$ implies that q tends to infinity when N tends to infinity.   QED

Definition. An irrational number α is diophantine if there exists a constant $C > 0$ and an exponent $r \geq 2$ such that for any rational p/q ($q > 0$) one has $|\alpha - p/q| > C/q^r$.

5. A diophantine number: the golden mean

The most famous example of a number which is badly approximated by the rationals is the golden mean $\phi = (1 + \sqrt{5})/2$.

Theorem. There exists a constant $C > 0$ such that for every rational p/q, we have $|\phi - p/q| > C/q^2$.

In fact, we could even prove that we may take $C = 1/\sqrt{5}$ and that ϕ is the irrational number which has the worst approximation by rationals (see [Niv56] for a precise statement and for further details on these questions of approximation by rationals).

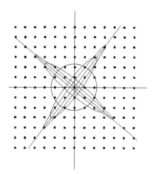

Fig. 10.7. Lattice and eigendirections

Proof. (Outline) Let us consider the matrix $\Phi = \begin{pmatrix} 0 & 1 \\ 1 & 1 \end{pmatrix}$. It has two eigenvalues: ϕ and $-\phi^{-1}$. The slopes of the eigen-directions are also ϕ and $-\phi^{-1}$ (see Fig. 10.7). The linear forms $\pi_1(x, y) = y - \phi x$ and $\pi_2(x, y) = y + \phi^{-1}x$ are eigenvectors of the transposed linear map, with eigenvalues $-\phi^{-1}$ and ϕ respectively. The matrix Φ acts linearly in the plane $\mathbb{R}^2$ and preserves the two eigen-lines as well as the lattice of integral points since its coefficients and those of its inverse are integers. Note that Φ dilates the first eigen-line ($\phi > 1$) and contracts the other one. We are seeking to measure the degree of approximation of ϕ by rationals. In other words, we are looking for points on the line of slope ϕ whose coordinates are "as integral as possible". Let us consider a disk D big enough in the plane whose center is the origin. In this disk, there is only a finite number of points with integral coordinates, so that there exists a constant $C_1 > 0$ such that for all integral points in D different from $(0,0)$, we have: $|\pi_1(q,p)\pi_2(q,p)| > C_1$. Let us study the effect of the action of the matrix Φ^n. The disk D is transformed in the interior D_n of an ellipse, laid down along the line of slope ϕ, and the estimate $|\pi_1(q,p)\pi_2(q,p)| > C_1$ for the integral points (q,p) different from 0 and located in D implies the same inequality for all

integral points of D_n different from 0. This is clear because the product $|\pi_1\pi_2|$ is invariant under the action of Φ. Thus the inequality $|\pi_1(q,p)\pi_2(q,p)| > C_1$ is valid for all integral points in all the D_n's. When n varies in $\mathbb{Z}$, these D_n's cover a whole "hyperbolic" neighborhood of the eigen-lines, of the form $|\pi_1(x,y)\pi_2(x,y)| < C_2$. To summarize, we have proven that there exists a constant $C_3 = \min(C_1, C_2)$ such that for any integral point (q,p) of the plane (different from $(0,0)$), we have $|\pi_1(q,p)\pi_2(q,p)| > C_3$. Now, let us distinguish two sets of rationals p/q according to whether $|\phi - p/q|$ is greater than or less than a fixed small enough quantity $C_4 > 0$. On the first set, the inequality $|\phi - p/q| \geq C_4$ implies in particular that $|\phi - p/q| \geq C_4/q^2$. On the second set, the inequality $|\phi - p/q| < C_4$ implies an inequality of the form $|\pi_2(q,p)| > C_5|q|$ (in fact $C_5 = \phi + \phi^{-1} - C_4 = \sqrt{5} - C_4$ is appropriate) so that we have $|\pi_1(q,p)| > C_3 C_5^{-1}/|q|$ and so $|\phi - p/q| > C_3 C_5^{-1}/q^2$. Thus indeed we have $|\phi - p/q| > C_6/q^2$ for all integral points different from 0 with $C_6 = \min(C_4, C_3 C_5^{-1})$.   QED

10.7 Solution to the stability problem "in the toy model"

Let us take up the problem again. Starting from a function u on the circle, whose integral is 0, and which is infinitely differentiable, we seek to know whether there exists a continuous function v whose Fourier coefficients are given for n different from 0 by

$$v_n = \frac{u_n}{(\exp(2i\pi n\alpha) - 1)}.$$

Since u is infinitely differentiable, the sequence of Fourier coefficients u_n is rapidly decreasing (see Box 3). The terms $(\exp(2i\pi n\alpha) - 1)$ which appear in the denominator are different from 0 but they may be arbitrarily small because α is irrational. This is the *small divisors phenomenon*. These denominators could be so small that the Fourier coefficients v_n may become very big and the Fourier series of v may diverge. Therefore, the difficulty is to know who is winning: is it the numerator which rapidly tends to zero or the denominator which may be very small? The answer, which the reader has already guessed, depends on the quality of the approximation of α by the rationals.

First of all let us assume that α satisfies a diophantine condition $|\alpha - p/q| > C/q^r$ (see Box 4). Let us note that $|(\exp(2i\pi n\alpha) - 1)|$ is nothing else than the euclidian distance between the points 1 and $\exp(2i\pi n\alpha)$ on the unit circle in the complex plane. Since the length of a chord is bigger than $2/\pi$ times the length of the arc which subtends it, we may write that $|\exp(2i\pi n\alpha) - 1|$ is $2/\pi$ times bigger than the length of the circular arc joining 1 to $\exp(2i\pi n\alpha)$, i.e $2/\pi \times 2\pi\times$ the distance between $n\alpha$ and the closest integer p. Thus, we obtain an estimate of the small divisor of the form:

$$|\exp(2i\pi n\alpha) - 1| > 4C/|n|^{r-1}.$$

Since u_n is rapidly decreasing, there exists for every k a constant C_k such that $|u_n| < C_k n^{-k}$. Thus, we obtain an estimate for the Fourier coefficients:

$$|v_n| < (C_k/4C)/|n|^{(k-r+1)}.$$

Since this is valid for every k, the sequence v_n is rapidly decreasing and hence the Fourier series converges to an infinitely differentiable function v. In other words, the continuous function v exists and the perturbed motion g is stable. In this case, we have obtained our justification of the Lagrange-Laplace method, at least under the diophantine condition and in the (naive) framework of our "toy model".

If the rotation angle of the non-perturbed motion is diophantine, the perturbed motion is always stable, whatever the perturbation u (assumed to have 0 integral and to be infinitely differentiable).

What happens if α is not diophantine, e.g. if it is the Liouville number we previously defined? *We may then construct unstable examples i.e. for which the averaging method does not work.* Let $\alpha = \lambda$ denote the Liouville number. We know that there exists a sequence of integers p_k such that $|\alpha - p_k/10^{k!}| < 2.10^{-(k+1)!}$. Thus, for every k, we have $|\exp(2i\pi 10^{k!}\alpha) - 1| < 2\pi.2.10^{k!-(k+1)!} = 4\pi.10^{-k.k!}$ (this time, note that a chord is smaller than the arc which subtends it). Let us construct a sequence u_n as follows. Let $u_0 = 0$ and $u_n = 0$ if $n > 0$ is not an integer of the form $10^{k!}$ and let $u_{10^{k!}} = k.(\exp(2i\pi 10^{k!}\alpha) - 1)$. Finally, let us define u_n for $n < 0$ by $u_n = \overline{u_{-n}}$ for $n < 0$. This sequence is evidently rapidly decreasing because $k.10^{-k.k!} = k.(10^{k!})^{-k}$. This defines the periodic function u (with real values) infinitely differentiable and with 0 integral. When we compute the corresponding coefficients v_n, we find, by their very construction, that $v_n = 0$ if n is not of the form $10^{k!}$ and $v_{10^{k!}} = k$ so that the v_n's are not bounded. Thus, there does not exist any continuous function v whose Fourier coefficients are the v_n's and our problem has no solution: there is no continuous function v such that $u(x) = v(x + \alpha) - v(x)$. We know that this means that the perturbed motion is not stable and that the averaging method does not apply.

The theorem of Kolmogorov-Arnold-Moser is analogous: it asserts that the averaging principle works if the frequencies which come into play are diophantine and if the perturbations are weak enough. A (slightly more) precise statement will be given in the following lines.

10.8 Are the irrational diophantine numbers rare or abundant?

We are all convinced that rational numbers are rare among real numbers, even if it took a lot of work from the mathematicians of the past to be clearly conscious of this fact. For a contemporary mathematician, who is used to the infinite sets *à la* Cantor, the explanation is easy: the rational numbers are countable whereas the real numbers are uncountable. For this reason, to assume that the ratio of the periods of two planets is irrational seems reasonable and the converse has very little chance of happening.

We saw in the previous paragraph that the "rational/irrational" distinction in celestial mechanics should better be replaced by a "non-diophantine/ diophantine" one. I have already explained that the Liouville number, although being mathematically irrational, is "physically rational" and we have just noted that if a frequency is equal to this Liouville number, the averaging method may fail.

Are the diophantine numbers abundant? There are essentially two possible mathematical definitions for abundance and it happens that the answer depends on the choice of the definition:

The first possible approach is that of *Lebesgue's measure*. Let us say that a subset X of $\mathbb{R}$ is *negligible in the sense of Lebesgue* or that it has *0 Lebesgue measure* if for every $\varepsilon > 0$, we may find a countable collection of intervals $I_n \subset \mathbb{R}$ whose sum of lengths is smaller than ε and whose union covers X. Let us say that $X \subset \mathbb{R}$ is of *full Lebesgue measure* if its complement is negligible in the sense of Lebesgue. One of the most interesting aspects of this concept is that the union of a countable collection of negligible sets is negligible. Of course, what is important for this theory to work is that a set cannot be both negligible and of full measure. This is an exercise left to the reader.

The second approach is due to Baire. Let us say that a subset X of $\mathbb{R}$ is *meager in the sense of Baire* if it is contained in a countable union of closed sets of empty interiors. Let us say that X is *residual in the sense of Baire* if its complement is meager. As with the previous definition, the countable union of meager sets is meager (easy) and a set cannot be both meager and residual (this is Baire's theorem).

Which notion of abundance is best adapted to our intuition? The question is delicate and sometimes generates violent polemics among mathematicians. For the case we are interested in, i.e. the abundance of diophantine numbers, the situation is caricatural.

Theorem. The set of irrational diophantine numbers is both meager in the sense of Baire and of full Lebesgue measure.

The proofs are not difficult but they are instructive. Let us write the definition of the set Dioph $\subset \mathbb{R}$ of diophantine numbers by using quantifiers:

$$\mathrm{Dioph} = \{\alpha \in \mathbb{R} \mid \exists r \in \mathbb{N} \, \exists n \in \mathbb{N} \, \forall (p,q) \in \mathbb{Z} \times \mathbb{N}^\star : |\alpha - p/q| \geq \frac{1}{nq^r}\}.$$

Thus Dioph is a countable union indexed by r and n of closed sets which are clearly of empty interiors: Dioph is meager in the sense of Baire.

In order to prove that Dioph is of full Lebesgue measure, let us fix a real $r > 2$ and let us consider the set

$$\text{Dioph}_r = \{\alpha \in \mathbb{R} \mid \exists C \in \mathbb{R}_+^* \; \forall (p, q) \in \mathbb{Z} \times \mathbb{N}^* : |\alpha - p/q| \geq C/q^r\}.$$

It suffices to prove that Dioph_r is of full Lebesgue measure because $\text{Dioph}_r \subset$ Dioph. In order to prove this, we show that its complement meets the interval $[0, 1]$ on a negligible set in the sense of Lebesgue (note that Dioph is invariant under integral translations). Indeed $[0, 1] \setminus \text{Dioph}_r$ is the intersection with $[0, 1]$ of the following sets defined for $C > 0$:

$$\text{NonDioph}_{r,C} = \bigcup_{q=1}^{+\infty} \bigcup_{p=0}^{q} \left] \frac{p}{q} - \frac{C}{q^r}, \frac{p}{q} - \frac{C}{q^r} \right[.$$

This is a countable union of intervals whose sum of lengths is smaller than $2C \sum_q \frac{q+1}{q^r}$. This sum converges because $r > 2$ and the sum is arbitrarily small if C is small enough. Thus, by definition, $\text{NonDioph}_{r,C}$ is negligible and this proves that Dioph is of full Lebesgue measure.   QED

Of course, the previous statement is not mathematically contradictory but it leaves us in an awkward situation. Which meaning will the physicist rather give to the concept of abundance? My personal experience seems to show that physicists do not either have any miraculous solution to suggest. I will come back to this question in the last section but for now let us do "as if" the good concept was that of Lebesgue.

We can therefore conclude that the set of rotation angles for which the perturbed motion is stable is of full Lebesgue measure and we should therefore be satisfied with this result since it covers most of the cases (but we should not forget that if we had preferred Baire to Lebesgue, we should have had the opposite conclusion).

10.9 A statement of the theorem of Kolmogorov-Arnold-Moser

It is difficult to give a clear-cut statement of the KAM theorem. I will first start by stating a precise theorem which is a special case and I will then try to describe the general theorem, but I will need to be much fuzzier then.

Let us consider a transformation f of the cylinder $\mathbb{R}/\mathbb{Z} \times [-1, 1]$ defined this time by $f(x, y) = (x + y, y)$. Again in this case, the circles $y = const$ are invariant and f induces a rotation on each one of them but contrarily to the "toy model", the angle of this rotation depends on the circle since it is equal to y. This map is often called a "twist" for obvious reasons. Now, let us perturb f, i.e. we consider a map g of the form

$$g(x, y) = (x + y + \varepsilon_1(x, y), y + \varepsilon_2(x, y)).$$

As a matter of fact, we ask that g maps the cylinder to itself, i.e. that $\varepsilon_2(x, \pm 1) = 0$ identically. We also assume that g preserves the area, i.e. that its jacobian is identically equal to 1. Let us fix an irrational number α in the interval $[-1, +1]$ and let us suppose that it is diophantine. *The KAM theorem asserts that if $\varepsilon_1, \varepsilon_2$ are small enough, then there exists a curve which is invariant by g, close to the curve $y = \alpha$, and on which the dynamics of g is conjugate to a rotation of angle α.*

We must first give a meaning to "$\varepsilon_1, \varepsilon_2$ small enough". The initial theorem was formulated in 1954 by Kolmogorov in the space of real analytic functions and it is with respect to this (exotic) topology that we may understand the smallness [Kol54]. Kolmogorov only gave global indications on the proof and it is Arnold who gave the rigorous proof of this theorem in 1961, still in the analytical case [Arn63a]. In 1962, Moser succeeded in accomplishing the feat of proving the theorem in the space of infinitely differentiable functions [Mos62]. In fact, Moser used functions which are 333 times differentiable and the topology of uniform convergence on these 333 derivatives... The mere fact that it is necessary to use as many derivatives shows the difficulty of the proof. Nowadays, it is known that the theorem is true with 4 derivatives and false with 3 [Her86].

I have to give up the idea of giving even a sketch of a proof of the theorem. I would simply like to explain that, contrarily to the toy model case, this is a *nonlinear* problem in the (infinite dimensional) space of curves. The linearization of this problem essentially leads to the problem we have already discussed. To switch from a nonlinear problem to a linear problem, the mathematician uses the implicit function theorem, which is correct in a Banach space but false in the Fréchet spaces which occur here. This is why this theorem requires quite formidable techniques of functional analysis (see about this point in the second part of [Her86]).

Each diophantine number α has a corresponding neighborhood in which the theorem applies. The more diophantine α is, i.e. the more difficulties it encounters to be approximated by rationals and the more the invariant circle of angle α is robust under the effect of the perturbation. Thus, given a perturbation $(\varepsilon_1, \varepsilon_2)$, we cannot apply the theorem to every diophantine number. Typically, given the perturbation, some invariant circles remain and the others "break down". Furthermore, the theorem warrants that for a small enough perturbation, the Lebesgue measure of the set of circles which remain is arbitrarily close to the full measure. Thus, we may say that if we perturb f a little, there is every chance that an orbit remains located on a circle and be almost periodic. The situation in the so-called *instability zone*, outside these invariant circles, is very complicated: a lot of problems remain open and research keeps being very active.

What is the link between this theorem and celestial mechanics? Let us consider the restricted three-body problem: two masses revolve one around the other in a Keplerian way and a third infinitely small mass orbits in the same plane. This third mass is attracted by the two others but does not perturb them. In order to describe the dynamics of the third mass, we introduce the phase space: two position coordinates and two velocity coordinates are needed, which gives a space of dimension 4. The conservation of total energy forces the third object to stay in a 3-dimensional submanifold. So, we have to study the dynamics of a vector field in a certain 3-dimensional manifold. For this purpose one can use the method of Poincaré's sections which consists in studying the successive returns of the orbit on a surface transverse to the vector field. This leads to iterate a transformation in dimension 2 of the type we previously considered. Without any detail, the KAM theorem we have cited allows to prove the stability of the system formed by these three bodies. Many more pages, formulae and pictures would be needed to justify this point.

When we consider a "real" solar system, with many planets, the phase space and Poincaré's sections are of higher dimension, and the invariant circles need to be replaced by invariant tori of higher dimensions. This complicates the statement of the theorem but the spirit remains the same: these invariant tori resist the perturbations if the frequency ratios in the initial system are diophantine enough. The general KAM theorem deals with this case.

Thus, the "physical" consequence of KAM is the following. *If we launch a system of planets of small enough masses around a Sun of big mass in initial conditions which are close to that of a Keplerian system, the dynamics which will result from this will be almost periodic, at least for a set of initial conditions whose Lebesgue measure becomes fuller and fuller as the masses of the planets tend to 0.* Outside this set of initial conditions, the theorem does not say anything, apart from the fact that they are rare (in the sense of Lebesgue measure).

This is the reason why our solar system "stands a good chance of being almost periodic"...

10.10 Is the KAM theorem useful in our solar system?

The KAM theorem and its proof are magnificent. From a certain view point, this may suffice to the mathematician. I have no intention of debating here in a few lines of the complex relationship between mathematics and physics but the KAM example could undoubtedly be used as a starting point.

Originating from Physics, the problem has generated a whole branch of mathematics which perfectly suffices to itself and which also generates some other problems which are often totally without any physical content. But it seems to me that even the "purest" mathematician has the duty to go back to the initial problem: has it been solved? Here are some elements of answer:

The KAM theorem applies in the case of "small enough" masses. If we closely study the proof we realize that it applies to very small masses, smaller by several orders of magnitude than what is observed in our solar system. It would clearly be useful to obtain efficient and effective versions of KAM, let us say for masses 1/1000 times the mass of the Sun. We are still very far away from this and, unfortunately, few colleagues find this mathematical issue to be fascinating.

The forces which act in the solar system are mostly gravitational but other forces are non-hamiltonian (e.g. the solar wind can "slow down" the planets). After several hundred thousands years, the effects are perhaps not negligible and the KAM theorem cannot help us to understand the situation. Indeed, is there an interest other than philosophical or mathematical to prove that the "theoretical" (= hamiltonian) solar system is stable or instable? The physicist wants to understand the situation for the near future (let us say that a few billion years would suffice him).

The union of the invariant tori given by the theorem has a large Lebesgue measure but it has an empty interior. Which is the good abundance concept in physics? As I have already explained earlier, mathematicians cannot answer this question and physicists have to show them the way.

Experience shows that many frequencies encountered in the solar system seem to be very rational. The following example, taken from [Bel86], is really impressive. Let us consider the angular frequencies ω_i^{obs} $(i = 1, ..., 9)$ of the 9 planets (measured in such a unit that the frequency of Jupiter equals 1). It turns out that when we modify *very slightly* these values, we can find "theoretical" frequencies ω_i^t which are *exactly* linked together with integral linear relations: the following table exhibits a 9×9 matrix with small integer entries, with a lot of zeros, which exactly anihilates the vector of theoretical frequencies. Note that the discrepancies $\Delta\omega/\omega = (\omega^{obs} - \omega^t)/\omega$ are extremely small.

	Planet	ω_i^{obs}	ω_i^t	$\Delta\omega/\omega$	n_1	n_2	n_3	n_4	n_5	n_6	n_7	n_8	n_9
1	Mercury	49.22	49.20	0.0004	1	-1	-2	-1	0	0	0	0	0
2	Venus	19.29	19.26	0.0015	0	1	0	-3	0	-1	0	0	0
3	Earth	11.862	11.828	0.0031	0	0	1	-2	1	-1	1	0	0
4	Mars	6.306	6.287	0.0031	0	0	0	1	-6	0	-2	0	0
5	Jupiter	1.000	1.000	0.0000	0	0	0	0	2	-5	0	0	0
6	Saturn	0.4027	0.4000	0.0068	0	0	0	0	1	0	-7	0	0
7	Uranus	0.14119	0.14286	-0.0118	0	0	0	0	0	0	1	-2	0
8	Neptune	0.07197	0.07143	0.0075	0	0	0	0	0	0	1	0	-3
9	Pluto	0.04750	0.04762	-0.0025	0	0	0	0	0	1	0	-5	1

The book [Bel86] contains a very interesting paragraph on these resonances which are observed in our solar system. It contains in particular a discussion on the "hypothesis of Moltchanov" according to which "every oscillatory system having been subject to an extended evolution is necessarily in resonance and is governed by a family of integers". Thus, for Moltchanov, the small non-hamiltonian forces keep the systems away from the diophantine frequencies and push them in the zone where the KAM theorem does not apply... It seems to me that justifying or invalidating this hypothesis remains a magnificent challenge for today's mathematicians.

References

[AA68]   Arnold, V.I., Avez, A.: Ergodic problems of classical mechanics. W. A. Benjamin, Inc., New York-Amsterdam (1968)

[And]   http://www-groups.dcs.st-andrews.ac.uk/history, a web site on the history of mathematics.

[Arn63a]   Arnold, V.I.: Proof of a theorem of A. N. Kolmogorov on the preservation of conditionally periodic motions under a small perturbation of the Hamiltonian (in Russian). Uspekhi Mat. Nauk, 18:F113, 13–40 (1963)

[Arn63b]   Arnold, V.I.: Small denominators and problems of stability of motion in classical and celestial mechanics (in Russian). Uspekhi Mat. Nauk, 18, 91–192 (1963)

[Arn83]   Arnold, V.I.: Geometric methods in the theory of ordinary differential equations. Springer, New York (1983)

[Arn89]   Arnold, V.I.: Mathematical Methods of Classical Mechanics, 2d ed. Springer (1989)

[Bar97]   Barrow-Green, J.: Poincaré and the three-body problem. History of Mathematics, 11. American Mathematical Society, Providence, RI; London Mathematical Society, London (1997)

[Bel86]   Béletski, V.: Essais sur le mouvement des corps cosmiques. Éditions Mir, French translation (1986)

[Cop92]   Copernicus, N.: On the Revolutions of the Heavenly Bodies, trans. E. Rosen. The Johns Hopkins University Press, Baltimore (1992). Originally published as volume 2 of Nicholas Copernicus' Complete Works, Jerzy Dobrzycki (Editor), Polish Scientific Publishers, Warsaw (1978)

[Gal01]   Gallavotti, G.: Quasi periodic motions from Hipparchus to Kolmogorov. Atti Accad. Naz. Lincei Cl. Sci. Fis. Mat. Natur. Rend. Lincei (9) Mat. Appl., 12, 125–152 (2001)

[Her86]   Herman, M.: Sur les courbes invariantes par les difféomorphismes de l'anneau. Astérique, 144 (1986)

[Kol41]   Kolmogorov, A.N.: The local structure of turbulence in incompressible viscous fluid for every large Reynold's numbers. C. R. (Dokl.) USSR Sci. Acad., 30, 301–305 (1941)

[Kol54]   Kolmogorov, A.N.: General theory of dynamical systems and classical mechanics. In: Proceedings of the International Congress of Mathematicians, Amsterdam, 1954. Erven P. Noordhoff N.V., Groningen (1957)

[Kör89]   Körner, T.: Fourier analysis. Cambridge University Press, Cambridge (1989)

[Mos62]   Moser, J.: On invariant curves of area-preserving mappings of an annulus. Nachr. Akad. Wiss. Göttingen Math.-Phys. Kl. II, 1–20 (1962)

[Niv56]   Niven, I.: Irrational numbers. The Carus Mathematical Monographs, n° 11. The Mathematical Association of America. Distributed by John Wiley and Sons, Inc., New York (1956)

[Pet93]   Peterson, I.: Newton's Clock: Chaos in the Solar System. W. H. Freeman & Co, N.Y. (1993)

[Poi90]   Poincaré, H.: Sur le problème des trois corps et les équations de la dynamique (1890). Œuvres, volume VII, Gauthier-Villars, Paris (1951)

[Ste69]   Sternberg, S.: Celestial Mechanics, parts I and II. W.A. Benjamin (1969)

[Yoc06]   Yoccoz, J.C.: Une erreur féconde du mathématicien Henri Poincaré [A fruitful error by the mathematician Henri Poincaré]. Soc. Math. France / Gazette Math., **107**, p. 19–26 (jan. 2006)

[Zee98]   Zeeman, C.: Gears from ancient Greeks (1998). The transparencies of this conference are available at http://www.math.utsa.edu/ecz/

A.N. Kolmogorov in his flat at Moscow University. With friendly permission of
Mathematisches Forschungsinstitut Oberwolfach/photo collection of Prof. Konrad Jacobs

The KAM Theorem

John H. Hubbard

Department of Mathematics, Cornell University, Ithaca, NY, USA
http://www.math.cornell.edu/People/Faculty/hubbard.html
jhh8@cornell.edu

Translated from the French by Thomas Ransford

I first heard about the KAM theorem when I was an undergraduate, around 1966. It seemed to me the most beautiful result in the world, but for many years my interests were engaged elsewhere. Around 1980, I came back to dynamical systems, and I quickly realized that the KAM theorem is indispensable.

Each year, for about fifteen years, I said to myself in September: this is the year that I am going to understand the proof. Each year, as March came around, I had to admit failure once again: I no longer knew the order of the quantifiers in the technical lemmas, and so was unable to apply them.

During these years, I tackled all the proofs that I knew: Arnold's [Arn63, AA68], Moser's [Mos62, Mos73], Sternberg's [Ste71], those based on the Nash–Hamilton implicit function theorem, those of Herman [FH83, Her86],... I did not succeed in mastering a single one. And I am far from being alone: I know numerous dynamicists who realize that they ought be able to prove the theorem, who even teach it sometimes, but who have never mastered the proof either.

After being pointed in the right direction by Pierre Lochak, I finally discovered the article of Bennettin, Galgani, Giorgilli and Strelcyn [BGGS84], which I found luminous. With the help of Yulij Ilyashenko, I discovered several improvements: this is the proof published in [HI02]. Ilyashenko gave an exposition of it at the Moscow mathematics seminar in 2002; in the audience were some participants from Kolmogorov's seminar in 1957; they told him that this proof was in fact the original proof.

One might wonder whether this is really true. At any rate, it is very hard to understand why Kolmogorov never published a proof of his most beautiful result.

The KAM theorem is so called in honour of Andreï Kolmogorov, Vladimir Arnold and Jürgen Moser. The theorem was announced in the Proceedings of the International Congress in Amsterdam [Kol57], but no detailed proof was published until that of Arnold [Arn63] in 1963. In the meantime, in 1962, Moser had published the article [Mos62], which establishes a different but related result, bringing to bear similar techniques.

These publications have generated countless other works, both in physics and in mathematics. In physics, the KAM theorem is a basic tool in the design of particle accelerators: indeed, accelerated particles fly round an accelerator billions of times, and the techniques guaranteeing the stability of the particle beam are the same as those which permit us to study the stability over billions of years of the solar system. In mathematics, the KAM theorem is at the origin of the small-divisor problem; it is the research area of scores of mathematicians.

I am not going to attempt to sketch the development of the subject over the last forty years; I lack the expertise and time for such a project. This article consists of two parts. The first part illustrates the KAM theorem in the context of two examples: the solar system and the forced pendulum. It requires only a moderate mathematical background.

In the second part, I give a rigorous statement of the theorem, and sketch the main ideas of the proof, the one written out in detail in [HI02]. The background required is more substantial: differentiable manifolds, symplectic forms, flows of vector fields, etc.

Part I. Two examples

11.1 The solar system

The KAM theorem yields a troubling answer to one of the oldest questions in celestial mechanics: is the solar system stable? Will it continue eternally more or less as we see it today? Or could it be that planetary interactions, between Jupiter and Saturn e.g. will eventually lead to catastrophes, where certain planets escape from the Sun, and others collide or fall into the Sun?

Kolmogorov's answer is that, in a system like the solar system, regular motion and chaotic motion are inextricably entwined, zones of chaos within zones of order and zones of order within zones of chaos, and all of this on every scale. Regular motion resembling Keplerian motion and motion which does not resemble it at all both appear with strictly positive probability. There is thus no reasonable answer to the question.

---

1. Newton's equations.

A system of n masses $m_1, \ldots, m_n$ with positions $\mathbf{x}_1, \ldots, \mathbf{x}_n$ satisfies Newton's law

$$m_i \mathbf{x}_i'' = \sum_{j \neq i} G m_i m_j \frac{\mathbf{x}_j - \mathbf{x}_i}{|\mathbf{x}_j - \mathbf{x}_i|^3}, \tag{11.1}$$

where $G \approx 6.662 \cdot 10^{-11}\ m^3/(kg\,s^2)$ is the universal gravitational constant. This equation follows the famous general law $\mathbf{F} = m\mathbf{a}$: *force equals mass times acceleration*. Indeed, on the left-hand side, we see $m_i \mathbf{x}_i''$, namely the i-th mass times its acceleration, and on the right-hand side we see the force acting on this mass: each mass m_j attracts m_i with a force proportional to the product of the masses and inversely proportional to the square of the distance between them. The numerator is the vector joining $\mathbf{x}_i$ to $\mathbf{x}_j$; it gives the direction of the force, but it already has length $|\mathbf{x}_j - \mathbf{x}_i|$, so we must divide by $|\mathbf{x}_j - \mathbf{x}_i|^3$ in order to obtain a force inversely proportional to the square of the distance.

A word of caution: we are speaking here of classical mechanics, namely the behaviour of the very good mathematical model of celestial mechanics given in Box 1. We are neglecting all physical phenomena other than gravitation: tides, the pressure of solar wind, not to mention the fact that the Sun is destined one day to become a red giant and swallow up several planets (including the Earth); we are also neglecting the relativistic corrections to Newton's equations. In the very long term (millions, even billions of years hence), these effects are certainly significant. Thus the word "eternally" used above does not apply to the real solar system; but we can still ask if the trajectories of the mathematical model are stable for ever.

Besides, the effects of attraction between planets, though small compared with the gravitational attraction of the Sun, are considerably greater than those of the other phenomena above, and produce measurable effects over much shorter periods (centuries, or merely years). It is of these that we shall speak.

11.1.1 The case of zero planetary masses

To understand how the KAM theorem applies to the solar system, it is essential to see that the equations governing planetary motion have a limit as the masses of the planets tend to zero, and that, moreover, the behaviour in this limit is not very different from the motion that we actually observe. If we were to replace all the planets by grains of sand, then these grains would follow very much the same trajectories as do the planets.

This phenomenon is nothing other than Galileo's experiment. He observed that a marble and a cannonball, released from the top of the Leaning Tower of Pisa, fall at the same speed; for cannonball and marble, read Jupiter and grain of sand.

2. The solar system with zero planetary masses.

The fundamental observation is that in the (11.1) one can simplify m_i. For $n+1$ masses m_0 (the Sun), $m_1, \ldots, m_n$ (the planets) in positions $\mathbf{x}_0, \mathbf{x}_1, \ldots, \mathbf{x}_n$, Newton's equations can be written

$$\mathbf{x}_0{}'' = G \sum_{j=1}^{n} m_j \frac{\mathbf{x}_j - \mathbf{x}_0}{|\mathbf{x}_j - \mathbf{x}_0|^3}$$

$$\mathbf{x}_i{}'' = Gm_0 \frac{\mathbf{x}_0 - \mathbf{x}_i}{|\mathbf{x}_0 - \mathbf{x}_i|^3} + \sum_{j=1,\ldots,n,\ j\neq i} Gm_j \frac{\mathbf{x}_j - \mathbf{x}_i}{|\mathbf{x}_j - \mathbf{x}_i|^3}.$$

As all the planetary masses $m_j, j = 1, \ldots, n$ tend to 0, but not m_0 (the Sun), these equations become

$$\mathbf{x}_0{}'' = \mathbf{0}$$

$$\mathbf{x}_i{}'' = Gm_0 \frac{\mathbf{x}_0 - \mathbf{x}_i}{|\mathbf{x}_0 - \mathbf{x}_i|^3}.$$

Since $\mathbf{x}_0{}'' = \mathbf{0}$, the mass m_0 travels in a straight line at constant speed; without changing the other equations, we can work in a system of coordinates where $\mathbf{x}_0 = \mathbf{0}$ (heliocentric). The other masses then satisfy the decoupled equations

$$\mathbf{x}_i{}'' = -Gm_0 \frac{\mathbf{x}_i}{|\mathbf{x}_i|^3}.$$

The difference between zero masses and small masses can be observed in the oscillation of the Sun. The centre of gravity of the Sun-Jupiter system is about one solar diameter from the centre, and as Jupiter rotates "around the Sun", the Sun itself orbits the centre of gravity. This is how we detect the existence of planets around other stars: except in a few recent cases, even the best telescopes do not see the planets; but they do see the oscillations that a big planet induces in its star.

Look at Box 2: the argument often heard that "if planets had no mass, then they would not experience the Sun's attraction" obviously does not apply to these equations.

The solar system with planets of zero mass is easy to understand: it is exactly what we all learned at school. Each planet obeys Kepler's laws.

1. The orbit of each planet is an ellipse with the Sun at one of the foci.
2. The radius vector from the Sun to each planet sweeps out equal areas in equal times.
3. The period of each planet is proportional to the length of its major axis raised to the power $3/2$.

In particular, the system is stable: its evolution over time will never lead it to diverge very far from its present state.

11.1.2 Irrational numbers and vectors

The KAM theorem states that if a "totally integrable" mechanical system admits a "sufficiently irrational" trajectory, then each "sufficiently small" perturbation of the system also admits "the same motion". (We shall see a more precise statement later.)

This notion of "sufficiently irrational" is the central concept of the KAM theorem, and, more generally, of all problems concerning "small divisors". It is a problem with a long history: Archimedes sought rational approximations of π, and the essence of his work consisted in finding good rational approximations of square roots. In the 19th century, Liouville used rational approximations of algebraic numbers to find the first transcendental numbers (see Box 4).

To say that a number θ is irrational is to say that $|\theta - p/q| \neq 0$, for every pair of integers p, q. In order to quantify this statement, we need not only that $|\theta - p/q| \neq 0$, but that $|\theta - p/q|$ is "big". An objection to this is that it is manifestly impossible: one can always approximate an irrational number by rational numbers. For example, the numbers

$$a_0 = 3, a_1 = 3.1 = \frac{31}{10}, a_2 = 3.14 = \frac{314}{100}, \dots$$

(which have been chosen suitably) approximate π more and more closely, and $|\pi - a_k|$ tends to 0 with k. So we cannot simply ask that $|\theta - p/q|$ be big; what we can ask, however, is that it can only be small *if the denominator is big*. (Certain irrational numbers are less easily approximated by rationals, in other words, the denominators in their rational approximations are forced to grow more rapidly—see Box 3.)

---

3. Some good approximations to π.

The decimal approximation

$$\left| \pi - \frac{314159}{100000} \right| < \frac{3}{1000000}$$

is not very good; we can do better with a much smaller denominator. Some good approximations are

$$\left| \pi - \frac{22}{7} \right| < \frac{2}{100}, \qquad \left| \pi - \frac{355}{113} \right| < \frac{3}{10\,000\,000}.$$

It is the rate of growth of the minimal denominators for finer and finer approximations which will measure good or bad "approximability" by rationals.

There are numerous ways of making this precise; the one we shall adopt will be to demand that, in order to find integers p, q such that

$$\left| \theta - \frac{p}{q} \right| < \varepsilon$$

it is necessary to take q at least of order $\varepsilon^{-1/3}$; this can be reformulated as saying that there exists a constant $\gamma > 0$ such that, for every pair of integers p, q, we have

$$|q\theta - p| \geq \gamma/q^2. \tag{11.2}$$

Consider π for example: to this day, the best we know is that $|q\pi - p| \geq \gamma/q^{7.0161}$, which is much less restrictive than (11.2). (It is nevertheless believed that in fact a lower bound of the form $|q\pi - p| \geq \gamma_r/q^r$ holds for each $r > 1$, which would be much stronger than $(11.2)^1$.)

---

4. Liouville and transcendental numbers.

The first quantitative statement stipulating that certain irrational numbers can only be approximated by rationals with "big" denominators was the (celebrated and elementary) result of Liouville (1844): if θ is an algebraic number of degree d, namely, if it satisfies an equation

$$a_d\theta^d + a_{d-1}\theta^{d-1} + \cdots + a_0 = 0$$

with integer coefficients (and with $a_d \neq 0$), then there exists a constant $C > 0$ such that, for every pair of coprime integers p, q (with $q \neq 0$), we have

$$\left| \theta - \frac{p}{q} \right| > \frac{C}{q^d}.$$

From this, one easily deduces that the number $\sum_{n=0}^{\infty} 10^{-n!}$ is transcendental, the first transcendental number to be discovered. The theory inspired by Liouville's result is now an important part of number theory.

The condition (11.2) is not unreasonably restrictive: it holds for *almost all* real numbers (in the mathematical sense of the term, namely, the exceptions form a set of Lebesgue measure zero), and this is true even for the condition that there exists a γ_r such that $|qa - p| > \gamma_r/q^r$ for an arbitrary $r > 1$ (and every pair of coprime integers p, q). On the other hand, it can be shown that the set of real numbers a satisfying $|qa - p| > \gamma/q$ (for every pair of integers p, q) is of measure zero[2].

The irrationality of numbers is not sufficient for stating KAM; we need a notion of irrationality of vectors. In the case of the solar system with zero

---

[1] Liouville proved that if an irrational number is "too well" approximable by rationals thus it is transcendental (see Box 4), but there also exist transcendental numbers which are "very badly" approximable by rationals: for example, for the number e we have a lower bound $|qe - p| \geq \gamma_r/q^r$ for every power $r > 1$. (Editor's note.)

[2] This last condition is satisfied, notably, by algebraic numbers of degree 2, but remarkably, to this day we do not know a single algebraic number of degree > 2 which is known to satisfy or not to satisfy the condition

planetary masses, the vector in question is $(P_1, \ldots, P_n)$, the vector of the planets' periods, or rather the vector of frequencies $(\omega_1, \ldots, \omega_n)$, where $\omega_i = 1/P_i$. Notice that an individual P_i means nothing: it depends on the choice of units of time. But the ratios of the P_i have a meaning: to say that Jupiter's year is 11.86 times as long as that of Earth obviously has a sense independent of time units. In this language, the irrationality of the number θ becomes the irrationality of the vector $(\theta, 1)$.

The analogue of the (11.2) for a vector of length n, is to demand that there exist $\gamma > 0$ such that for every vector with integer coefficients $(k_1, \ldots, k_n)$ we have

$$|k_1\omega_1 + \cdots + k_n\omega_n| \geq \frac{\gamma}{(k_1^2 + \cdots + k_n^2)^{n/2}}. \qquad (11.3)$$

This condition is indeed "sufficiently irrational". It is not the weakest condition under which KAM can be proved, but it is the easiest to use.

An important question is to know to what extent the vectors $(\omega_1, \ldots, \omega_n) \in \mathbb{R}^n$ satisfying the condition (11.3) are not too exceptional: we should like ordinary motions to be preserved—not merely certain exceptional motions. And this is indeed the case: the vectors $(\omega_1, \ldots, \omega_n) \in \mathbb{R}^n$ satisfying condition (11.3) are of full measure, in other words their complement is of measure zero. If we select a vector at random, for example by throwing a ten-sided die to choose successive digits of the coordinates, we will obtain a vector satisfying condition (11.3) with probability 1, that is to say "almost surely" in the mathematical sense of the term (and thus "with certainty" in practice).

For the sake of clarity, let us make a precise statement.

Definition 1. *Denote by Ω_γ^n the subset of $\mathbb{R}^n$ formed by vectors $(\omega_1, \ldots, \omega_n) \in \mathbb{R}^n$ such that, for every vector with integer coefficients $(k_1, \ldots, k_n)$, we have*

$$|k_1\omega_1 + \cdots + k_n\omega_n| \geq \frac{\gamma}{(k_1^2 + \cdots + k_n^2)^{n/2}}.$$

Theorem 1. *The subset of $\mathbb{R}^n$*

$$\mathbb{R}^n - \bigcup_{\gamma > 0} \Omega_\gamma^n$$

is of (Lebesgue) measure zero.

11.1.3 Linear windings on a torus

There is a more geometric way of describing the motion of the solar system under the hypothesis of zero planetary masses: as a trajectory winding around an n-dimensional torus, where n is the number of the planets. Each planet travels along an ellipse, which is topologically a circle, and metrically a circle on which the planet orbits at constant speed if we use time as parameter, starting e.g. from the apogee of each ellipse.

We shall think of the circle as the quotient space $\mathbb{R}/\mathbb{Z}$, where we identify numbers which differ by an integer, and of the n-dimensional torus as $(\mathbb{R}/\mathbb{Z})^n$. Still parametrised by time, the trajectory whose frequencies are $\omega = (\omega_1, \ldots, \omega_n)$ is at time t at the point

$$(a_1 + t\omega_1, \ldots, a_n + t\omega_n)$$

if at time 0 the system is at $(a_1, \ldots, a_n)$, in other words, if a time a_k/ω_k has elapsed since the k-th planet's last visit to its apogee.

In order to extract from this parametrisation the real positions of the planets in space, we have to calculate the parametrisation of the ellipses as a function of time; this is quite feasible, thanks to Kepler's second law, but it requires the use of *elliptic functions*, and will be of no use to us.

It is very instructive to consider the motions

$$t \mapsto \mathbf{a} + t\omega$$

on the torus $(\mathbb{R}/\mathbb{Z})^n$; we shall call these motions the *linear flow on* $(\mathbb{R}/\mathbb{Z})^n$ *in the direction* ω. These motions have an interesting geometry, which depends in detail on the vector ω. For example, the trajectory is dense on the torus if and only if ω is irrational, namely the relation $k_1\omega_1 + \cdots + k_n\omega_n = 0$ with all the k_i integers implies $k_1 = \cdots = k_n = 0$. If, on the contrary, all the frequences are multiples of a common frequency, the winding is around a circle embedded in the torus.

Figure 11.1 shows a dense winding on the torus, and another corresponding to $\omega = (5, 2)$.

One can visualize things as in Fig. 11.1 above; it is easier (but less pretty) to imagine the torus as a cube (a square when $n = 2$) with opposite sides stuck together, as in Fig. 11.2.

An irrational winding has lots of combinatorics, depending in a delicate way on the direction vector ω; e.g. on the right of Fig. 11.2, the first seven intersections s_i of the trajectory with the circle on the torus corresponding to the vertical side of the square occur in circular order $s_1, s_7, s_2, s_6, s_3, s_4, s_5$.

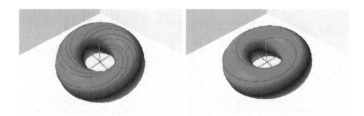

Fig. 11.1. Two windings on a torus: on the right, the direction is $(2, 5)$ and the trajectory is a circle embedded in the torus, which makes two turns in one sense while making five in the other; for the left-hand figure, the direction is $(1, \sqrt{2})$ and the winding is dense, but we have only drawn the segment $0 \leq t \leq 10$

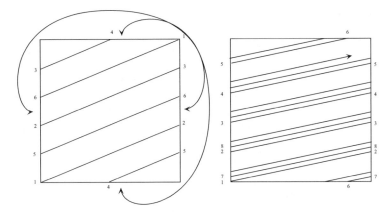

Fig. 11.2. It is much simpler to draw a winding in a square in the plane, where opposite sides are identified

To say that a motion $\mathbf{x}(t)$ of the perturbed system is "the same" as a motion of the unperturbed system $\mathbf{x}_1(t)$ dense on a torus T_1, means that the orbit $\mathbf{x}(t)$ is dense on a torus T, and that $\mathbf{x}(t)$ fills T in the same way combinatorially; more precisely, there exists a homeomorphism $\Phi : T \to T_1$ such that $\Phi(\mathbf{x}(t)) = \mathbf{x}_1(t)$.

We finally arrive at a rigorous statement in the particular case of the solar system.

Theorem 2 (KAM for the solar system). *Let $\mathbf{x}_1(t)$ be a motion of the zero-masses system, for which the frequency vector satisfies (11.3) (and thus $\mathbf{x}_1(t)$ is dense on a torus T_1). Then there exists $\varepsilon > 0$ such that, if the planets are given masses $m_i < \varepsilon$, there exists a trajectory $\mathbf{x}(t)$ of the system thus perturbed, dense on a torus T, and a homeomorphism $\Phi : T \to T_1$, such that $\Phi(\mathbf{x}(t)) = \mathbf{x}_1(t)$.*

Moreover, the set of trajectories of this kind is a set of positive measure in the set of all trajectories, and the probability of being on such a trajectory tends to 1 as ε tends to zero.

Should we believe that the solar system is on one of these nice stable trajectories described by the KAM theorem? Surely not, for the periods of the orbits of Jupiter and Saturn are in a ratio 5 : 2, thus rational, and the hypotheses are not satisfied. This is not (either figuratively or literally) the end of the world. A refinement of the KAM theorem guarantees that there also exist stable motions where such ratios may be rational: second-order zones of order. According to Jacques Laskar (Paris Observatory), the motions of the solar system are not compatible with this second order, and we need to seek zones of even higher order to describe the solar system.

11.2 The forced pendulum

The solar system is of course of vital importance, as much in physics as in history, astronomy and philosophy. But it is difficult to illustrate with pictures showing zones of order and chaos; it can be managed for the "restricted three-body problem"[3] but this requires some good will. The mathematical and graphical study of the forced pendulum is much simpler.

We shall consider the forced pendulum governed by the differential equation

$$x'' + \sin x = \varepsilon \cos 2\pi t;$$

the case $\varepsilon = 0$ is the unperturbed case. Following standard procedure in the study of differential equations, we introduce a speed variable y and write our equation as a system

$$x' = y \qquad y' = -\sin x + \varepsilon \cos 2\pi t. \tag{11.4}$$

When $\varepsilon = 0$ we can draw the *phase plane*, namely the (x, y)-plane with certain trajectories drawn in; this is illustrated in Fig. 11.3.

The picture corresponding to $\varepsilon > 0$ is much more complicated, since the trajectories cross each other all over the place. But there does exist a picture which really gives us information: it is the one where we do not draw an entire trajectory but only the points

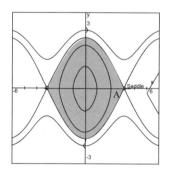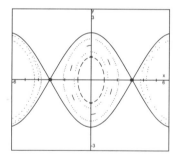

Fig. 11.3. On the left, the more-or-less circular trajectories are the pendulum's oscillations about its equilibrium position $(0, 0)$. They fill a certain region A in the plane, shaded light grey in the figure, which is of area exactly $2\sqrt{2} \int_{-\pi}^{\pi} \sqrt{1 + \cos x} \, dx = 16$. The trajectories above and below the figure (outside A) represent the motions where the pendulum makes complete turns, in one direction (above the figure) or the other (below). These two types of motion are separated by trajectories which take an infinite time to fall from the position of unstable equilibrium, and an infinite time to return there; in the picture, these bound A. On the right, we see the same picture under a stroboscope, flashing with period 2π

---

[3] Namely, the three-body problem with one body of zero mass: it is the trajectory of the latter which is then interesting (and complicated): the two massive bodies follow the (Keplerian) trajectories of the standard two-body problem

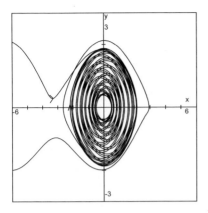 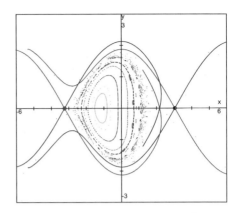

Fig. 11.4. On the left, two trajectories of the forced pendulum for $\varepsilon = .2$. One can convince oneself that one of them is the projection of a curve that winds around a torus. On the right, similar orbits seen under a stroboscope

$$\begin{pmatrix} x(0) \\ y(0) \end{pmatrix}, \begin{pmatrix} x(1) \\ y(1) \end{pmatrix}, \begin{pmatrix} x(2) \\ y(2) \end{pmatrix}, \ldots;$$

you might say that we illuminate the pendulum with a stroboscope, adjusted to be in phase with the force $\cos 2\pi t$ driving the pendulum, which has period 1.

This programme, applied to the unforced pendulum, gives the picture on the right of Fig. 11.3, and applied to the forced pendulum (with $\varepsilon = 0.2$) yields the image on the right of Fig. 11.4, much more comprehensible than that on the left.

On the right of Fig. 11.3, we see some points which "fill in" the trajectories in the picture on the left, but the "fillings-in" are different from one trajectory to the next because each has its own period, which may be rational or irrational. If ever we were to hit a trajectory with a rational period, we would see there only a finite number of points, but as this only occurs with probability zero it essentially never happens.

In Fig. 11.4, certain trajectories seem to want to be closed curves. Notice that if we work in $\mathbb{R}/\mathbb{Z} \times \mathbb{R}^2$, with the first coordinate corresponding to time (viewed with period 1), then the discrete trajectories filling in the simple curves in Fig. 11.4 (right) correspond to continuous trajectories filling in tori.

In the context of the forced pendulum, the KAM theorem gives the following result.

Theorem 3. *For every period $\alpha > 2\pi$, there exists C such that if $|\varepsilon| < C$, then the differential equation 11.4 has a periodic solution of period α. The set of these solutions is of positive measure, and the ratio of their measure to the area of the region A occupied by the periodic motions of unforced motion tends to 1 as $\varepsilon \to 0$.*

But there are trajectories other than those described by Theorem 3, which we can already glimpse on the right in Fig. 11.4, and which become predom-

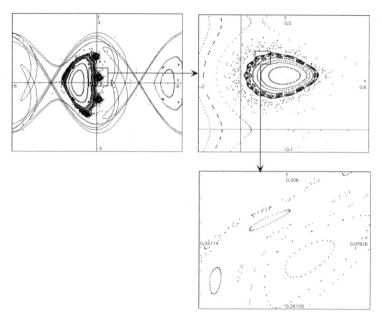

Fig. 11.5. This figure shows three successive blow-ups of the phase plane for the forced pendulum with $\varepsilon = .3$. We see regions of order, then regions of chaos, which, magnified, reveal regions of order and regions of chaos, and a second enlargement reveals the same structure repeated

inant as ε grows (see Fig. 11.5). There are good reasons for calling some of these trajectories "chaotic", but in the middle of these chaotic regions we observe zones of order, then in the zones of order further zones of chaos, which themselves contain zones of order, etc.

Figure 11.5 gives three successive blow-ups when $\varepsilon = 0.3$.

Part II. Precise statement and sketch of the proof

What distinguishes integrable systems from those obtained by perturbing them slightly, is that the former have a multitude of conservation laws, which disappear following the perturbation. For example, in the solar system, the energy and angular momentum of each planet are conserved: as soon as the planets are given non-zero mass, they can exchange energy and angular momentum, and all that remains is global conservation of energy and angular momentum. In the case of the unforced pendulum, the energy $y^2/2 - \cos x$ is conserved, but this is no longer true once $\varepsilon \neq 0$.

According to a "sociological" view of mathematics, a system, in general, should be able to do whatever is permitted by the laws governing it: the normal state of anarchy is chaos! From this point of view, we should expect that, in the absence of conservation laws, typical motions should be dense in the space available to them; Kolomogorov's theorem denies this, saying that *when the laws are relaxed a bit, the majority of motions stay "pretty much" where they were*, as if in fear of a non-existent police force.

Why did Kolmogorov think that his statement was true? Before him, I imagine that everyone believed the contrary: that as soon as conservation laws were lost, trajectories would go anywhere.

The only motivation I can see is that the solar system is indeed there, and that it would take almost a divine miracle for this to be true if chaos were really the generic state of mechanical systems. Moreover, Kolmogorov was not the first to ponder this: it appears explicitly in the work of Weierstrass, and implicitly in what I have read of Lagrange.

I have never seen any intuitive argument, however fuzzy, which makes plausible the statement of KAM. However, if one believes in it, then one can see that to say KAM is true is to say that a certain equation has a solution. One can then try to tackle this equation; this is what Kolmogorov and his successors did.

We shall see how to write the statement in the form of an equation, and describe what Kolmogorov did to solve it. For this project, we need to use a much more sophisticated vocabulary: we shall take for granted the notions of differentiable manifold, differential form, symplectic form, flow of a vector field, and Lie bracket.

11.3 Summary of Hamiltonian mechanics

Everything appearing in this section is treated, in much greater detail, in [Arn85].

Let (X, σ) be a symplectic manifold, in other words, X is a differentiable manifold and σ is a nowhere-vanishing 2-form on X such that $d\sigma = 0$. Then every function H on X has a symplectic gradient $\nabla_\sigma H$, which is the unique vector field such that

$$\sigma(\xi, \nabla_\sigma H) = dH(\xi) \qquad (11.5)$$

for every vector field ξ. We can then consider the Hamiltonian differential equation

$$\dot{x} = (\nabla_\sigma H)(x). \qquad (11.6)$$

5. The gradient and the symplectic gradient.

Very early on in courses on differential and integral calculus, one meets the gradient of a function $f : \mathbb{R}^n \to \mathbb{R}$: it is the vector field

$$\nabla f = \left(\frac{\partial f}{\partial x_1}, \ldots, \frac{\partial f}{\partial x_n} \right).$$

The gradient is often thought of as being the derivative Df of f, which measures to first order the change in f at a point $\mathbf{x}$ in the direction ξ:

$$f(\mathbf{x} + t\xi) = f(\mathbf{x}) + tDf(\mathbf{x})(\xi) + \text{terms smaller than } t.$$

But this is wrong. The derivative is a map from $\mathbb{R}^n$ into $\mathbb{R}$, the gradient is an element of $\mathbb{R}^n$, and like the symplectic gradient it is only defined with the help of a geometric structure on the ambient space, in this case, the scalar product $\langle \cdot, \cdot \rangle$: ∇f is the unique vector field such that

$$df(\xi) = \langle \xi, \nabla f \rangle$$

in close analogy with the formula (11.5). One can perfectly well consider the differential equation ("gradient equation")

$$\dot{\mathbf{x}} = \nabla f(\mathbf{x}), \tag{11.7}$$

which is indeed much used in optimization, since the solutions have a tendency to converge towards local maxima of f. The gradient formalism and that of the Hamiltonian gradient are thus almost indentical. But the gradient equation is infinitely less interesting than the Hamiltonian equation (11.6), because its solutions are non-recurrent. Indeed, because f increases along the solutions, they can never, as they evolve, return to their starting point. All the interest in dynamical systems lies in recurrence, which is prohibited by the differential equation (11.7), but which is allowed by, and, according to the Poincaré recurrence theorem, almost imposed by, Hamilton's equation.

If $X = \mathbb{R}^{2n}$, with coordinates $\mathbf{q} = (q_1, \ldots, q_n), \mathbf{p} = (p_1, \ldots, p_n)$ and 2-form $\sigma = \sum_i dp_i \wedge dq_i$, then the (11.6) can be re-written in form of the celebrated *equations of Hamiltonian mechanics*

$$
\begin{aligned}
\dot{q}_i &= \frac{\partial H}{\partial p_i} \\
\dot{p}_i &= -\frac{\partial H}{\partial q_i}
\end{aligned}
\tag{11.8}
$$

Example 1. Let us see how the solar system fits into this picture. We have met, in the case of a single body of zero mass, the equation $\mathbf{x}'' = -\mathbf{x}/|\mathbf{x}|^3$. We shall see (in the planar case, where $\mathbf{x} = \begin{pmatrix} q_1 \\ q_2 \end{pmatrix} \in \mathbb{R}^2$) that this is just Hamilton's equation for the manifold $X = \mathbb{R}^2 \times \mathbb{R}^2$ whose points we denote by

$$\left(\begin{pmatrix} q_1 \\ q_2 \end{pmatrix}, \begin{pmatrix} p_1 \\ p_2 \end{pmatrix} \right)$$

with the standard symplectic form $\sigma = dp_1 \wedge dq_1 + dp_2 \wedge dq_2$ and Hamiltonian

$$H\left(\begin{pmatrix} q_1 \\ q_2 \end{pmatrix}, \begin{pmatrix} p_1 \\ p_2 \end{pmatrix} \right) = \frac{1}{2}(p_1^2 + p_2^2) - \frac{1}{(q_1^2 + q_2^2)^{1/2}}.$$

Indeed, in this case we have

$$\nabla_\sigma H = \left(\begin{pmatrix} p_1 \\ p_2 \end{pmatrix}, -\frac{1}{(q_1^2 + q_2^2)^{3/2}} \begin{pmatrix} q_1 \\ q_2 \end{pmatrix} \right),$$

because

$$(dp_1 \wedge dq_1 + dp_2 \wedge dq_2) \left(\left(\begin{pmatrix} \xi_1 \\ \xi_2 \end{pmatrix}, \begin{pmatrix} \eta_1 \\ \eta_2 \end{pmatrix} \right), \left(\begin{pmatrix} p_1 \\ p_2 \end{pmatrix}, -\frac{1}{(q_1^2 + q_2^2)^{3/2}} \begin{pmatrix} q_1 \\ q_2 \end{pmatrix} \right) \right)$$

$$= \eta_1 p_1 + \frac{\xi_1 q_1}{(q_1^2 + q_2^2)^{3/2}} + \eta_2 p_2 + \frac{\xi_2 q_2}{(q_1^2 + q_2^2)^{3/2}}$$

$$= \left[DH\left(\begin{pmatrix} q_1 \\ q_2 \end{pmatrix}, \begin{pmatrix} p_1 \\ p_2 \end{pmatrix} \right) \right] \left(\begin{pmatrix} \xi_1 \\ \xi_2 \end{pmatrix}, \begin{pmatrix} \eta_1 \\ \eta_2 \end{pmatrix} \right).$$

The differential equation $\mathbf{x}' = \nabla_\sigma H(\mathbf{x})$ is thus

$$q_1' = p_1 \qquad p_1' = -\frac{q_1}{(q_1^2 + q_2^2)^{3/2}} \qquad (11.9)$$

$$q_2' = p_2 \qquad p_2' = -\frac{q_2}{(q_1^2 + q_2^2)^{3/2}} \qquad (11.10)$$

If we differentiate the equations from the first group and substitute the expressions from the second group, we recover the equation $\mathbf{x}'' = -\mathbf{x}/|\mathbf{x}|^3$ from which we started.

Example 2. Is the forced pendulum Hamiltonian? One might think not, first of all because energy is not conserved and the system is thus not "conservative" in the naive sense of the term, and secondly because there are 3 variables x, y, t, and a symplectic manifold is always of even dimension. But there is a trick which allows us to put this equation (and moreover any forced Hamiltonian system) into Hamiltonian form. Let us invent a new variable s (the dual variable of t), and consider $\mathbb{R}^4$ with coordinates t, s, x, y and symplectic form $ds \wedge dt + dy \wedge dx$. Let us take the Hamiltonian function

$$H\left(\begin{pmatrix} t \\ s \end{pmatrix}, \begin{pmatrix} x \\ y \end{pmatrix} \right) = s + \frac{1}{2}y^2 - \cos x + \varepsilon x \cos t.$$

The Hamiltonian equations then become

$$t' = \frac{\partial H}{\partial s} = 1 \qquad s' = -\frac{\partial H}{\partial t} = \varepsilon x \sin t \tag{11.11}$$

$$x' = \frac{\partial H}{\partial y} = y \qquad y' = -\frac{\partial H}{\partial x} = -\sin x + \varepsilon \cos t \tag{11.12}$$

As s does not appear on the right-hand side, the equations are just those of the forced pendulum.

Let us return to (11.6): $\dot{x} = (\nabla_\sigma H)(x)$. We shall denote by ϕ_f^t the flow of the vector field $\nabla_\sigma f$. It has two essential properties

- ϕ_f^t preserves f, in other words $f \circ \phi_f^t = f$;
- ϕ_f^t preserves σ, in other words $(\phi_f^t)^*\sigma = \sigma$.

The central construction in Kolmogorov's theorem is that of a certain symplectic diffeomorphism, which we shall construct as the flow of a Hamiltonian function.

We shall need the *Poisson bracket* in order to state the theorem and to understand the method of proof. The Lie bracket $[\nabla_\sigma f, \nabla_\sigma g]$ of the vector fields is well-defined, as for any vector fields, and symplectic because ϕ_f, ϕ_g are. One might wonder whether it is the symplectic gradient of a function: this turns out indeed to be the case.

We define the Poisson bracket $\{f, g\}$ of two functions on X by the equivalent formulas

$$\{f, g\} = \sigma(\nabla_\sigma g, \nabla_\sigma f) = df(\nabla_\sigma g) = -dg(\nabla_\sigma f). \tag{11.13}$$

This corresponds to the the Lie bracket:

$$\nabla_\sigma \{f, g\} = [\nabla_\sigma f, \nabla_\sigma g].$$

We shall say that the functions f, g commute if $\{f, g\} = 0$. This certainly implies that the flows ϕ_f, ϕ_g commute:

$$\phi_f(s) \circ \phi_g(t) = \phi_g(t) \circ \phi_f(s).$$

In the standard case of (11.8), the Poisson bracket is calculated via the formula

$$\{f, g\} = \sum_{i=1}^{n} \left(\frac{\partial f}{\partial q_i} \frac{\partial g}{\partial p_i} - \frac{\partial f}{\partial p_i} \frac{\partial g}{\partial q_i} \right). \tag{11.14}$$

Let $\mathbb{T} = \mathbb{R}/\mathbb{Z}$. A *totally integrable system* will be the symplectic manifold $X = \mathbb{T}^n \times \mathbb{R}^n$, with variables $(\mathbf{q} \in \mathbb{T}^n, \mathbf{p} \in \mathbb{R}^n)$, symplectic form $\sum_i dp_i \wedge dq_i$ as above, and with Hamiltonian function $H(\mathbf{p})$ depending only on $\mathbf{p}$.

In this case, it is very easy to integrate the (11.8): the solution with initial value $(\mathbf{q}_0, \mathbf{p}_0)$ is simply

$$\mathbf{q}(t) = \mathbf{q}_0 + t\frac{\partial H}{\partial \mathbf{p}}(\mathbf{p}_0) = \mathbf{q}_0 + t\,\omega(\mathbf{p}_0),$$

$$\mathbf{p}(t) = \mathbf{p}_0$$

In particular, each coordinate $p_1, \ldots, p_n$ is conserved, and the trajectory is a linear motion on the torus $\mathbb{T}^n \times \{\mathbf{p}_0\}$.

A celebrated theorem of Liouville (the same Liouville as the one of transcendental numbers, see Box 4) states that this situation occurs whenever one has any mechanical system with n degrees of freedom[4] and n commuting conservation laws. More precisely, let us suppose that X is a symplectic manifold of dimension $2n$, and that $f_1, \ldots, f_n$ are n functions on X. Then, if

1. the functions $f_1, \ldots, f_n$ commute; set $F = (f_1, \ldots, f_n)$;
2. $F^{-1}(0)$ has a compact component Y, and
3. the function $F = (f_1, \ldots, f_n) : X \to \mathbb{R}^n$ is a submersion on Y, in other words $DF(\mathbf{y})$ is surjective for all $\mathbf{y} \in Y$,

then $F^{-1}(0)$ is a torus, and there exists a neighbourhood U of Y in X and coordinate functions

$$\mathbf{q} : U \to \mathbb{T}^n \qquad \mathbf{p} : U \to \mathbb{R}^n$$

with respect to which σ can be written as $\sum dp_i \wedge dq_i$, and f_i as p_i.

---

6. Liouville and the solar system.

In Example 1, we saw that the planar solar system fits into the framework of Hamiltonian mechanics; we now want to show that Liouville's theorem applies in the case of zero masses. Once again, it suffices to consider the case of one planet, because the planets clearly behave independently of one another. It is a system with two degrees of freedom, so we need two conservation laws. We already have one: the function

$$H\left(\begin{pmatrix} q_1 \\ q_2 \end{pmatrix}, \begin{pmatrix} p_1 \\ p_2 \end{pmatrix}\right) = \frac{1}{2}(p_1^2 + p_2^2) - \frac{1}{(q_1^2 + q_2^2)^{1/2}}.$$

For the second, we shall take the function

$$M = q_1 p_2 - q_1 p_2.$$

---

[4] We are talking here about n mechanical degrees of freedom, in other words the dimension of the set of positions of the system: for example the degree of freedom corresponding to one coordinate of position corresponds to the two dimensions *position* and *speed* in phase space, and n degrees of freedom correspond to a phase space of dimension $2n$

One easily checks that

$$M' = q_1'p_2 + q_1 p_2' - q_2' p_1 - q_2 p_1' = p_1 p_2 + q_1 \frac{q_2}{(q_1^2 + q_2^2)^{3/2}} - p_2 p_1 - q_2 \frac{q_1}{(q_1^2 + q_2^2)^{3/2}} = 0;$$

so it is indeed a conserved quantity. It remains to see that $\{H, M\} = 0$. The Poisson bracket is given by the formula (11.14):

$$\{H, M\} = \frac{\partial H}{\partial q_1} \frac{\partial M}{\partial p_1} - \frac{\partial H}{\partial p_1} \frac{\partial M}{\partial q_1} + \frac{\partial H}{\partial q_2} \frac{\partial M}{\partial p_2} - \frac{\partial H}{\partial p_2} \frac{\partial M}{\partial q_2}$$

$$= \frac{q_2}{(q_1^2 + q_2^2)^{3/2}} (-q_1) - (p_1)(p_2) + \frac{q_1}{(q_1^2 + q_2^2)^{3/2}} (q_2) - (p_2)(-p_1) = 0.$$

Liouville's theorem thus applies to the solar system with planets of zero mass.

11.4 A precise statement of Kolmogorov's theorem

As the KAM theorem mentions a "sufficiently small perturbation", we need a means of measuring the size of perturbations; in practice, this means choosing a function space, or rather, choosing a norm $\|f\|$ for our functions.

Kolmogorov saw that the proof of KAM is much simpler if one works with real-analytic functions, and if one uses as norm their upper bound on compact neighbourhoods of the origin in $\mathbb{C}^n$ and of $\mathbb{R}/\mathbb{Z}$ in $(\mathbb{C}/\mathbb{Z})^n$. For one thing, this yields a multitude of different norms, corresponding to different neighborhoods, and when one solves a problem with data bounded on a certain compact set, even if one is unable to bound the solutions on that set, one can sometimes bound them on smaller compact sets.

There is another reason: looking at Box 7 on Newton's method, one sees a constant M which essentially measures a least upper bound (a sup) of second derivatives. The evaluation of M is often difficult, and it is almost always the main obstacle to applications of the theorem 5. But for analytic functions, things are simpler: if an analytic function f is bounded on an open set U and if V is relatively compact in U, then one can bound the second derivatives of f on V in terms of $\sup_U |f|$, thanks to Cauchy's inequalities.

The important domains are

$$B_\rho = \{\mathbf{p} \in \mathbb{C}^n \mid |\mathbf{p}| \leq \rho\}$$
$$C_\rho = \{\mathbf{q} \in \mathbb{C}^n/\mathbb{Z}^n \mid |\operatorname{Im}(\mathbf{q})| \leq \rho\}$$
$$A_\rho = C_\rho \times B_\rho = \{(\mathbf{q}, \mathbf{p}) \in \mathbb{C}^n/\mathbb{Z}^n \times \mathbb{C}^n \mid |\mathbf{p}| \leq \rho, \ |\operatorname{Im}(\mathbf{q})| \leq \rho\}.$$

We shall denote by $\mathcal{B}_\rho, \mathcal{C}_\rho, \mathcal{A}_\rho$ the Banach algebras of functions continuous on these compact sets and analytic on the interiors, with sup-norm $\|f\|_\rho$. The elements of $\mathcal{B}_\rho$ can be expanded as power series, and those of $\mathcal{C}_\rho$ can be expanded as Fourier series

$$f(\mathbf{z}) = \sum_{\mathbf{k} \in \mathbb{Z}^n} f_{\mathbf{k}} e^{2\pi i \, \mathbf{k} \cdot \mathbf{z}}.$$

We are finally in a position to give a rigorous statement of Kolmogorov's theorem.

Theorem 4. *Let $\rho, \gamma > 0$ be two numbers, and let $h(\mathbf{q}, \mathbf{p}) = h_0(\mathbf{p}) + h_1(\mathbf{q}, \mathbf{p})$ be a Hamiltonian with $h_0 \in \mathcal{B}_\rho$, $h_1 \in \mathcal{A}_\rho$ and $\|h_0\|_\rho \leq 1$. Let us write*

$$h_0(\mathbf{p}) = a + \omega \cdot \mathbf{p} + \frac{1}{2}\mathbf{p} \cdot C\mathbf{p} + o(|\mathbf{p}|^2),$$

the expansion of order 2 of $\mathbf{h}_0$, with $\omega \in \Omega_\gamma$ and C symmetric and invertible. Then for every $\rho_* < \rho$, there exists $\varepsilon > 0$, depending on C and γ but not on the remainder term $o(|\mathbf{p}|^2)$, such that when $\|h_1\|_\rho < \varepsilon$, there exists a symplectic map $\Phi : A_{\rho_*} \to A_\rho$ such that, setting $(\mathbf{q}, \mathbf{p}) = \Phi(\mathbf{Q}, \mathbf{P})$ and $H = h \circ \Phi$, we have

$$H(\mathbf{Q}, \mathbf{P}) = A + \omega \cdot \mathbf{P} + R(\mathbf{Q}, \mathbf{P}) \tag{11.15}$$

with $R(\mathbf{Q}, \mathbf{P}) \in O(|\mathbf{P}|^2)$.[5]

In particular, the torus $\mathbf{P} = 0$ is invariant under the flow $\nabla_\sigma H$, and on this torus the flow ϕ_H is linear with direction ω.

11.5 Strategy of the proof

Equation (11.15) is an equation for a diffeomorphism Φ, which we have to solve. Moreover, the solution should be symplectic, which adds the equation $\Phi^* \sigma = \sigma$. As is so often the case when solving a non-linear equation, this is done using Newton's method: we obtain Φ as a limit of symplectic diffeomorphisms Φ_i, each one calculated from the preceding one as the solution of a linearized equation.

---

7. Newton's method and Kantorovitch's theorem.

Solving systems of non-linear equations is nearly always a difficult problem, especially when the number of unknowns becomes large, or even infinite as in Theorem 5 below. There are not many theoretical tools available, and even fewer practical tools. In certain problems there are particular approaches that arise naturally, but in general the first idea that comes to mind is to apply Newton's method, or one of its numerous variants; Kantorovitch's theorem is just about the only result

---

[5] $O(|\mathbf{P}|^2)$ denotes the class of functions of the form $|\mathbf{P}|^2 O(1)$ where $O(1)$ is a function bounded in a neighbourhood of $\mathbf{P} = 0$. (Likewise, in what follows $o(|\mathbf{p}|)$ will denote the class of functions of the form $|\mathbf{p}|o(1)$ where $o(1)$ will be a function tending to 0 as $\mathbf{p} \to \mathbf{0}$.) As far as possible, we prefer to write $R(\mathbf{Q}, \mathbf{P}) \in O(|\mathbf{P}|^2)$ rather than $R(\mathbf{Q}, \mathbf{P}) = O(|\mathbf{P}|^2)$ (except perhaps in some finite-order expansions)

which guarantees us convergence. Newton's algorithm for solving an equation $f(\mathbf{x}) = 0$ consists of choosing a point $\mathbf{x}_0$, and then defining

$$\mathbf{x}_{i+1} = \mathbf{x}_i - [Df(\mathbf{x}_i)]^{-1} f(\mathbf{x}_i).$$

Of course, this makes sense only if $[Df(\mathbf{x}_i)]$ is invertible; in finite dimensions, this requires that the number of unknowns be the same as the number of equations. Notice that the calculation of $\mathbf{x}_{i+1}$ starting from $\mathbf{x}_i$ is the solution of a *linear* equation; from the practical point of view, a computer trying to solve an equation by Newton's method spends most of its time solving linear equations.

Theorem 5 (Kantorovitch). *Let E and F be Banach spaces, $U \subset E$ be an open set and $f : U \to F$ a map of class C^1. Suppose that $\mathbf{x}_0 \in U$ is a point where $[Df(\mathbf{x}_0)] : E \to F$ is an isomorphism. Put $h_0 = -[Df(\mathbf{x}_0)]^{-1}(f(\mathbf{x}_0))$, $\mathbf{x}_1 = \mathbf{x}_0 + h_0$, and define the ball $U_0 = B_{\|h_0\|}(\mathbf{x}_1)$. Then if*

1. $U_0 \subset U$
2. $\|[Df(\mathbf{y}_1)] - [Df(\mathbf{y}_2)]\| \leq M$ *for all* $\mathbf{y}_1, \mathbf{y}_2 \in U_0$, *and*
3. $\|f(\mathbf{x})\| \ \|[Df(\mathbf{x}_0)]^{-1}\|^2 M \leq \frac{1}{2}$,

then the equation $f(\mathbf{x}) = 0$ has a unique solution in U_0, and Newton's method starting from $\mathbf{x}_0$ is defined for all i and converges to this solution. If moreover we have $\|f(\mathbf{x})\| \ \|[Df(\mathbf{x}_0)]^{-1}\|^2 M = k < \frac{1}{2}$, then the method is "superconvergent": if we put

$$C = \frac{1-k}{2(1-2k)} \|[Df(\mathbf{x}_0)]^{-1}\| M, \quad \mathbf{x}_{i+1} = \mathbf{x}_i + h_i \quad \text{and} \quad h_{i+1} = -[Df(\mathbf{x}_i)]^{-1} f(\mathbf{x}_i),$$

then $\|h_{i+1}\| \leq C\|h_i\|^2$. "Superconvergence" corresponds to doubling the number of correct digits at each iteration, which is quite different from geometric convergence, which, though rapid, only adds a fixed number of significant figures at each iteration.

For a proof, see [HH02].

A diffeomorphism is a complicated object and difficult to control. In practice, we shall write

$$\Phi_i = \phi_i \circ \phi_{i-1} \circ \cdots \circ \phi_1, \tag{11.16}$$

where each ϕ_i is the Hamiltonian flow ϕ_{g_i} at time 1 for a certain "Hamiltonian" function g_i, which will be the unknown in our linearized problem. This has two important advantages:

- the unknown g_i is a (numerical) function, and functions are much easier to handle than diffeomorphisms;
- the map ϕ_{g_i} is automatically a diffeomorphism, and it is automatically symplectic.

As in any construction based on successive approximations, the proof is by recurrence: at the i-th step we have constructed a Hamiltonian $\tilde{h} = \Phi_i^* h$, which we expand up to order 2 in $\mathbf{p}$, whose coefficients will be Fourier series in $\mathbf{q}$. More precisely, we shall write $\tilde{h} = \tilde{h}_0 + \tilde{h}_1$, where

- $\tilde{h}_1$ contains the terms that are constant or linear in $\mathbf{p}$, except for the term constant in $\mathbf{q}$, and
- $\tilde{h}_0$ is all the rest.

We would like to eliminate the term $\tilde{h}_1$, but we shall not succeed in doing so in one go, at least not by solving a linear equation. Instead, we shall suppose that $\tilde{h}_1$ is "of order ε_i", a notion which needs to be defined. We shall solve a linear equation for a function g such that $\phi_g^* \tilde{h}$ is "better" than $\tilde{h}$. We would like the problem term $(\phi_g^* \tilde{h})_1$ to be of order $\varepsilon_{i+1} \sim \varepsilon_i^2$; this would be the quadratic convergence of Newton's method. We shall not succeed: the "Newton method" used is not quite the standard one, but we can still manage to make ε_{i+1} small enough to guarantee convergence.

We expand $\phi_g^* \tilde{h}$ to first order in g:

$$\phi_g^* \tilde{h} = \tilde{h} + \{g, \tilde{h}\} + o(|g|) = \tilde{h}_0 + \tilde{h}_1 + \{g, \tilde{h}_0\} + \{g, \tilde{h}_1\} + o(|g|).$$

Our hope is to eliminate the terms which are not $O(|\mathbf{p}|)^2$, except the term constant in $\mathbf{q}$. In order to apply the standard Newton method, we would need to solve the equation

$$\tilde{h}_1 + \{g, \tilde{h}_0\} + \{g, \tilde{h}_1\} \in o(|\mathbf{p}|).$$

We shall attack this in a slightly different way: let us suppose that $\{g, \tilde{h}_1\}$ is of order 2, since g and $\tilde{h}_1$ are both small. We can declare anything we like to be "small"; the problem is to know whether the inequalities obtained at the end justify our choice. Thus the linear equation to be solved is

$$\tilde{h}_1 + \{g, \tilde{h}_0\} \in o(|\mathbf{p}|). \tag{11.17}$$

Equation (11.17) is a system of "Diophantine partial differential equations".

11.5.1 Diophantine partial differential equations

Let $g \in \mathcal{C}_\rho$, namely a function of $\mathbf{q} \in \mathbb{C}^n / \mathbb{Z}^n$. The linear equation to be solved for our "Newton method" is the equation

$$Df(\omega) = \sum_{i=1}^{n} \omega_i \frac{\partial f}{\partial q_i} = g, \tag{11.18}$$

where the unknown f is an element of $\mathcal{C}_{\rho'}$ for a certain $\rho' < \rho$.

This equation can be solved with the help of Fourier series if we set

$$f(\mathbf{q}) = \sum_{\mathbf{k} \in \mathbb{Z}^n} f_{\mathbf{k}} e^{2\pi i \, \mathbf{k} \cdot \mathbf{q}}, \quad g(\mathbf{q}) = \sum_{\mathbf{k} \in \mathbb{Z}^n} g_{\mathbf{k}} e^{2\pi i \, \mathbf{k} \cdot \mathbf{q}},$$

then the solution of the problem is

$$f_{\mathbf{k}} = \frac{1}{2\pi i \, (\mathbf{k} \cdot w)} g_{\mathbf{k}}. \tag{11.19}$$

One sees immediately that for the problem to have a solution, it is necessary that $g_0 = 0$, and then f_0 is arbitrary; except for this choice, the series giving f is unique. The two properties are essential.

The formula (11.19) shows that the convergence properties of f depend on the Diophantine properties of w. If there exists an integer vector $\mathbf{k} \neq \mathbf{0}$ such that $\mathbf{k} \cdot w = 0$, then the Fourier series for f simply does not exist, because we have divided by zero to find the coefficients. The $\mathbf{k} \cdot w$ are the fearsome *small divisors*, and when they are too small, they prevent the series for f from converging. Our Diophantine condition is sufficient to guarantee that such horrors do not occur, and even that the series for f also converges in the interior of the same domain C_ρ on which g is defined. But this does not give that f is bounded on C_ρ, so f does not belong to C_ρ. We shall have to be satisfied with $\|f\|_{\rho'}$ for $\rho' < \rho$.

We have to solve an equation of this type at each step of Newton's method, and must therefore take an infinite number of radii $\rho = \rho_0 > \rho_1 > \rho_2 \dots$. The ρ_i need to be chosen with care, large enough that the domain of definition of the limit is non-empty, but small enough to guarantee convergence. The tool we use for this is the following statement.

Proposition 1. *If $g \in C_\rho$ and $w \in \Omega_\gamma$, then for δ satisfying $0 < \delta < \rho$ we have*

$$\|f\|_{\rho-\delta} \leq \frac{\kappa_n}{\gamma \delta^{2n}} \|g\|_\rho \quad and \quad \|Df\|_{\rho-\delta} \leq \frac{\kappa_n}{\gamma \delta^{2n+1}} \|g\|_\rho,$$

where κ_n is a constant depending only on n.

11.5.2 The essential construction of the proof

Let us return to the (11.17):

$$h_1 + \{g, h_0\} \in O(|\mathbf{p}|^2),$$

where we have eliminated the tildes and replaced $o(|\mathbf{p}|)$ by $O(|\mathbf{p}|^2)$, which is apparently stronger, but is in fact equivalent because we are working with analytic functions. The unknown in this equation is g, which we may take to be of degree 1 in $\mathbf{p}$:

$$g = \lambda \, \mathbf{q} + X(\mathbf{q}) + \sum_{i=1}^{n} Y_i(\mathbf{q})p_i; \tag{11.20}$$

indeed, only the linear terms of g can contribute to the linear terms of $\{g, h_0\}$. The new unknowns are thus λ, X and Y_i; let us note that X and Y_i are functions of $\mathbf{q}$ only, and can therefore be expanded as Fourier series.

We write

$$h_0(\mathbf{q}, \mathbf{p}) = a + \omega \, \mathbf{p} + \frac{1}{2}\mathbf{p} \cdot C(\mathbf{q})\mathbf{p} + R(\mathbf{q}, \mathbf{p})$$

$$\text{with } a \in \mathbb{R}, \ \omega \in \Omega_\gamma, \text{ and } R(\mathbf{q}, \mathbf{p}) \in O(|\mathbf{p}|^3) \qquad (11.21)$$

$$h_1(\mathbf{q}, \mathbf{p}) = A(\mathbf{q}) + B(\mathbf{q}) \, \mathbf{p} \text{ with } \overline{A} = 0,$$

where $\overline{A}$ is the average of A on the torus $\mathbf{p} = 0$, which is already written in a.
If we substitute the expression (11.20) for g into $h_1 + \{g, h_0\}$, we find

$$(h_1 + \{g, h_0\})(\mathbf{q}, \mathbf{p}) = \omega \cdot \lambda + A(\mathbf{q}) + DX(\mathbf{q})(\omega)$$
$$+ \Big(B(\mathbf{q}) + \big(\lambda + DX(\mathbf{q}) \big) C(\mathbf{q}) + \omega DY(\mathbf{q}) \Big) \cdot \mathbf{p} + O(|\mathbf{p}|^2).$$

It therefore comes down to solving the equations

$$DX(\mathbf{q})(\omega) = -A(\mathbf{q}) \qquad (11.22)$$
$$DY(\mathbf{q})(\omega) = -B(\mathbf{q}) - \big(\lambda + DX(\mathbf{q}) \big) C(\mathbf{q}). \qquad (11.23)$$
$$(11.24)$$

for X and $Y_i, i = 1, \ldots n$.

These are equations of the form (11.18), which we know how to solve. The hypothesis $\overline{A} = 0$ ensures that we can solve (11.22) and find $X \in \mathcal{C}_{\rho'}$ for all $\rho' < \rho$. Once this has been found, we substitute it into the (11.23). The vector λ is then determined by the condition that the averages of the right-hand sides in the n (11.23) are all zero. With this λ, we can solve the (11.23) and find $Y_i \in \mathcal{C}_{\rho''}$ for all $\rho'' < \rho'$.

For us to be able to repeat the operation, ρ' and ρ'' need to be chosen with care. We refer to [HI02] for the details.

References

[AA68]   Arnold, V.I., Avez, A.: Ergodic problems of classical mechanics. Benjamin, NY (1968)

[Arn63]   Arnold, V.I.: Proof of a theorem of A.N. Kolmogorov on the preservation of conditionally periodic motions under a small perturbation of the Hamiltonian. Uspekhi Mat. Nauk, **18**, 13-40 (1963)

[Arn85]   Arnold, V.I.: Mathematical methods of classical mechanics. Springer Verlag (1985)

[BGGS84]   Benettin, G., Galgani, L., Giorgilli, A., Strelcyn, J.-M.: A proof of Kolmogorov's theorem on invariant tori using canonical transformations defined by the Lie method. Il Nuovo Cimento, **79**B (1984)

[FH83]   Fathi, A., Herman, M.: Sur les courbes invariantes par les difféomorphismes de l'anneau. Astérisque, **103-104** (1983)

[Her86]   Herman, M.: Sur les courbes invariantes par les difféomorphismes de l'anneau. Astérisque, **144** (1986)

[HH02]   Hubbard, B., Hubbard, J.: Vector Calculus, Linear Algebra and Differential forms, 2nd ed. Prentice Hall, Upper Saddle River, New Jersey 07458 (2002)

[HI02]   Hubbard, J., Ilyashenko, Y.: A proof of Kolmogorov's theorem. Discrete and Continuous Dynamical systems, **4**, 1-20 (2003)

[HW91]   Hubbard, J., West, B.: Differential Equations: a dynamical systems approach. Springer, Berlin, Heidelberg, New York (1991)

[Kol57]   Kolmogorov, A.N.: General theory of dynamical systems and classical mechanics. In: Proceedings of the International Congress of Mathematicians, Amsterdam, 1954. Erven P. Noordhoff N.V., Groningen (1957)

[Mos62]   Moser, J.: On the invariant curves of area-preserving mappings of an annulus. Nachr. Akad. Wiss. Gttingen Math-Phys. Kl. II, 1-20 (1962)

[Mos73]   Moser, J.: Stable and random motions in dynamical systems. Princeton Univ. Press, Princeton, NJ (1973)

[Ste71]   Sternberg, S.: Celestial mechanics, parts I and II. Benjamin, NY (1971)

From Kolmogorov's work on entropy
of dynamical systems to non-uniformly
hyperbolic dynamics

Denis V. Kosygin[1] and Yakov G. Sinai[2]

[1] Courant Institute of Mathematical Sciences, New York University, USA
   kosygin@cims.nyu.edu
[2] Department of Mathematics, Princeton University, USA
   sinai@Math.Princeton.EDU

12.1 General dynamical systems

Assume that $(X, \mathcal{T}, \mu)$ is a probability space. By a dynamical system we understand in this paper a group or a semi-group of measure-preserving transformations $T^t : X \to X$ with the invariant measure μ, such that the map $(x, t) \to T^t x$ is measurable. If $t \in \mathbb{N}$ or $\mathbb{Z}$, one deals with a *discrete time dynamical system*. The generator corresponding to $t = 1$ is called an automorphism if T is invertible and an endomorphism if it is not. If $t \in \mathbb{R}$ or $\mathbb{R}^+$, then $\{T^t\}$ is a *continuous time dynamical system* which is called a flow or a semi-flow, respectively.

One of the most popular examples of a dynamical system is a Hamiltonian flow when X is a symplectic manifold and μ is the Liouville measure.

Two dynamical systems $(X, \mathcal{T}, \mu, (T^t))$ and $(Y, \mathcal{S}, \nu, (S^t))$ are (metrically) *isomorphic* if there exist zero measure subsets $N_1 \subset X$, $N_2 \subset Y$ and an isomorphism ϕ of measure spaces $\phi \colon X \setminus N_1 \to Y \setminus N_2$ such that $\forall t$, $\phi \circ T^t \circ \phi^{-1} = S^t$.

12.1.1 Bernoulli shifts

An important example of a discrete time dynamical system is provided by a *Bernoulli shift*. Consider an infinite sequence of independent Bernoulli trials with outcomes 0 and 1, and with probabilities of outcomes $P(0) = 1 - p$ and $P(1) = p$. The probability space X is the space of binary sequences $\{0, 1\}^{\mathbb{Z}}$ with the product measure $\mu = P^{\otimes \mathbb{Z}}$. The transformation $S \colon X \to X$ which translates a sequence $(e_n) \in X$ by one unit to the left $S(e_n) = (e'_n)$, where $e'_n = e_{n+1}$, is called the *left shift*, or simply the *shift* in X. The triplet $(X, \mu, \{S^n\}_{n \in \mathbb{Z}})$ is called the *Bernoulli shift*. More general Bernoulli shifts are obtained when one considers Bernoulli trials with outcomes in an arbitrary probability space.

Two Bernoulli shifts with spaces of outcomes of different cardinalities may nevertheless be isomorphic. (See a famous example due to Meshalkin in [CFS82], Chap. 8, n° 1, p. 181.)

12.1.2 Spectral properties of dynamical systems

With every dynamical system $\{T^t\}$ one can relate the *adjoint* group or semi-group $\{U^t\}$ which acts on the space of measurable functions: $U^t f(x) = f(T^t x)$. Elements of the semigroup $\{U^t\}$ are unitary operators of the space $L^2(X, \mu)$. Adjoint semigroups of isomorphic dynamical systems are conjugate and there-fore have the same spectral properties. Until 1958, basic invariants of dynam-ical systems came from the spectral theory.

12.2 Kolmogorov's paper on entropy

Kolmogorov's paper [Kol58] which introduced the notion of entropy of a dynamical system appeared in 1958. There are all reasons to believe that it can be considered as a starting point of the important theory which is now called deterministic chaos. Before [Kol58] the main method of study of dynamical systems was the spectral one. This was the influence of the paper by von Neumann [vonN32] where he gave a complete metrical clas-sification of dynamical systems with pure point spectrum[3]. Quite soon after [vonN32] the problem of isomorphism of two different Bernoulli shifts emerged and became very popular. Several leading Soviet mathematicians including Pontryagin tried without any success to attack it. On the other hand the spectral theory of dynamical systems had some success. Kolmogorov himself understood quite well spectral properties of dynamical systems generated by Gaussian stationary processes. This case provides many examples of systems with singular spectra. The complete theory was developed in the works of Fomin [Fom50], Girsanov [Gir58], and Maruyama [Mar49]. At the same time new examples of systems with countable (multiplicity) Lebesgue spectrum[4] started to appear. It was a sensational discovery by Gelfand and Fomin that

---

[3] A dynamical system is said to have a *pure point spectrum* if the eigenvectors of operators U^t (adjoint group) span a dense linear subspace of $L^2(X, \mathcal{T}, \mu)$. Recall, more generally, that the *spectrum* of an operator A is the set of values λ such that $A - \lambda\mathrm{Id}$ is not *invertible*, and that its *point spectrum* (the set of its eigenvalues) is the set of values λ such that $A - \lambda\mathrm{Id}$ is not *injective*. The part of the spectrum, which is not pure point is said *continuous spectrum*, and it is decomposed (*via* the Lebesgue decomposition theorem) into *absolutely continuous spectrum* and *singular spectrum*

[4] That means essentially that the associated unitary operator U^t is a shift of count-able multiplicity. See the precise definition in Appendix 2 of [CFS82]

the geodesic flows on compact manifolds of constant negative curvature have countable Lebesgue spectrum. Their proof [GF52] used methods from the representation theory. However there was no understanding why this type of spectrum was so abundant.

In spite of being a probabilist Kolmogorov always stressed the importance of study of mixing[5] in classical dynamical systems, i.e. systems generated by ODEs and PDEs (as opposed to probabilistic systems, generated by stochastic processes). This can be already seen in his famous talk at the Amsterdam Mathematical Congress [Kol54a]. Probably the main motivation came from the turbulence theory where Kolmogorov already had classical results (see [Kol41a], [Kol41b], [Kol42]). But at that time it was not clear at all what are the mechanisms of mixing in classical dynamical systems and whether they are the same as in the probability theory. In [Kol58] Kolmogorov was motivated by his intention to provide a general metrical invariant of dynamical system which would allow to distinguish these two types of systems. For this purpose he proposed the concept of entropy of a dynamical system.

Today the common definition of the entropy of a dynamical system $(X, \mathcal{T}, \mu, (T^n))$ is done in three steps. If $\tau = \{A_1, \ldots, A_n\}$ is a finite measurable partition of X, its entropy is $H(\tau) = -\sum_{j=1}^{n} \mu(A_j) \ln \mu(A_j)$ (of course $x \ln x = 0$ if $x = 0$). If $\tau' = \{A'_1, \ldots, A'_m\}$ is another finite measurable partition of X, let us denote by $\tau \vee \tau'$ the partition generated by τ and τ' (i.e. the partition defined by the intersections $A_i \cap A'_j$). The entropy of the couple (T, τ) is the limit

$$h(T, \tau) = \lim_{N \to \infty} \frac{1}{N} H(\tau \vee T\tau \vee \cdots \vee T^{N-1}\tau).$$

Finally, the entropy of the transformation T is $h(T) = \sup_\tau h(T, \tau)$, with the supremum taken over the set of finite measurable partitions of X. It is obvious that the entropy is a metric invariant of dynamical systems.

Initially it was believed that for probabilistic dynamical systems the entropy can be positive while for classical dynamical systems it should be zero. It is not hard to see that the entropy of the Bernoulli shift with the outcomes in $E = \{a_1, \ldots, a_n\}$ and with the probability law $P(a_i) = p_i$ is $h = -\sum_{i=1}^{n} p_i \ln p_i$. The invariant proposed by Kolmogorov showed that two Bernoulli shifts with different values of entropy are non-isomorphic which was a great success in the theory of isomorphism of dynamical systems. The complete solution of the isomorphism problem of Bernoulli shifts was given by Ornstein in [Orn70] (see also [Orn74]): two Bernoulli shifts are isomorphic if and only if they have the same entropy.

After the paper [Kol58] appeared Rokhlin proposed to compute the entropy of a linear automorphism of the torus. According to the explained above point of view one of us (Ya. G. S.) tried to prove that it is zero. When the

---

[5] A dynamical system is mixing if for each pair of measurable subsets A, B, $\mu(T^{-t}A \cap B) \to \mu(A)\mu(B)$ as $t \to \infty$

corresponding drawings were shown to Kolmogorov he immediately realized
that the entropy should be positive. After that it was not so difficult to get the
final result (see [Sin59]). Now the expression for entropy follows from a much
more general statement (see [Sin66]), but at that time any new example of a
classical dynamical system with positive entropy appeared as a big surprise.

12.3 The notion of hyperbolicity

All the examples suggested that the positivity of entropy of a classical dynamical system was connected with the existence of invariant foliations which
initially were called expanding and contracting (see [Sin66]). Later, following
Smale, people started to call them unstable and stable foliations. Anosov in
[Ano67] introduced a class of dynamical systems which now bear his name.
He proved that Anosov systems are topologically stable and are ergodic[6] if
they have an absolutely continuous invariant measure. The geodesic flows on
compact manifolds of negative curvature are the main examples of Anosov
flows. Anosov's results provided a far going generalization of Hopf's results:
indeed, in [Hop39], [Hop40] Hopf had proved the ergodicity of the geodesic
flows on n-dimensional manifolds of constant negative curvature with $n > 2$.
For $n = 2$ Hopf proved ergodicity for the case of variable negative curvature.

During the same years Smale worked on various problems in the topological
theory of dynamical systems (see [Sma67]). In particular, he considered problems of structural stability and topological classification of multi-dimensional
dynamical systems.

One of the outcomes of Anosov's and Smale's works was the notion of
hyperbolicity of a dynamical system and a general statement that hyperbolic
dynamical systems have stable and unstable foliations. If they preserve an absolutely continuous invariant measure then they are ergodic, have positive entropy and, under some additional assumptions, countable Lebesgue spectrum.
The corresponding theory covered the majority of examples of smooth (i.e.
sufficiently regular) dynamical systems with positive entropy and explained
the appearance of this type of spectrum.

We shall give the definition of what is called now a uniformly hyperbolic
(or Anosov) diffeomorphism.

Let M be a compact smooth manifold. The class of smoothness plays
no role. By this reason we assume that M is C^∞ and consider a C^∞-
diffeomorphism T of M.

Definition 1. T *is called a uniformly hyperbolic (or Anosov) map if the*
tangent bundle $\mathcal{T}(M)$ *can be decomposed onto two sub-bundles* $\mathcal{T}^{(s)}(M)$,
$\mathcal{T}^{(u)}(M)$, *i.e.* $\mathcal{T}(M) = \mathcal{T}^{(s)}(M) \oplus \mathcal{T}^{(u)}(M)$ *so that both* $\mathcal{T}^{(s)}(M)$, $\mathcal{T}^{(u)}(M)$

---

[6] A dynamical system is ergodic if each of its measurable invariant subsets differs
from the empty set or the entire space X by a set of zero measure

*are invariant under the tangent map dT and for some constants $C < \infty$,
$\lambda < 1$ and all $n > 0$*

$$\|dT^n e\| \leq C\lambda^n \|e\|, \qquad e \in T^{(s)}(M), \tag{12.1}$$

$$\|dT^{-n} e\| \leq C\lambda^n \|e\|, \qquad e \in T^{(u)}(M). \tag{12.2}$$

It follows from the invariance of $T^{(s)}(M)$, $T^{(u)}(M)$ that $dT^n e \in T^{(s)}_{T^n x}(M)$ if $e \in T^{(s)}_x(M)$, and $dT^{-n} \in T^{(u)}_{T^{-n}x}(M)$ if $e \in T^{(u)}_x(M)$. It is also important that at each point $x \in M$ the subspaces $T^{(s)}_x(M)$, $T^{(u)}_x(M)$ are transversal to each other. Conditions (12.1) and (12.2) are called conditions of uniform hyperbolicity. The term *uniform* reflects the fact that the numbers C and λ, as well as the estimate from below of the distance between $T^{(s)}_x(M)$ and $T^{(u)}_x(M)$, do not depend on x. Similar conditions can hold along individual trajectories which in this case are called hyperbolic (see [Pes77]). Analogous definitions can be given for continuous time.

The number of examples of uniformly hyperbolic or Anosov systems is not so big. According to hypothesis by Smale each uniformly hyperbolic map is topologically isomorphic to an automorphism of a nil-manifold[7] or its finite covering. In the case of continuous time the main examples of Anosov systems are the geodesic flows on manifolds of negative curvature.

12.4 Lorenz system

Frequently, in applications, some of the conditions of hyperbolicity are violated. If we take billiard systems inside domains with strictly concave boundaries then these dynamical systems are discontinuous but almost every trajectory is uniformly hyperbolic. Discontinuity makes the whole theory much more complicated (see [Szá00]).

An interesting theory is connected with the so-called Lorenz system

$$\frac{dx}{dt} = -\sigma(x - y),$$

$$\frac{dy}{dt} = -xz + rx - y, \tag{12.3}$$

$$\frac{dz}{dt} = xy - bz,$$

which was proposed by Lorenz in 1962 (see [Lor63]). In (12.3) σ, r, b are numerical coefficients. Paper [Lor63] became popular only in the middle of seventies and is considered as another corner-stone of chaos theory. It gave a remarkable example of a strange attractor. The concept of such attractors was proposed by Ruelle and Takens in [RT71] and has spread also among physicists.

---

[7] Compact quotient space of a simply connected nilpotent Lie group by a discrete subgroup

The theories explaining the behavior of trajectories in (12.3) were proposed by Guckenheimer and Williams (see [GW79]) and by Afraimovich, Bykov, Shilnikov (see [ABS77]). Only recently Tucker [Tuc99] gave a computer-assisted proof of the fact that for some domain of parameters σ, r, b Lorenz system yields a hyperbolic discontinuous map.

12.5 Hyperbolicity in one-dimensional systems

We shall discuss below another way of weakening the condition of uniform hyperbolicity. Let us start from the one-dimensional dynamics. The analogue of the uniform hyperbolicity is the condition of uniform expansion. It is realized for maps $Tx = \{f(x)\}$ with $|f'(x)| \geq \Lambda > 1$ for $0 \leq x \leq 1$. Here $\{y\}$ denotes the fractional part of y. Under a simplifying assumption $\{f(0)\} = \{f(1)\} = 0$ and $f \in C^2([0,1])$ Rényi proved in [Rén57] that such maps have an absolutely continuous invariant measure which is a manifestation of strong statistical properties.

Expandings are always discontinuous (in a compact space). Soon after [Rén57] several attempts were made to prove the existence of an absolutely continuous invariant measure for smooth maps $Tx = \{f(x)\}$. The most popular example is the family of parabolic maps T_α where $f_\alpha(x) = \alpha x(1-x)$, $0 \leq \alpha \leq 4$: in this case $\{f_\alpha(x)\} = f_\alpha(x)$ thus $T_\alpha = f_\alpha$ is smooth. The value $\alpha = 4$ was considered by von Neumann and Ulam. They proved in [UV47] that an absolutely continuous invariant measure for f_4 exists and its density has the form $1/(\pi\sqrt{x(1-x)})$. The next step was done by Ruelle in [Rue77] where he found the value $\alpha \in [0,4]$ satisfying $(\alpha-2)^2(\alpha+2) = 16$, for which the trajectory of the critical point $x = 1/2$ is eventually periodic and proved that in this case the map also has an absolutely continuous invariant measure. Similar result for the family $g_\alpha(x) = \alpha \sin 2\pi x$ was proven by Bunimovich in [Bun70]. Later these results were generalized by Misiurewicz (see [Mis81]) and Ognev (see [Ogn81]). A great breakthrough was made by Jakobson in [Jak81]. He proved that the set of values of α for which T_α possesses an absolutely continuous invariant measure has positive Lebesgue measure. Later studies showed that each value of α considered earlier by Ruelle, Misiurewicz and Ognev is a density point of this set[8]. Recent progress in works by Graczyk, Lyubich, Świątek and many others provides practically complete understanding of the dynamics of the quadratic family f_α. In particular, for almost every α in the interval $[0,4]$ f_α has an invariant measure which is either absolutely continuous, or is concentrated on an orbit of an attractive periodic point. Survey [GS99] contains extensive references and we refer the reader to it.

In the one-dimensional case the hyperbolicity of an individual trajectory means that the product of derivatives along this trajectory grows exponentially. Collet and Eckmann in [CE83] gave a formal definition of this property.

---

[8] That is a point at which the density (in the sense of Lebesgue measure) of this set is 1

Definition 2. *Let f be a C^1-transformation of the unit interval $[0,1]$. The forward trajectory of a point $x \in [0,1]$ is hyperbolic if for some $C > 0$, $\Lambda > 1$*

$$\left| \left(\frac{d}{dx} f^n \right) (x) \right| \geq C \Lambda^n, \qquad n = 1, 2, \ldots . \tag{12.4}$$

The map f is called unimodal, if it has exactly one point $c \in (0,1)$ such that $f'(c) = 0$ and f is strictly increasing on one of the sub-intervals $[0,c)$, $(c,1]$ and is strictly decreasing on the other. Collet and Eckmann proved that a unimodal map f whose critical point c satisfies the condition

$$\left| \left(\frac{d}{dx} f^n \right) (f(c)) \right| \geq C \Lambda^n, \qquad n = 1, 2, \ldots , \tag{12.5}$$

possesses an absolutely invariant continuous measure. Condition (12.5) is called the Collet-Eckmann condition. It means that the trajectory of the critical value $f(c)$ is forward hyperbolic.[9]

It is possible to show that if the critical point satisfies Collet-Eckmann condition, then the forward trajectory of almost every point is hyperbolic. However, the constant C in (12.4) may depend on the initial point and the hyperbolicity is not uniform.

Clearly, exponential growth in (12.5) is impossible if a trajectory goes through a critical point c where the derivative is zero. Jakobson in [Jak81] proved that if the trajectory of the critical point during the dynamics does not come too close to itself then an absolutely continuous invariant measure exists. In another statement he showed that the set of α for which this condition holds for quadratic family f_α introduced above has positive Lebesgue measure.

In [Jak81] Jakobson proved the non-uniform hyperbolicity of maps f_α for some parameters α using the method of *parameter exclusion*. In this method one begins by verifying that the trajectory of the critical value $f_\alpha(c)$ is hyperbolic for finite number of iterates $n = 1, 2, \ldots, N = N_0$ and initial interval Δ_0 of parameters. Then N is increased to $N_1 > N_0$ and those parameters from Δ_0 for which the hyperbolicity conditions are violated for any $n \leq N_1$ are excluded from consideration and thus new set $\Delta_1 \subset \Delta_0$ is constructed. The same procedure is applied to N_1, Δ_1 and we obtain a new pair $N_2 > N_1$, $\Delta_2 \subset \Delta_1$, and so on. Parameters in the set $\Delta = \cap \Delta_k$ specify non-uniformly hyperbolic maps for which Collet-Eckmann condition is fulfilled. Jakobson has shown that Δ has positive Lebesgue measure.

There were several proofs of Jakobson's theorem. Especially we would like to mention the proof by Benedicks and Carleson [BC85], because the authors used later the main ideas of their proof for the two-dimensional case. The proof in [BC85], as in [Jak81], also uses the method of parameter exclusion.

---

[9] In addition to unimodality Collet and Eckmann in [CE83] used some technical assumptions about f. Later it was shown by Nowicki and van Strien in [NV88] and by Kozlovski in [Koz00] how these additional assumptions can be removed and the existence proof of absolutely continuous invariant measure can be extended to maps with finitely many non-flat critical points

12.6 Two-dimensional systems

In [BC91] Benedicks and Carleson considered Hénon map of the two-dimensional plane

$$T(x, y) = (1 - ax^2 - y, bx)$$

which depends on two real parameters a and b. This map was introduced by Hénon in 1976 (see [Hén76]). Computer simulations indicated T may possess a strange attractor similar to Lorenz attractor mentioned above, that for $a \approx 1.4$, $b \approx 0.3$. Benedicks and Carleson proved that for a set of parameters a, b of positive Lebesgue measure T possesses a strange attractor which contains a dense hyperbolic trajectory. Later several authors (Mora and Viana [MV93], Wang and Young [WY01]) proved that for the same set of parameters the restriction of T to this attractor is a non-uniform hyperbolic system with strong statistical properties. They also showed the existence of such attractors in more general families of maps, which include Hénon map as a particular case.

In non-uniformly hyperbolic cases stable and unstable tangent sub-spaces $\mathcal{T}_x^{(s)}(M)$, $\mathcal{T}_x^{(u)}(M)$ can exist only for some points $x \in M$ and the constants C, λ in (12.1), (12.2) may depend on x. The distance between $\mathcal{T}_x^{(s)}(M)$ and $\mathcal{T}_x^{(u)}(M)$ may be also arbitrarily small. Furthermore, for some points $\mathcal{T}_x^{(s)}(M)$ and $\mathcal{T}_x^{(u)}(M)$ have common directions, along which there is no expansion under both forward and backwards iterates of T. For two-dimensional systems, such as Hénon map, stable and unstable directions at such points coincide and leaves of stable and unstable foliations are tangent to each other. We shall call these tangency points as critical points of Hénon map, because they play a role similar to the role of critical points of one-dimensional maps. Benedicks and Carleson related the existence of the attractor to the hyperbolic behaviour of this set of critical points which must satisfy conditions similar to the Collet-Eckmann conditions (12.5).

Existence proofs for attractors of Hénon type are intrinsically difficult due to the fact that exact locations of critical points are unknown until the attractor, whose entire existence depends on the hyperbolic behaviour of trajectories of critical points, is constructed. The proof is also complicated because after the construction of the attractor is finished the set of critical points turns out to be uncountable.

Benedicks and Carleson overcame the difficulties mentioned above with the help of the method of parameter exclusion which in some sense is close to the method used by Jakobson for one-dimensional maps.

When $b = 0$ the transformation T maps the entire plane to the x-axis and its restriction to the x-axis after suitable change of coordinates is given by the quadratic map f_α. In particular, $(a, b) = (2, 0)$ corresponds to f_α with $\alpha = 4$ which is a hyperbolic map. Benedicks and Carleson consider maps with small b, $a \leq 2$ and close to 2. The smallness of b implies the strong contraction by T in the y-direction. Thus for small values of parameter b trajectories of T have the same hyperbolic properties as the corresponding trajectories of f_α for long intervals of time. Even though actual positions of critical points of Hénon map are not known they can be approximated by the orbit of the critical point c of the corresponding one-dimensional map.

Let Ω_0 be an invariant domain of T, containing point $(c, 0)$. Such a domain always exists for small values of b. For $n \geq 1$ put $\Omega_n = T^n \Omega_0$. Denote by $\mathcal{C}_0$ a small neighborhood in Ω_0 of $(c, 0)$. For small values of b the set $\mathcal{C}_0$ serves as the initial approximation to the set $\mathcal{C}$ of genuine critical point, which is yet unknown. If f_α satisfies the Collet-Eckmann condition one can use (12.5) to establish the hyperbolicity of an initial segment of the trajectory of $\mathcal{C}_0$ for an open set Δ_0 of parameters a, b. This in turn allows to refine the approximation to $\mathcal{C}$ and to construct a smaller set $\mathcal{C}_1 \subset \mathcal{C}_0 \cap \Omega_1$; by further exclusion of parameters from the set Δ_0 it is possible to extend the initial segment of the trajectory of $\mathcal{C}_1$ with the hyperbolic behaviour. The remaining set of parameters is denoted Δ_1 and the entire process is repeated for the pair $\mathcal{C}_1$, Δ_1. And so on. We obtain decreasing sequences of sets Ω_n, $\mathcal{C}_n$ and Δ_n, $n = 1, 2, \ldots$. Sets Ω_n converge to Hénon attractor $\Omega = \cap \Omega_n$. The set $\Delta = \cap \Delta_n$ has positive Lebesgue measure and for $(a, b) \in \Delta$ the critical set $\mathcal{C} = \cap \mathcal{C}_n$ is well defined. The set $\mathcal{C}$ is a zero-measure set containing points of the attractor Ω whose trajectories are forward hyperbolic and where stable and unstable directions coincide.

Note that all known proofs by the method of the parameter exclusion require b to be very small and the existence of Hénon attractor for parameters $a \approx 1.4$, $b \approx 0.3$, considered by Hénon, is not proven. The existence of the Hénon attractor for the conservative Hénon maps (i.e.[10] $b = 1$) is also unknown.

12.7 Conservative systems

The problem of establishing non-uniform hyperbolicity for conservative dynamical systems presents additional difficulties. As an example consider a transformation of the two-dimensional torus

$$T(\phi, z) = (\phi', z'), \quad z' = z + kV(\phi), \quad \phi' = \phi + z' \pmod 1,$$

where $V(\phi)$ is a smooth function on the unit circle. T is the standard map with parameter k. Case $V(\phi) = 1/(2\pi) \sin 2\pi \phi$ was considered by Chirikov in [Chi79]. It preserves the Lebesgue measure. We shall assume that V has finitely many critical points on the unit circle, and all these points are non-degenerate.

When $k = 0$ the two-dimensional torus is partitioned onto invariant invariant circles $z = const$, and the restriction of T to each circle acts as a rotation by z. According to the Kolmogorov-Arnold-Moser theory most of the invariant circles persist for small values of k and the restriction of T to each

---

[10] The Jacobian of T is b

circle is conjugated to a rotation by an irrational angle (see [Kol54b], [Mos62], [Arn63]). As k increases, the area filled by invariant circles decreases. When k exceeds a critical value k_c there are no invariant curves whose projection onto the z-axis is the whole interval $[0, 1)$. The remains of the invariant circles for $k > k_c$ are described by the Aubry-Mather theory (see [Per79], [AL83], [Mat84]). They form invariant Cantor sets - cantori, on which the action of T is still conjugate to a rotation.

When k is large the domain of T may be partitioned in two regions. One region, whose area is close to 1, is the *hyperbolic* region $\mathcal{H}$. The restriction of T to $\mathcal{H}$ has strong hyperbolic properties. In order to describe them it is convenient to make a linear change of coordinates $(\phi, z) = (x, x - y)$ (mod 1). Then T takes the form

$$T(x, y) = (2x + kV(x) - y, x) \quad (\text{mod } 1).$$

For large values of k the transformation T expands in x-direction and its inverse expands in y-direction. The remaining part $\mathcal{S}$ consists of several strips. Each strip in $\mathcal{S}$ corresponds to a critical point of function V and the restriction of T to it resembles Hénon map. The set $\mathcal{S}$ is in some sense an *a priori* approximation to the set of critical points $\mathcal{C}$ of T. It is expected that just as for the one-dimensional quadratic family, and for Hénon map, T is non-uniformly hyperbolic for a set of parameters of positive measure.

The preservation of the area brings additional features in the description of the critical set. It also presents new technical difficulties in the proof of non-uniform hyperbolicity by parameter elimination. On the other hand it also allows to conjecture the existence of leaves of invariant stable and unstable foliations for T for almost every point of the torus. Tangency points of these foliations are the critical points of T. As in case for Hénon map they are not known in advance and have to be constructed together with the invariant foliations by successive approximations. Consider the partition ξ_0 of the torus onto the circles $y = const$ and put $\xi_n = T^n \xi_0$ for $n = 1, 2 \ldots$. For the partition η_0 onto the circles $x = const$ put $\eta_n = T^{-n} \eta_0$, $n = 1, 2 \ldots$. If k is such that T is non-uniformly hyperbolic it is natural to expect that the partitions η_n, ξ_n converge as $n \to \infty$ to invariant stable and unstable foliations $\xi^{(s)}$, $\xi^{(u)}$, respectively. Tangency points of η_n and ξ_n converge to tangency points of $\xi^{(s)}$ and $\xi^{(u)}$.

Similar to the set $\mathcal{C}_0$ for Hénon maps, the critical region $\mathcal{S}$ produces tangency points of stable and unstable foliations. In fact, tangency points of the first order form continuous curves. When images of these curves visit $\mathcal{S}$ again there appear points of second order of tangency of elements of approximating partitions η_n and ξ_n. Points of second order of tangency persist under small parameter perturbations. Points of higher order of tangency are non-generic, after a small perturbation of parameter they split into finitely many tangency points of the second order. Thus in the measure-preserving case the role of critical points is played by the points of the second order of tangency.

The trajectory of each point can be decomposed into "hyperbolic segments" between two consecutive visits to S. On any such segment the images of the point remain in $\mathcal{H}$ where T has strong hyperbolic properties. After a visit to S the expanding direction may turn and become a contracting direction, which results in some loss of the hyperbolicity along the trajectory of a point of tangency. Furthermore, each visit to S of a tangency point of the second order to creates additional tangency points of the second order surrounding the original tangency point. The frequency of returns to S must be controlled in order to control the structure of the foliations ξ_n and η_n in the vicinity of tangency points. For Hénon map the rate of return of approximations to critical points to the critical region was controlled by making parameter b small. For standard maps the entire hierarchical structure of foliations after multiple visits to S has to be described.

An additional difficulty in the process of parameter elimination is presented by the rapid growth of the number of tangency points of the second order in approximations $\mathcal{C}_n$. In both cases of Hénon map and the standard map the number of tangency points in $\mathcal{C}_n$ grows exponentially in n. For the standard map the base of the exponent is proportional to k whereas for Hénon map it can be made as close to 1 as necessary by choosing b sufficiently small.

12.8 Conclusion

Kolmogorov believed and stressed it several times that general measure-preserving dynamical systems are mixtures of quasi-periodic motions and motions with positive entropy. We are still very far from elucidating the situation.

References

[ABS77]   Afraĭmovič, V.S., Bykov, V.V., Sil'nikov, L.P.: The origin and structure of the Lorenz attractor. Dokl. Akad. Nauk SSSR, **234**, 336–339 (1977)

[AL83]   Aubry, S., Le Daeron, P.Y.: The discrete Frenkel-Kontorova model and its extensions. I. Exact results for the ground-states. Physica D, **8**, 381–422 (1983)

[Ano67]   Anosov, D.V.: Geodesic flows on closed Riemann manifolds with negative curvature. Proc. Steklov Inst. Math., **90**, 235 p. (1967)

[Arn63]   Arnol'd, V.I.: Proof of a theorem of A. N. Kolmogorov on the preservation of conditionally periodic motions under a small perturbation of the Hamiltonian. Uspehi Mat. Nauk, **18**, 13–40 (1963)

[BC85]   Benedicks, M., Carleson, L.: On iterations of $1 - ax^2$ on $(-1, 1)$. Ann. of Math., **122**, 1–25 (1985)

[BC91]   Benedicks, M., Carleson, L.: The dynamics of the Hénon map. Ann. of Math., **133**, 73–169 (1991)

[Bun70]   Bunimovič, L.A.: On a transformation of the circle. Akad. Nauk SSSR Mat. Zametki, **8**, 205–216 (1970)

[CE83]   Collet, P., Eckmann, J.-P.: Positive Liapunov exponents and absolute
           continuity for maps of the interval. Ergodic Theory Dynam. Systems, **3**,
           13–46 (1983)
[CFS82]   Cornfeld, I.P., Fomin, S.V., Sinaǐ, Ya.G.: Ergodic Theory. Grund. math.
           Wiss. n° 245, Springer (1982)
[Chi79]   Chirikov, B.V.: A universal instability of many-dimensional oscillator sys-
           tems. Phys. Rep., **52**, 264–379 (1979)
[Fom50]   Fomin, S.: On dynamical systems in a space of functions. Ukrain. Mat.
           Žurnal, **2**, 25–47 (1950)
[GF52]   Gel'fand, I.M., Fomin, S.V.: Geodesic flows on manifolds of constant neg-
           ative curvature. Uspehi Matem. Nauk (N.S.), **7**, 118–137 (1952)
[GF55]   Gel'fand, I.M., Fomin, S.V.: Geodesic flows on manifolds of constant neg-
           ative curvature. Amer. Math. Soc. Transl., **1**, 49–65 (1955)
[Gir58]   Girsanov, I.V.: Spectra of dynamical systems generated by stationary
           Gaussian processes. Dokl. Akad. Nauk SSSR (N.S.), **119**, 851–853 (1958)
[GS99]   Graczyk, J., Świątek, G.: Smooth unimodal maps in the 1990s. Ergodic
           Theory Dynam. Systems, **19**, 263–287 (1999)
[GW79]   Guckenheimer, J., Williams, R.F.: Structural stability of Lorenz attrac-
           tors. Publ. Math. Inst. Hautes Études Sci. Publ. Math., **50**, 59–72 (1979)
[Hén76]   Hénon, M.: A two-dimensional mapping with a strange attractor. Comm.
           Math. Phys., **50**, 69–77 (1976)
[Hop39]   Hopf, E.: Statistik der geodätischen Linien in Mannigfaltigkeiten nega-
           tiver Krümmung. Ber. Verh. Sächs. Akad. Wiss. Leipzig, **91**, 261–304
           (1939)
[Hop40]   Hopf, E.: Statistik der Lösungen geodätischer Probleme vom unstabilen
           Typus. II. Math. Ann., **117**, 590–608 (1940)
[Jak81]   Jakobson, M.V.: Absolutely continuous invariant measures for one-
           parameter families of one-dimensional maps. Comm. Math. Phys., **81**,
           39–88 (1981)
[Kol41a]   Kolmogorov, A.N.: The local structure of turbulence in incompressible
           viscous fluid for very large Reynold's numbers. C. R. (Doklady) Acad.
           Sci. URSS (N.S.), **30**, 301–305 (1941)
[Kol41b]   Kolmogorov, A.N.: Dissipation of energy in the locally isotropic turbu-
           lence. C. R. (Doklady) Acad. Sci. URSS (N.S.), **32**, 16–18 (1941)
[Kol42]   Kolmogorov, A.N.: Equations of turbulent motion of an incompressible
           fluid. Bull. Acad. Sci. URSS. Ser. Phys. [Izvestia Akad. Nauk SSSR], **6**,
           56–58 (1942)
[Kol54a]   Kolmogorov, A.N.: Théorie générale des systèmes dynamiques et
           mécanique classique. In: Proceedings of the International Congress of
           Mathematicians, vol. 1, 315–333, Amsterdam (1954), Erven P. Noordhoff
           N.V., Groningen (1957)
[Kol54b]   Kolmogorov, A.N.: On conservation of conditionally periodic motions for
           a small change in Hamilton's function. Dokl. Akad. Nauk SSSR (N.S.),
           98, 527–530 (1954)
[Kol58]   Kolmogorov, A.N.: A new metric invariant of transient dynamical systems
           and automorphisms in Lebesgue spaces. Dokl. Akad. Nauk SSSR (N.S.),
           119, 861–864 (1958)
[Kol91a]   Kolmogorov, A.N.: The local structure of turbulence in incompressible
           viscous fluid for very large Reynolds numbers. Proc. Roy. Soc. London
           Ser. A, **434**, 9–13 (1991)

[Kol91b] Kolmogorov, A.N.: Dissipation of energy in the locally isotropic turbu-
          lence. Proc. Roy. Soc. London Ser. A, **434**, 15–17 (1991)
[Koz00]   Kozlovski, O.S.: Getting rid of the negative Schwarzian derivative condi-
          tion. Ann. of Math., **152**, 743–762 (2000)
[Lor63]   Lorenz, E.N.: Deterministic Nonperiodic Flow. J. Atmospheric Sci., **20**,
          130–141 (1963)
[Mar49]   Maruyama, G.: The harmonic analysis of stationary stochastic processes.
          Mem. Fac. Sci. Kyūsyū Univ. A, **4**, 45–106 (1949)
[Mat84]   Mather, J.N.: Nonexistence of invariant circles. Ergodic Theory Dynam.
          Systems, **4**, 301–309 (1984)
[Mis81]   Misiurewicz, M.: Absolutely continuous measures for certain maps of an
          interval. Publ. Math. Inst. Hautes Études Sci., **53**, 17–51 (1981)
[Mos62]   Moser, J.: On invariant curves of area-preserving mappings of an annulus.
          Nachr. Akad. Wiss. Göttingen Math.-Phys. Kl. II, **1962**, 1–20 (1962)
[MV93]   Mora, L., Viana, M.: Abundance of strange attractors. Acta Math., **171**,
          1–71 (1993)
[NV88]   Nowicki, T., van Strien, S.: Absolutely continuous invariant measures
          for C^2 unimodal maps satisfying the Collet-Eckmann conditions. Invent.
          Math., **93**, 619–635 (1988)
[NV90]   Nowicki, T., van Strien, S.: Hyperbolicity properties of C^2 multi-modal
          Collet-Eckmann maps without Schwarzian derivative assumptions. Trans.
          Amer. Math. Soc., **321**, 793–810 (1990)
[Ogn81]   Ognev, A.I.: Metric properties of a class of mappings of a segment. Akad.
          Nauk SSSR Mat. Zametki, **30**, 723–736, 797 (1981)
[Orn70]   Ornstein, D.S.: Bernoulli shifts with the same entropy are isomorphic.
          Adv. in Math., **4**, 337–352 (1970)
[Orn74]   Ornstein, D.S.: Ergodic theory, randomness, and dynamical systems. Yale
          University Press, New Haven, Conn. (1974)
[Per79]   Percival, I.C.: Variational principles for invariant tori and cantori. In:
          Nonlinear dynamics and the beam-beam interaction, AIP Conf. Proc.
          n° **57**, 302–310, Sympos. Brookhaven Nat. Lab., New York (1979)
[Pes77]   Pesin, Ja.B.: Characteristic Ljapunov exponents, and smooth ergodic the-
          ory. Uspehi Mat. Nauk, **32**, 55–112, 287 (1977)
[Rén57]   Rényi, A.: Representations for real numbers and their ergodic properties.
          Acta Math. Acad. Sci. Hungar, **8**, 477–493 (1957)
[RT71]   Ruelle, D., Takens, F.: On the nature of turbulence. Comm. Math. Phys.,
          20, 167–192 (1971)
[Rue77]   Ruelle, D.: Applications conservant une mesure absolument continue par
          rapport à dx sur $[0,1]$. Comm. Math. Phys., **55**, 47–51 (1977)
[Sin59]   Sinaĭ, Ya.G.: On the concept of entropy for a dynamic system. Dokl.
          Akad. Nauk SSSR, **124**, 768–771 (1959)
[Sin61]   Sinaĭ, Ya.G.: Dynamical systems with countable Lebesgue spectrum. I.
          Izv. Akad. Nauk SSSR Ser. Mat., **25**, 899–924 (1961)
[Sin66]   Sinaĭ, Ya.G.: Classical dynamic systems with countably-multiple
          Lebesgue spectrum. II. Izv. Akad. Nauk SSSR Ser. Mat., **30**, 15–68 (1966)
[Sma67]   Smale, S.: Differentiable dynamical systems. Bull. Amer. Math. Soc., **73**,
          747–817 (1967)
[Szá00]   Szász, D. (ed.): Hard ball systems and the Lorentz gas. Encyclopaedia of
          Mathematical Sciences n° **101**, Springer, Berlin (2000)

[Tuc99]   Tucker, W.: The Lorenz attractor exists. C. R. Acad. Sci. Paris Sér. I
           Math., **328**, 1197–1202 (1999)
[UV47]   Ulam, S., von Neumann, J.: On combinations of stochastic and determin-
           isitc processes. Bull. Amer. Math. Soc., **53**, 1120 (1947)
[vonN32]   von Neumann, J.: Zur Operatorenmethode in der klassischen Mechanik.
           Ann. of Math., **33**, 587–642 (1932)
[WY01]   Wang, Q., Young, L.S.: Strange attractors with one direction of instabil-
           ity. Comm. Math. Phys., **218**, 1–97 (2001)

From Hilbert's 13th Problem to the theory of neural networks: constructive aspects of Kolmogorov's Superposition Theorem

Vasco Brattka

Laboratory of Foundational Aspects of Computer Science, Department of
Mathematics and Applied Mathematics, University of Cape Town, South Africa
http://cca-net.de/vasco
BrattkaV@maths.uct.ac.za

13.1 Hilbert's 13th problem

In the year 1900, in his famous lecture in Paris, Hilbert presented twenty-three problems which he considered as a challenge for future generations of mathematicians [Hil00]. As we know today, many of these problems actually attracted considerable attention and some of them even led to important developments in mathematics (see [Gra00] for a recent discussion). The 13th problem is concerned with algebraic equations

$$a_n x^n + a_{n-1} x^{n-1} + \ldots + a_1 x + a_0 = 0$$

and it is entitled *"Impossibility of the solution of the general equation of the 7-th degree by means of functions of only two arguments"*. Hilbert presented his question in terms of *nomography* (the theory of *nomograms*, which are graphical presentations of continuous functions depending on several arguments for computational purposes), a field which is almost forgotten as a mathematical discipline today but which was popular at the time of Hilbert's talk (the success of this discipline was very much due to Maurice d'Ocagne [d'Oc99]). It is easy to see that nomograms can be employed in order to compute values of functions which depend only on two variables. By iterating such calculations one can obviously evaluate functions which can be represented as superpositions of functions of at most two variables. Given nomograms for the corresponding solution functions the described method allows to compute the solutions of algebraic equations up to degree six. This is because Tschirnhausen Transformations allow to bring the general equations of degree five to seven into the following normal forms [Hil27]:

$$x^5 + ax + 1 = 0,$$
$$x^6 + ax^2 + bx + 1 = 0,$$
$$x^7 + ax^3 + bx^2 + cx + 1 = 0.$$

The Tschirnhausen Transformations themselves can be represented by superpositions of addition and functions which depend only on one variable. Moreover, it is known that the solutions of algebraic equations are given, at least locally, by continuous functions of their coefficients (although they are not given by globally defined continuous functions). Hence, locally the solutions of the equations of degree five, six and seven above can be considered as continuous functions depending on a, (a, b) and (a, b, c), respectively, and the equation of degree seven is the one of lowest degree for which it is not obvious whether its solutions can be obtained by superpositions of continuous functions of only two variables and it was Hilbert's conjecture that this is not possible in general.[1]

Behind this specific problem of solving the general septic equation, there is the more general question whether addition is the only function which inherently depends on two variables. It is a basic observation that at least any arithmetic operation can be represented by superpositions of functions of only one variable and addition:

$$a - b = a + (-b),$$
$$a \cdot b = \frac{1}{4}\left((a + b)^2 - (a - b)^2\right),$$
$$\frac{a}{b} = a \cdot \frac{1}{b}.$$

For short, let us say in the following that addition is *universal* for a class $\mathcal{F}$ of real-valued functions, if any function $f \in \mathcal{F}$ in this class can be represented by a finite number of superpositions of addition and functions in $\mathcal{F}$ which do only depend on one variable. In general, by a *superposition* of functions we mean a substitution of these functions and projections. For instance, the function $h : \mathbb{R}^3 \to \mathbb{R}$, defined by

$$h(x, y, z) := g(x, f(z, y))$$

is defined as a superposition of $g : \mathbb{R}^2 \to \mathbb{R}$ and $f : \mathbb{R}^2 \to \mathbb{R}$. Especially, our consideration above shows that addition is universal for the class $\mathcal{F}$ of arithmetic operations (depending on one or more variables). In contrast to this, addition is not universal for the class of analytic functions as Hilbert already pointed out in his description of his 13th problem [Hil00] (see also the example of Ostrowski's function mentioned in [Hil27].) In particular a positive solution of Hilbert's conjecture about the impossibility of the solution of the general septic equation would imply that addition is not universal for the class of continuous functions (say, on the unit interval $[0, 1]$). It were Arnol'd and Kolmogorov who disproved this conjecture in 1957 by showing that addition

---

[1] It is worth noticing that Maurice d'Ocagne showed in [d'Oc00] that the septic equation can be solved with the help of nomograms. This does not solve the conjecture but it relativizes its original motivation. Hilbert added a remark mentioning this result in a later publication of his talk [Hil01]

is universal for the class of continuous functions and finally Kolmogorov even provided a strong normal form for the corresponding superpositions. General presentations of Hilbert's 13th problem and Kolmogorov's Superposition Theorem can be found in [Vit69, Lor76, Kah82], algebraic aspects of Hilbert's 13th problem are discussed in [Dix93].

13.2 Kolmogorov's Superposition Theorem

In a series of papers Kolmogorov and his student Arnol'd studied representations of continuous functions by superpositions of continuous functions with a smaller number of variables. While Kolmogorov's results in [Kol56] already imply that any continuous function depending on several variables can be obtained as a superposition of continuous functions of only three variables, it was Arnol'd's result [Arn57] which finally refuted Hilbert's conjecture and showed that addition is universal for continuous functions (on the unit interval). In the same issue of the same journal Kolmogorov presented the following normal form for such a superposition [Kol57]: for every integer $n \geq 2$ there exist continuous functions $\varphi_{pq} : [0,1] \to \mathbb{R}$ such that every continuous function $f : [0,1]^n \to \mathbb{R}$ is representable in the form

$$f(x_1, ..., x_n) = \sum_{q=1}^{2n+1} g_q \left(\sum_{p=1}^{n} \varphi_{pq}(x_p) \right),$$

where the functions $g_q : \mathbb{R} \to \mathbb{R}$ are continuous. Instead of a detailed proof Kolmogorov presented a brief sketch of a direct construction which is elementary in the sense that it does not employ any advanced tools.

By refining this construction Sprecher proved in 1963 that the functions φ_{pq} can be represented as $\varphi_{pq}(x) = \lambda_p \varphi_q(x)$ with certain constants $0 < \lambda_p \leq 1$ and continuous functions $\varphi_q : [0,1] \to \mathbb{R}$ (see [Spr65]). Additionally, Lorentz proved that the functions g_q can be chosen all the same [Lor66]. Combining all these normal forms we arrive at our final version of the Kolmogorov Superposition Theorem (which, more appropriately, might also be called the *Kolmogorov-Lorentz-Sprecher Theorem*):

Theorem 1 (Kolmogorov's Superposition Theorem). *For each $n \geq 2$ there exist continuous functions $\varphi_q : [0,1] \to \mathbb{R}$, $q = 0, ..., 2n$ and positive constants $\lambda_p \in \mathbb{R}$, $p = 1, ..., n$ such that the following holds true: for each continuous function $f : [0,1]^n \to \mathbb{R}$ there exists a continuous function $g : [0,1] \to \mathbb{R}$ such that*

$$f(x_1, ..., x_n) = \sum_{q=0}^{2n} g \left(\sum_{p=1}^{n} \lambda_p \varphi_q(x_p) \right).$$

It is worth noticing that the functions φ_q and the constants λ_p are independent from the represented function f and do only depend on the dimension n. More than stated, one can choose $\lambda_p > 0$ such that $\sum_{p=1}^{n} \lambda_p \leq 1$. Typically, the construction is done by choosing some *rationally independent* constants λ_p (which are constants such that $\sum_{p=1}^{n} r_p \lambda_p = 0$ with rational r_p implies $r_p = 0$ for all $p = 1, ..., n$). Moreover, the functions φ_q can be chosen to be Lipschitz continuous of a certain order but they cannot be continuously differentiable in general (see [LGM96]).

Much of the work following Kolmogorov's original construction was devoted to the question what kind of smoothness conditions could be imposed on the functions φ_q and g. Already in 1954 Vitushkin had proved that addition is not universal for the class of r-times continuously differentiable functions for any $r \geq 1$ [Vit54]. A very readable presentation of Kolmogorov's Superperition Theorem, its proof and a discussion of further results along the forementioned line can be found in [LGM96, Vit77]. Instead of discussing any smoothness conditions, we will focus on computability conditions.

Kolmogorov's original proof sketch of his theorem was rather constructive. However, he has not defined the functions φ_q and g explicitly but he only showed that such functions exist (without considering the technical details of the concrete construction). Meanwhile, some possible alternatives to turn Kolmogorov's considerations into a more detailed proof have been studied from different perspectives. On the one hand, a very explicit definition of the functions φ_q and g has recently been presented by Sprecher [Spr96, Spr97]. In particular, Sprecher defined a specific continuous function $\varphi : [0, 1] \rightarrow \mathbb{R}$, which we will call *Sprecher's function* in the following. One can use this function in order to define

$$\varphi_q(x) := c\varphi(x + aq) + bq$$

with certain constants a, b, c. On the other hand, rather abstract but very brief and elegant proofs of Kolmogorov's Superposition Theorem have been presented by Hedberg [Hed71] and Kahane [Kah75]. These proofs are based on the Baire Category Theorem and we will discuss the computational implications of such proofs in a later section.

For a long time the Kolmogorov Superposition Theorem was considered as a side result in Approximation Theory without any applications. In 1976 Lorentz mentioned the deduction of the multi-variable case of the Weierstraß Approximation Theorem from the single variable case as the only application of the Kolmogorov Superposition Theorem known to him [Lor76] and he writes: *"Perhaps Kolmogorov's theorem is of the nature of a pathological example whose main purpose is to disprove hopes that are too optimistic."* It was only 11 years later when Hecht-Nielsen realized that Kolmogorov's Superposition Theorem allows to characterize the power of feedforward neural networks[2] [Hec87] and since then the theorem has established itself as

---

[2] See the definition p. 273

applicable. In recent years other applications of Kolmogorov's Theorem to topics such as Radon transform and topological groups have been considered [Fri95, LMP97]. We will briefly discuss the applications to neural networks in a later section. Last but not least, these applications to computer science related topics increase the desire to understand the computational nature of Kolmogorov's Superposition Theorem.

In the following sections we will start to employ Sprecher's recent ideas in order to show that a computational version of Kolmogorov's Superposition Theorem can be proved. It is very natural to ask whether addition is universal for the class of computable functions (these are those functions which admit an algorithm that in principle allows a Turing machine to evaluate the function up to any prescribed precision). Such functions are studied in a field called *computable analysis* (which is the theory of computability on real numbers and functions) and it is known that they have to be continuous [Wei00, Ko91, PR89]. Hence, computability is another strengthening of the notion of continuity, logically independent of the smoothness conditions which have been mentioned above. Probably the first computational version of the Kolmogorov Superposition Theorem has been presented by Nakamura, Mines and Kreinovich [NMK93]. We will revisit this topic in the sound and rigorous framework offered by the representation based approach to computable analysis.

13.3 Computability of Sprecher's function

In this section we will briefly recall the notion of a computable real-valued function as it is used in computable analysis and we will show that Sprecher's function φ is computable in this sense. Roughly speaking, a partial real-valued function[3] $f :\subseteq \mathbb{R}^n \to \mathbb{R}$ is *computable*, if there exists a Turing machine M which in the long run transforms each infinite sequence $p \in \Sigma^\omega$ representing some input $x \in \mathbb{R}^n$ into an infinite sequence $r \in \Sigma^\omega$ representing the corresponding output $f(x)$. Precise definitions can be found in Weihrauch [Wei00]; other equivalent approaches have been presented by Pour-El and Richards [PR89] and Ko [Ko91].

It is easy to see that the given definition sensitively relies on an appropriate choice of the representation. Here, we will use the so-called *Cauchy representation* $\rho :\subseteq \Sigma^\omega \to \mathbb{R}$ of the real numbers, where roughly speaking, $\rho(p) = x$ if p is a sequence of rational numbers $(q_i)_{i \in \mathbb{N}}$ (encoded over the alphabet Σ) which rapidly converges to x, i.e. $|q_k - q_i| \leq 2^{-k}$ for all $i \geq k$. A corresponding representation ρ^n for the n-dimensional Euclidean space $\mathbb{R}^n$ can be derived easily. More generally, a *representation* of a set X is a surjective mapping

---

[3] The notation $f :\subseteq E \to F$ recalls that f needs only to be defined on a subset of E (*partial function*). When we write $f : E \to F$, that means that f is defined at *every* point of E (*total function*)

$\delta :\subseteq \Sigma^\omega \to X$. Using this notion we can define the computability concept of computable analysis precisely.

Definition 1 (Computable functions). Let δ and δ' be representations of X and Y, respectively. A partial function $f :\subseteq X \to Y$ is called (δ, δ')-*computable*, if there exists a Turing machine M with one-way output tape which computes a function $F_M :\subseteq \Sigma^\omega \to \Sigma^\omega$ such that $\delta' F_M(p) = f\delta(p)$ for all $p \in \text{dom}(f\delta)$.

The definition can be generalized to functions of higher arity[4] straightforwardly. The situation of the previous definition can be visualized by the commutative diagram of Fig. 13.1.

We can generalize this computability concept even to (partial) multivalued operations $f :\subseteq X \rightrightarrows Y$, which are operations where the image $f(x)$ is a (not necessarily single-valued) subset of Y. In this case we write $\delta' F_M(p) \in f\delta(p)$, i.e. "$\in$" instead of "$=$" in the definition above, in order to define computability.

We will call the (ρ^n, ρ)-computable functions $f :\subseteq \mathbb{R}^n \to \mathbb{R}$ *computable* for short. Computability can analogously be defined for other types of functions, as total functions $f : [a, b] \to \mathbb{R}$ and sequences $f : \mathbb{N} \to \mathbb{R}$, using representations $\rho|^{[a,b]}$ of $[a, b]$ and $\delta_\mathbb{N}$ of $\mathbb{N}$. There exists a well-known characterization of computable functions $f : [0, 1] \to \mathbb{R}$ which we will use in order to prove the computability of Sprecher's function φ. Roughly speaking, a function $f : [0, 1] \to \mathbb{R}$ is computable, if and only if $f|_D$ restricted to some "effectively dense" subset $D \subseteq [0, 1]$ is computable and f admits some computable *modulus of continuity* m, which is a function $m : \mathbb{N} \to \mathbb{N}$ such that

$$|x - y| < 2^{-m(n)} \implies |f(x) - f(y)| < 2^{-n}$$

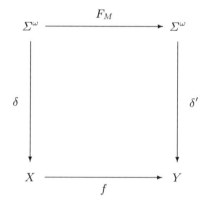

Fig. 13.1. Computable functions

---

[4] Arity is the number of variables on which the function depends : $f(x)$ is unary (arity 1), $f(x, y)$ is binary (arity 2),..., $f(x_1, ..., x_k)$ is k-ary (arity k)

holds for all $x, y \in [0,1]$ and $n \in \mathbb{N}$. As a dense subset we will use in the following the set $\mathbb{Q}_\gamma$ of rational numbers between 0 and 1 in expansion with respect to base γ. This set can be defined by $\mathbb{Q}_\gamma := \bigcup_{k=1}^\infty \mathbb{Q}_{\gamma k}$ and

$$\mathbb{Q}_{\gamma k} := \left\{ \sum_{r=1}^k i_r \gamma^{-r} : i_1, ..., i_k \in \{0, 1, ..., \gamma - 1\} \right\}$$

for all integers $k \geq 1$ and $\gamma \geq 2$. To make the characterization of computable functions $f : [0,1] \to \mathbb{R}$ more precise we assume that $\nu : \mathbb{N} \to \mathbb{Q}_\gamma$ is some effective standard numbering of the rational numbers in γ-expansion (cf. [Wei00] for the discussion of numberings).

Proposition 1 (Characterization of computable functions). *Let $\gamma \geq 2$. A function $f : [0,1] \to \mathbb{R}$ is computable, if and only if the following holds:*

1. *$f \circ \nu : \mathbb{N} \to \mathbb{R}$ is a computable sequence of real numbers,*
2. *f admits a computable modulus of continuity $m : \mathbb{N} \to \mathbb{N}$.*

For a proof cf. [Wei00, Bra03a, Ko91, PR89]. The main purpose of the remaining part of this section is to prove that there exist computable functions $\varphi_q : [0,1] \to \mathbb{R}$ which can be employed for a proof of the computable version of Kolmogorov's Superposition Theorem. In Kolmogorov's original proof the existence of such functions φ_q has been proved without any explicit construction. Recently, Sprecher defined a concrete function $\varphi : [0,1] \to \mathbb{R}$ [Spr96, Spr97] which can be used to construct the functions φ_q. We will prove that φ is a computable function. More precisely, for each dimension $n \geq 2$ and each base $\gamma \geq 2n + 2$ Sprecher has defined a separate function $\varphi : [0,1] \to \mathbb{R}$ as follows.

Definition 2 (Sprecher's function). Let $n \geq 2$ and $\gamma \geq 2n + 2$. Let $\varphi : [0,1] \to \mathbb{R}$ be the unique continuous function which fulfills the equation

$$\varphi\left(\sum_{r=1}^k i_r \gamma^{-r} \right) = \sum_{r=1}^k \tilde{i}_r 2^{-m_r} \gamma^{-\frac{n^{r-m_r}-1}{n-1}} \tag{13.1}$$

where

$$\tilde{i}_r := i_r - (\gamma - 2)\langle i_r \rangle,$$

$$m_r := \langle i_r \rangle \left(1 + \sum_{s=1}^{r-1} \prod_{t=s}^{r-1} [i_t] \right)$$

for all $r \geq 1$, $\langle i_1 \rangle := 0$, $[i_1] := 0$ and

$$\langle i_r \rangle := \begin{cases} 0 \text{ if } i_r = 0, 1, ..., \gamma - 2 \\ 1 \text{ if } i_r = \gamma - 1 \end{cases}$$

$$[i_r] := \begin{cases} 0 \text{ if } i_r = 0, 1, ..., \gamma - 3 \\ 1 \text{ if } i_r = \gamma - 2, \gamma - 1 \end{cases}$$

for $r \geq 2$.

Sprecher has proved that there exists such a unique continuous function φ on $[0,1]$, mainly because φ (as defined on $\mathbb{Q}_\gamma$ by (13.1)) is strictly increasing on $\mathbb{Q}_\gamma$ and

$$|\varphi(d) - \varphi(d')| \leq \frac{1}{\gamma \cdot 2^{k-1}} \qquad (13.2)$$

holds for all consecutive numbers $d, d' \in \mathbb{Q}_{\gamma k}$ and all $k \geq 1$. Especially, φ is strictly increasing and $\varphi[0,1] = [0,1]$. We will use the same idea and the previous proposition in order to prove:

Proposition 2 (Computability of Sprecher's function). *For each $n \geq 2$ and $\gamma \geq 2n + 2$ Sprecher's function $\varphi : [0,1] \to \mathbb{R}$ is computable.*

Proof. Let $n \geq 2$ and $\gamma \geq 2n + 2$. Since a standard numbering ν allows to extract the digits $i_1, ..., i_k$ of the γ-expansion of each number $\nu(l) \in \mathbb{Q}_{\gamma k}$ from l, we can directly deduce from the definition of φ that $\varphi \circ \nu$ is a computable sequence of real numbers.

Now, let $c \in \mathbb{N}$ be a number such that $c \geq \log_2 \gamma$. Then $m : \mathbb{N} \to \mathbb{N}$ with $m(k) = c \cdot k$ is a computable function. Let $x, y \in [0,1]$ and $k \in \mathbb{N}$ with $|x-y| < 2^{-m(k)}$ where without loss of generality $x \leq y$. Then $|x-y| < \gamma^{-k}$ and there exist consecutive numbers $d < d'$ in $\mathbb{Q}_{\gamma k}$ such that $d \leq x \leq y \leq d'$, where at least the first or the last inequality is strict. Since φ is strictly increasing by definition, we obtain by (13.2)

$$|\varphi(x) - \varphi(y)| < |\varphi(d) - \varphi(d')| \leq \frac{1}{\gamma \cdot 2^{k-1}} \leq \frac{1}{2^k}.$$

Thus, the function m is a computable modulus of continuity of φ and by Proposition 1 φ is a computable function.

The proof especially shows that Sprecher's function φ is Lipschitz continuous of order $1/\log_2 \gamma$, i.e.

$$|\varphi(x) - \varphi(y)| \leq 2 \cdot |x - y|^{1/\log_2 \gamma}.$$

But on the other hand, it is easy to see that φ is not differentiable. Figure 13.2 shows the graph of Sprecher's function for dimension $n = 2$ and base $\gamma = 10$ together with a part of the graph magnified by factor 10.

The important property of Sprecher's function is that certain intervals I which correspond to the "plateaux" in the graph of φ are mapped to very small intervals $\varphi(I)$. To be more precise, define for each fixed $n \geq 2$ and $\gamma \geq 2n + 1$

$$I_i^k := \left[i \cdot \gamma^{-k}, i \cdot \gamma^{-k} + \delta_k\right]$$

for all $k \geq 1$, and $i = 1, ..., \gamma^k - 1$, where

$$\delta_k := \frac{\gamma - 2}{(\gamma - 1)\gamma^k}.$$

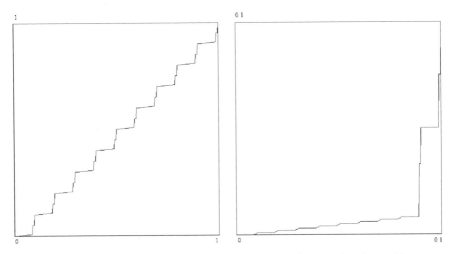

Fig. 13.2. The graph of Sprecher's function φ for $n = 2$ and $\gamma = 10$

Sprecher has proved [Spr96] that disjoint intervals I_i^k of each fixed precision k are mapped to small disjoint intervals $\varphi(I_i^k)$. This "separation property" will implicitly be used to prove the Fundamental Lemma 1 below.

It is well-known that one can define computable functions by case distinction, provided that both functions coincide at the border. More precisely, if $f : [0,1] \to \mathbb{R}$ and $g : [1,2] \to \mathbb{R}$ are computable functions with $f(1) = g(1)$, then $h : [0,2] \to \mathbb{R}$ defined by

$$h(x) := \begin{cases} f(x) \text{ if } 0 \le x \le 1 \\ g(x) \text{ if } 1 < x \le 2 \end{cases}$$

is a computable function too (see [Wei00]). Since $\varphi(1) = 1$ we can extend Sprecher's function to the interval $[0,2]$ by $\varphi(x+1) := \varphi(x)+1$ for all $x \in (0,1]$. Thus, from now on we can assume without loss of generality that Sprecher's function is a computable function $\varphi : [0,2] \to \mathbb{R}$, if necessary. Especially, we will use this fact in the following definition where we define functions $\xi_q : [0,1]^n \to \mathbb{R}$ with the help of Sprecher's function φ. We will call ξ_q the *Kolmogorov maps.*

Definition 3 (Kolmogorov maps). Let $n \ge 2$ and $\gamma \ge 2n + 2$. Define

1. $\lambda_p \in \mathbb{R}$ for $p = 1, 2, ..., n$ and $\lambda \in \mathbb{R}$ by

$$\lambda_1 := \frac{1}{2}, \quad \lambda_{p+1} := \frac{1}{2}\sum_{r=1}^{\infty}\gamma^{-p\frac{n^r-1}{n-1}}, \quad \lambda := \sum_{p=1}^{n}\lambda_p,$$

2. $\varphi_q : [0,1] \to \mathbb{R}$ for $q = 0, 1, ..., 2n$ by

$$\varphi_q(x) := \frac{1}{2n+1}\left(\frac{1}{2}\varphi\left(x + \frac{q}{\gamma(\gamma-1)}\right) + \frac{1}{\lambda}q\right),$$

3. $\xi_q : [0,1]^n \to \mathbb{R}$ for $q = 0, 1, ..., 2n$ by

$$\xi_q(x_1, ..., x_n) := \sum_{p=1}^{n} \lambda_p \varphi_q(x_p).$$

It should be noticed that it is the definition of ξ_q which includes the reduction of the number of variables. Let us consider an example: for instance for $n = 2$, $\gamma = 10$ and $q = 0$ we obtain

$$\xi_0(x, y) = \frac{1}{20}\varphi(x) + \frac{\lambda_2}{10}\varphi(y)$$

with constant

$$\lambda_2 = \frac{1}{2}\sum_{r=1}^{\infty} 10^{-(2^r - 1)} = 0.05050005000000050000000000000005...$$

Figure 13.3 displays the graph of the function ξ_0 together with the squares S_{0ij}^1 defined below.

Since we will employ the functions φ_q and λ in order to prove a computable version of Kolmogorov's Superposition Theorem, we have to prove that these functions and constants are computable.

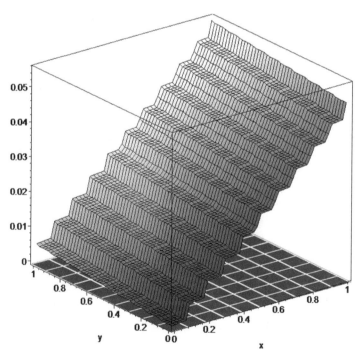

Fig. 13.3. The function ξ_0 together with the squares S_{0ij}^1 for $n = 2$ and $\gamma = 10$

Proposition 3. *Let $n \geq 2$ and $\gamma \geq 2n + 2$. Then:*

1. $\lambda_p \in \mathbb{R}$ *is computable for all $p = 1, 2, ..., n$ and $\lambda \in \mathbb{R}$ is computable,*
2. $\varphi_q : [0,1] \to \mathbb{R}$ *is computable for all $q = 0, 1, ..., 2n$,*
3. $\xi_q : [0,1]^n \to \mathbb{R}$ *is computable for all $q = 0, 1, ..., 2n$.*

Proof. Let $n \geq 2$ and $\gamma \geq 2n + 2$.

1. Obviously, λ_1 is computable. We note that for $p \geq 1$ and $\gamma > 2$ we obtain

$$\left| \sum_{r=1}^{k} \gamma^{-p \frac{n^r - 1}{n-1}} - \sum_{r=1}^{i} \gamma^{-p \frac{n^r - 1}{n-1}} \right| < \sum_{r=k+1}^{i} 2^{-\frac{n^r-1}{n-1}} \leq \sum_{r=k+1}^{i} 2^{-r} < 2^{-k}$$

for all $i > k$. Thus, $(\sum_{r=1}^{k} \gamma^{-p \frac{n^r-1}{n-1}})_{k \in \mathbb{N}}$ is a rapidly converging and computable sequence of real numbers and the limit $2\lambda_{p+1}$ is a computable real number. Hence, λ is computable too.

2. Since $\frac{q}{\gamma(\gamma-1)} < 1$ for $q = 0, 1, ..., 2n + 1$ the functions φ_q are well-defined, if we assume that Sprecher's function φ is defined on $[0, 2]$. Moreover, it is straightforward to prove that φ_q is computable for $q = 0, 1, ..., 2n + 1$ using some standard closure properties of computable functions [Wei00] and the fact that λ is computable.

3. Follows directly from (1) and (2).

The next step is to transfer the nice separation property of Sprecher's function to the Kolmogorov maps ξ_q. Essentially, this transfer has already been performed by Sprecher [Spr96]. We use his results to prove the Fundamental Lemma 1 below. Roughly speaking, the idea is that ξ_q maps certain cubes S_{qi} to very small and disjoint intervals $\xi_q(S_{qi})$. We will define these cubes with the help of the intervals I_i^k defined above. Since later on we will need coverings of $[0,1]^n$, it is not sufficient to consider purely the intervals I_i^k, but we will need $2n + 1$ families I_{qi}^k with $q = 0, ..., 2n$ of such intervals where $I_{qi}^k := I_i^k - aq$ is a slight translation of I_i^k defined precisely as follows: for each fixed $n \geq 2$ and $\gamma \geq 2n + 1$ let

$$I_{qi}^k := \left[i \cdot \gamma^{-k} - aq, i \cdot \gamma^{-k} + \delta_k - aq \right] \cap [0,1]$$

for all $k \geq 1$, $q = 0, 1, ..., 2n$ and $i = 1, ..., \gamma^k$, where δ_k is defined as above and

$$a := \frac{1}{\gamma(\gamma - 1)}.$$

Figure 13.4 visualizes the translated intervals I_{qi}^1 in case of dimension $n = 2$ and base $\gamma = 10$ (row q displays the intervals $I_{q1}^1, ..., I_{q10}^1$).

Now the following lemma states the separation property of the Kolmogorov maps ξ_q which can be deduced from the separation properties of Sprecher's function φ on corresponding intervals.

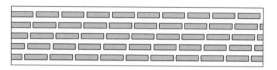

Fig. 13.4. The intervals I_{qi}^1 for $n = 2$ and $\gamma = 10$

Lemma 1 (Fundamental Lemma). *For each fixed $n \geq 2$, $\gamma \geq 2n + 2$ and $k \geq 1$, the intervals $\xi_q(S_{qi_1...i_n}^k)$ with*

$$S_{qi_1...i_n}^k := I_{qi_1}^k \times I_{qi_2}^k \times ... \times I_{qi_n}^k, \ i_1, ..., i_n \in \{1, ..., \gamma^k\}, \ q = 0, ..., 2n$$

are pairwise disjoint. Moreover, for each $k \geq 1$ each point $x \in [0,1]^n$ is covered for at least $n + 1$ values of $q = 0, ..., 2n$ by some cube $S_{qi_1...i_n}^k, i_1, ..., i_n \in \{1, ..., \gamma^k\}$.

Proof. Let $q \in \{0, ..., 2n\}$. Sprecher has proved that the intervals $\Xi(S_{qi_1...i_n}^k)$ with $i_1, ..., i_n \in \{1, ..., \gamma^k\}$ are pairwise disjoint (see Lemma 4 and the following in [Spr96]), where $\Xi : [0,1]^n \to \mathbb{R}$ is the function defined by

$$\Xi(x_1, ..., x_n) := \sum_{p=1}^{n} 2\lambda_p \varphi(x_p + aq).$$

Since $2\lambda = \sum_{p=1}^{n} 2\lambda_p < 2$ and $\varphi[0,2] = [0,2]$, we obtain range$(\Xi) \subseteq [0,4)$ and

$$\xi_q(x_1, ..., x_n) = \sum_{p=1}^{n} \lambda_p \frac{1}{2n+1} \left(\frac{1}{2}\varphi(x_p + aq) + \frac{1}{\lambda}q \right)$$

$$= \frac{1}{2n+1} \left(\frac{1}{4} \sum_{p=1}^{n} 2\lambda_p \varphi(x_p + aq) + q\frac{1}{\lambda} \sum_{p=1}^{n} \lambda_p \right)$$

$$= \frac{1}{2n+1} \left(\frac{1}{4}\Xi(x_1, ..., x_n) + q \right).$$

Thus range$(\xi_q) \subseteq \left[\frac{q}{2n+1}, \frac{q+1}{2n+1} \right)$ and $\xi_q(S_{qi_1...i_n}^k)$ is just an affine transformation of range(Ξ). Hence the intervals $\xi_q(S_{qi_1...i_n}^k)$ with $i_1, ..., i_n \in \{1, ..., \gamma^k\}, q = 0, ..., 2n$ are disjoint.

Since the length of the "gaps", i.e. the distance of any two consecutive intervals I_{qi}^k and I_{qi+1}^k is $\delta := 1/((\gamma - 1)\gamma^k)$ we can conclude that the length $\delta_k = (\gamma - 2)\delta$ of I_{qi}^k as well as the "shift length" $a = \gamma^{k-1}\delta$ are integral multiples of δ. Thus, either two shifted gaps overlap or they are disjoint. In order to overlap, the shift $aq = q/(\gamma(\gamma - 1))$ has to be an integral multiple of $1/\gamma^k$ but this implies $q \geq \gamma - 1 \geq 2n + 1$ which is impossible. It follows that each point $x \in [0,1]$ is covered for at least $2n$ of the $2n + 1$ values $q = 0, ..., 2n$ by some interval $I_{qi}^k, i = 1, ..., \gamma^k$. This implies that each point $x \in [0,1]^n$

is covered for at least $n + 1$ of the $2n + 1$ values $q = 0, ..., 2n$ by some cube $S^k_{qi_1...i_n}$ with $i_1, ..., i_n \in \{1, ..., \gamma^k\}$.

Figure 13.3 shows the squares S^k_{qij} with delay $q = 0$, precision $k = 1$, dimension $n = 2$ and base $\gamma = 10$.

In the next section we will use the Fundamental Lemma to prove the computational version of Kolmogorov's Superposition Theorem.

13.4 A computable Kolmogorov Superposition Theorem

In this section we will use the functions defined in the previous section in order to prove a computable version of Kolmogorov's Superposition Theorem. For each $n \geq 2$, $\gamma \geq 2n + 2$ and each function $g : [0, 1] \to \mathbb{R}$ we define a function $h_g : [0, 1]^n \to \mathbb{R}$ by $h_g := \sum_{q=0}^{2n} g \circ \xi_q$, i.e.

$$h_g(x_1, ..., x_n) := \sum_{q=0}^{2n} g \circ \xi_q(x_1, ..., x_n) = \sum_{q=0}^{2n} g \left(\sum_{p=1}^{n} \lambda_p \varphi_q(x_p) \right),$$

where $\xi_q, \varphi_q, \lambda_p$ are the functions and constants, respectively, as defined in the previous section. Then the statement of Kolmogorov's Superposition Theorem could be reformulated as follows: for each continuous f there exists some continuous g such that $f = h_g$. As a preparation of the proof of this Theorem (and of a computable version of it) we start with an approximation Lemma, which has been originally proved by Lorentz [Lor66]. For completeness, we adapt the proof to our setting. If X is a topological space then we denote by $\mathcal{C}(X)$ the set of continuous functions $f : X \to \mathbb{R}$ and by

$$||f|| := \sup_{x \in [0,1]^n} |f(x)|$$

we denote the supremum norm for functions $f \in \mathcal{C}([0, 1]^n)$.

Lemma 2 (Lorentz' Lemma). *Let $n \geq 2$, $\gamma \geq 2n + 2$ and consider $\theta := (2n + 1)/(2n + 2)$. For each continuous function $f : [0, 1]^n \to \mathbb{R}$ there exists some continuous function $g : [0, 1] \to \mathbb{R}$ such that*

$$||f - h_g|| < \theta ||f|| \text{ and } ||g|| < \frac{1}{n} ||f||.$$

Proof. Let $\varepsilon > 0$ be such that $\frac{n}{n+1} + \varepsilon < \theta$. Since f is uniformly continuous on $[0, 1]^n$, we can choose some $k \geq 1$ such that the oscillation of f on each cube $S^k_{qi_1...i_n}$ is less than $\varepsilon ||f||$, i.e.

$$\max_{\substack{i_1, ..., i_n \in \{1, ..., \gamma^k\} \\ q \in \{0, ..., 2n\}}} \text{diam} f(S^k_{qi_1...i_n}) < \varepsilon ||f|| .$$

Let $c_{qi_1...i_n}^k$ be the center of the cube $S_{qi_1...i_n}^k$. Now we can define $g : [0,1] \to \mathbb{R}$ by

$$g(y) := \tfrac{1}{n+1} f(c_{qi_1...i_n}^k) \text{ for all } y \in \xi_q(S_{qi_1...i_n}^k)$$

and $q = 0,...,2n$ and $i_1,...,i_n \in \{1,...,\gamma^k\}$ and by linearization for all other $y \in [0,1]$. By the Fundamental Lemma 1 g is well-defined and $||g|| < \frac{1}{n}||f||$. Moreover, for each $x \in [0,1]^n$ there are at least $n+1$ values of $q = 0,...,2n$ such that $x \in S_{qi_1...i_n}^k$ for some $i_1,...,i_n \in \{1,...,\gamma^k\}$. For these $n+1$ values of q we obtain $g \circ \xi_q(x) = \tfrac{1}{n+1} f(c_{qi_1...i_n}^k)$ and $|f(c_{qi_1...i_n}^k) - f(x)| < \varepsilon||f||$. For the remaining n values of q we obtain $|g \circ \xi_q(x)| \leq \frac{1}{n+1}||f||$. Altogether,

$$||f - h_g|| = \left\| f - \sum_{q=0}^{2n} g \circ \xi_q \right\| < \frac{n+1}{n+1}\varepsilon||f|| + \frac{n}{n+1}||f|| < \theta||f||.$$

In the following we will conclude the effective version of Kolmogorov's Superposition Theorem from Lorentz' Lemma. To this end, we have to use some further notions from computable analysis. We want to show that the operator, which maps f to a corresponding function g such that $f = h_g$, is computable. To express computability of such operators $\mathcal{C}[0,1]^n \to \mathcal{C}[0,1]$ we need a representation of $\mathcal{C}^n := \mathcal{C}([0,1]^n)$ (for short we write $\mathcal{C} := \mathcal{C}^1$). We will use the so-called *Cauchy representation* $\delta^n :\subseteq \Sigma^{\mathbb{N}} \to \mathcal{C}^n$ (cf. [Wei00]). Roughly speaking, $\delta^n(p) = f$, if and only if p is an (appropriately encoded) sequence of rational polynomials $(p_i)_{i \in \mathbb{N}}$, $p_i \in \mathbb{Q}[x_1,...,x_n]$ which converges rapidly to f, i.e. $||p_i - p_k|| \leq 2^{-k}$ for all $i > k$. Computability on the space $\mathcal{C}^n$ will from now on be understood with respect to the representation δ^n. An important property of this representation is that it allows evaluation and type conversion: the function $\mathcal{C}^n \times [0,1]^n \to \mathbb{R}, (f,x) \mapsto f(x)$ (evaluation of f at x) is computable (i.e. (δ^n, ρ^n, ρ)-computable) and for any representation δ of X a function $F : X \times [0,1]^n \to \mathbb{R}$ is (δ, ρ^n, ρ)-computable, if and only if the function $G : X \to \mathcal{C}^n$ with $G(x)(y) := F(x,y)$ is (δ, δ^n)-computable (cf. [Wei00]). For the proof of the computable Kolmogorov Superposition Theorem we will additionally use a notion of effectivity for subsets.

Definition 4 (Recursively enumerable open subsets). Let δ, δ' be representations of X, Y, respectively. An open subset $U \subseteq X \times Y$ is called (δ, δ')-r.e., if there is a (δ, δ', ρ)-computable function $F : X \times Y \to \mathbb{R}$ such that $(X \times Y) \setminus U = F^{-1}(\{0\})$.

For instance the set $\{(x,y) \in \mathbb{R}^2 : x < y\}$ is an r.e. open subset (usually we will not mention the representations if they are clear from the context). It is straightforward to generalize this definition to subsets of higher arity. For spaces Y like $\mathbb{N}, \mathbb{R}, \mathcal{C}^n$ we have the following uniformization property: *if $U \subseteq X \times Y$ is an r.e. open subset such that for any $x \in X$ there exists some $y \in Y$ such that $(x,y) \in U$, then there is a computable multi-valued operation $S : X \rightrightarrows Y$ such that* graph$(S) \subseteq U$, *i.e. $(x,y) \in U$ for any $x \in X$ and $y \in S(x)$* (cf. Theorem 12.4 in [Bra03a]).

Theorem 2 (Computable Kolmogorov Superposition Theorem). *Let $n \geq 2$ and $\gamma \geq 2n + 2$. There exists a computable multi-valued operation $K : \mathcal{C}^n \rightrightarrows \mathcal{C}$ such that $f = \sum_{q=0}^{2n} g \circ \xi_q$, i.e.*

$$f(x_1, ..., x_n) = \sum_{q=0}^{2n} g \left(\sum_{p=1}^{n} \lambda_p \varphi_q(x_p) \right)$$

for all $f \in \mathcal{C}^n$ and $g \in K(f)$.

Proof. Using the concepts of evaluation and type conversion and Proposition 3 we can conclude that the function $H : \mathcal{C} \rightarrow \mathcal{C}^n, g \mapsto h_g$ is computable. Thus, Lorentz' Lemma 2 and the fact that the supremum norm $\mathcal{C}^n \rightarrow \mathbb{R}, f \mapsto ||f||$ is computable, do directly imply that the set

$$M := \left\{ (f, g) \in \mathcal{C}^n \times \mathcal{C} : ||f - h_g|| < \theta ||f|| \text{ and } ||g|| < \frac{1}{n} ||f|| \right\}$$

with $\theta := (2n + 1)/(2n + 2)$ is an r.e. open set, i.e. there exists a computable function $F : \mathcal{C}^n \times \mathcal{C} \rightarrow \mathbb{R}$, such that $F^{-1}\{0\} = (\mathcal{C}^n \times \mathcal{C}) \setminus M$. This again implies by uniformization and Lorentz' Lemma 2 that there exists a computable multi-valued operation $L : \mathcal{C}^n \rightrightarrows \mathcal{C}$ such that $(f, g) \in M$ for all $f \in \mathcal{C}^n$ and $g \in L(f)$.

Now let $f \in \mathcal{C}^n$. We define a sequence $(g_i)_{i \in \mathbb{N}}$ of functions in $\mathcal{C}$ inductively as follows: we choose some $g_0 \in L(f)$ and if g_i is already defined, we choose some

$$g_{i+1} \in L(f - h_{g_0} - h_{g_1} - ... - h_{g_i}).$$

By Lorentz' Lemma 2 and induction we obtain

$$\left\| f - \sum_{i=0}^{r} h_{g_i} \right\| < \theta^{r+1} ||f|| \text{ and } ||g_r|| < \frac{1}{n} \theta^r ||f||.$$

These inequalities imply that the series $g := \sum_{i=0}^{\infty} g_i$ and $\sum_{i=0}^{\infty} h_{g_i}$ converge uniformly and effectively. By continuity and linearity of H we obtain

$$f = \sum_{i=0}^{\infty} h_{g_i} = h_{\sum_{i=0}^{\infty} g_i} = h_g = \sum_{q=0}^{2n} g \circ \xi_q .$$

We still have to show that the procedure which maps f to g defines a computable operation $K : \mathcal{C}^n \rightrightarrows \mathcal{C}$. This can be deduced from the previous considerations using some recursive closure schemes for multi-valued operations as presented in [Bra03a].

The proof presented here shows that the map $H : \mathcal{C} \rightarrow \mathcal{C}^n, g \mapsto h_g$ is surjective, which is a reformulation of the statement of the classical Kolmogorov Superposition Theorem 1, and it shows that H admits a computable multi-valued right inverse $K : \mathcal{C}^n \rightrightarrows \mathcal{C}$. In particular, we obtain the following corollary.

Corollary 1. *For each* $n \geq 2$ *there exist some computable functions* $\varphi_q : [0,1] \to \mathbb{R}$, $q = 0, ..., 2n$ *and computable constants* $\lambda_p \in \mathbb{R}$, $p = 1, ..., n$ *such that the following holds true: for each computable function* $f : [0,1]^n \to \mathbb{R}$ *there exists a computable function* $g : [0,1] \to \mathbb{R}$ *such that*

$$f(x_1, ..., x_n) = \sum_{q=0}^{2n} g \left(\sum_{p=1}^{n} \lambda_p \varphi_q(x_p) \right).$$

Especially, this shows that addition is universal for the class of computable functions $f : [0,1]^n \to \mathbb{R}$. However, Theorem 2 above states much more than the previous corollary: given an algorithm for f, we can *effectively* find an algorithm for some g such that $h_g = f$. Thus, there exists an algorithm (a program for K) which transfers a description of f into a description of a suitable g. And this algorithm does not only work for computable f, but even for continuous f.

13.5 Aspects of dimension

There is a straightforward interpretation of the Kolmogorov Superposition Theorem in terms of embeddings. The proof of Theorem 2 shows that

$$\xi = (\xi_0, ..., \xi_{2n}) : [0,1]^n \to \mathbb{R}^{2n+1}$$

is a computable embedding of $[0,1]^n$ into $\mathbb{R}^{2n+1}$. Here, ξ has to be injective, since the Kolmogorov Superposition Theorem especially holds for all non-injective functions $f : [0,1]^n \to \mathbb{R}$. Moreover, range($\xi$) $\subseteq [0,1]^{2n+1}$. Thus, we are in the situation of the commutative diagram of Fig. 13.5.

Here $\times_{i=0}^{2n} g : [0,1]^{2n+1} \to \mathbb{R}^{2n+1}$, $(x_0, ..., x_{2n}) \mapsto (g(x_0), ..., g(x_{2n}))$ denotes the function g, applied to $2n + 1$ variables in parallel and $\Sigma : \mathbb{R}^{2n+1} \to \mathbb{R}$

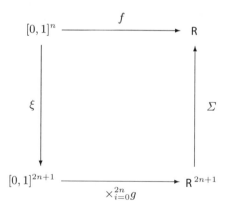

Fig. 13.5. Kolmogorov's Superposition Theorem

denotes the ordinary addition (with $2n + 1$ input variables). We formulate a corollary of our Theorem 2 in terms of this diagram.

Corollary 2. *There exists a computable embedding* $\xi : [0, 1]^n \to \mathbb{R}^{2n+1}$ *such that for any continuous function* $f : [0, 1]^n \to \mathbb{R}$ *there is a continuous function* $g : [0, 1] \to \mathbb{R}$ *such that* $f = \Sigma \circ \times_{i=0}^{2n} g \circ \xi$.

Here "continuous" might be simultaneously replaced by "computable" in both occurrences. Kolmogorov's Superposition Theorem, reformulated in this way, has a striking similarity to well-known embedding theorems in dimension theory (see [Eng78]). For instance, by a result of Nöbeling, any n-dimensional topological space can be embedded into a (very special) subspace of the $2n + 1$-dimensional Euclidean space $\mathbb{R}^{2n+1}$. Now the question arises whether the number $2n + 1$ of functions φ_q in the Kolmogorov Superposition Theorem is optimal. By results of Doss [Dos63] it was known that this is true in case of $n = 2$ if the functions φ_q are assumed to be increasing. For the general case it has been proved by Sternfeld [Ste85] that the number of functions cannot be reduced and his result even holds for much more general forms of superposition. Let X be a compact metric space and let us call a tuple $(\varphi_0, ..., \varphi_k)$ of functions $\varphi_q \in \mathcal{C}(X)$ a *basic family* for X, if every $f \in \mathcal{C}(X)$ admits a representation

$$f(x) = \sum_{q=0}^{k} g_q(\varphi_q(x))$$

for all $x \in X$ with some $g_q \in \mathcal{C}(\mathbb{R})$. In this terminology, Kolmogorov proved that $X = [0, 1]^n$ admits a basic family $(\varphi_0, ..., \varphi_{2n})$ of a very special form. Ostrand generalized this result to arbitrary compact metric spaces [Ost65, Ste79]. Sternfeld proved the following theorem which characterizes the (topological) dimension of a compact metric space in terms of the size of basic families [Ste85, Ste89, Lev90].

Theorem 3 (Sternfeld). *Let* X *be a compact metric space. Then* X *has dimension* $\dim(X) \leq n$ *if and only if it admits a basic family with at most* $2n + 1$ *functions. In case of* $\dim(X) = n$ *each such basic family has at least* $2n + 1$ *functions.*

As for non-compact spaces, Doss [Dos77] proved that the original Kolmogorov Superposition Theorem can be extended to the case of unbounded continuous functions $f : \mathbb{R}^n \to \mathbb{R}$, if one allows $4n$ functions φ_i (while $2n + 1$ functions are enough in case of compact spaces). Doss has also studied other variants of the Kolmogorov Superposition Theorem where addition is replaced by multiplication [Dos76].

13.6 Aspects of constructivity

As mentioned in the beginning, our focus here is on aspects of constructivity of the Kolmogorov Superposition Theorem. However, the notion of "constructivity" is not mathematically defined and comes associated with a whole variety of meanings. One possible interpretation is to specify that constructive mathematics (including analysis) means mathematics with intuitionistic logic [BB85] (see [Kus84] for Markov's school of constructive analysis). An alternative is to consider computability just as another classical property of classical objects as we did in the previous sections. Computable analysis, understood in this sense [Wei00], is a model of constructive analysis in the forementioned sense (but implicitly employs certain principles, such as Markov's principle[5], which are not admissible in the pure intuitionistic philosophy).

With respect to the Kolmogorov Superposition Theorem constructivity has been interpreted in different rather informal ways [Nee94, NMK93]. The proof of the computable version of the Kolmogorov Superposition Theorem which we have presented follows closely the ideas of Sprecher and his very explicit construction [Spr96]. Now one might think that much less explicit proofs of the Theorem do not allow for such a constructivization. Very elegant and brief proofs of the Kolmogorov Superposition Theorem typically employ the Baire Category Theorem [Hed71, Kah75]. Often, this principle is counted as "non-constructive" since the existence of certain objects is guaranteed without presenting a specific representative. In many cases of applications of the Baire Category Theorem [Jon97] this is actually true (for instance this is the case for the Open Mapping Theorem, the Closed Graph Theorem and Banach's Inverse Mapping Theorem [Bra03b]). However, in general one has to analyze carefully in which sense the Baire Category Theorem is applied [Bra01]. Baire's Category Theorem states that a complete metric space X cannot be decomposed into a countable union of closed subsets A_n with empty interior[6] (cf. [GP65]). Classically, we can bring this statement into the following two equivalent logical forms:

1. For all sequences $(A_n)_{n\in\mathbb{N}}$ of closed subsets $A_n \subseteq X$ with empty interior, there exists some point $x \in X \setminus \bigcup_{n=0}^{\infty} A_n$,
2. for all sequences $(A_n)_{n\in\mathbb{N}}$ of closed subsets $A_n \subseteq X$ with $X = \bigcup_{n=0}^{\infty} A_n$, there exists some $k \in \mathbb{N}$ such that A_k has a non-empty interior.

From the computational point of view the content of both logical forms of the theorem is different. While the first statement can be proved constructively

---

[5] By Markov's principle we can perform an unbounded search: whenever a recursively enumerable set A in $\mathbb{N}^2$ contains the graph of a total function $f : \mathbb{N} \to \mathbb{N}$, then it contains even the graph of a total computable function f. This has been implicitly used for the uniformization property mentioned before Theorem 2

[6] Recall that a subset is said to have *empty interior* if it does not contain a whole non-empty open ball

and computationally (i.e. there exists an algorithm which, given the sequence of subsets, determines a corresponding point x), the second statement does not allow for such an algorithmic version (actually, one can employ so-called simple sets in order to prove that there is no algorithm which can determine a corresponding index k in a reasonable sense [Bra01]).

Now, let us take a look at Kahane's version of the proof of the Kolmogorov Superposition Theorem [Kah75] (we will closely follow the exposition in [LMP97]). Let $\Phi \subseteq C$ denote the metric subspace of C (endowed with the ordinary maximum metric) of all increasing functions $\varphi : [0,1] \to \mathbb{R}$ with $\varphi[0,1] \subseteq [0,1]$. Let Φ^n denote the corresponding metric product space $(n \in \mathbb{N})$.

Theorem 4 (Kahane). *Let $\lambda_p > 0$ with $\sum_{p=1}^{n} \lambda_p \leq 1$ be rationally independent numbers. The set $K \subseteq \Phi^{2n+1}$ of all such tuples $(\varphi_0, ..., \varphi_{2n}) \in \Phi^{2n+1}$ which have the property that for any $f \in C^n$ there is some $g \in C$ satisfying*

$$f(x_1, ..., x_n) = \sum_{q=0}^{2n} g\left(\sum_{p=1}^{n} \lambda_p \varphi_q(x_p)\right)$$

is fat (i.e. a set of second Baire category)[7].

Altogether, Kahane's Theorem even states more than the Kolmogorov Superposition Theorem 1. Not only the specific functions $\varphi_0, ..., \varphi_n$ (defined with the help of Sprecher's function φ) are suitable for the proof of the Kolmogorov Superposition Theorem, but quasi all (increasing continuous functions) $\varphi_0, ..., \varphi_n$ could be employed for this purpose. However, one has to recall that "quasi all" is to be understood in a topological sense (as a synonym of "for a fat[8] set of points"). Now, the question arises whether this automatically guarantees that there are computable functions among these "quasi all" functions $\varphi_0, ..., \varphi_n$. Actually, this is the case since Kahane's proof uses the constructive version (1) of the Baire Category Theorem above and can be turned into a computable version. To this end, we define for the moment $h_{g,\varphi} : [0,1]^n \to \mathbb{R}$ by

$$h_{g,\varphi}(x_1, ..., x_n) := \sum_{q=0}^{2n} g\left(\sum_{p=1}^{n} \lambda_p \varphi_q(x_p)\right),$$

for each given $\varphi = (\varphi_0, ..., \varphi_{2n}) \in \Phi^{2n+1}$. We assume that $n \geq 2$, $\lambda_p > 0$ with $\sum_{p=1}^{n} \lambda_p \leq 1$ are some fixed rationally independent computable numbers and $\varepsilon := 1/(4n+3)$.

---

[7] We recall that a set $K \subseteq \Phi^{2n+1}$ is said *meager* (or of first Baire category) if it can be included in a countable union $K = \bigcup_{i=0}^{\infty} A_i$ of closed subsets $A_i \subseteq \Phi^{2n+1}$ with empty interior, and said *fat* (or of second Baire category) otherwise

[8] In the sense of the footnote above, i.e. of second Baire category

Lemma 3. *For any $f \in C^n$ with $f \neq 0$ the set*

$$\Omega(f) = \{\varphi \in \Phi^{2n+1} : (\exists g \in C)$$

$$\left(\|g\| < \|f\| \text{ and } \left\| f - \sum_{q=0}^{2n} h_{g,\varphi} \right\| < (1 - \varepsilon)\|f\| \right) \}$$

is an r.e. open subset of Φ^{2n+1} and $\bigcap_{f \in \mathbb{Q}[x_1,\ldots,x_n]\setminus\{0\}} \Omega(f)$ is a computable and countable intersection of such subsets (i.e. a computable G_δ-set).

From a proof which is similar to the proof of Lorentz' Lemma 2 (see Lemma 2.2 in [LMP97]) it follows that $\Omega(f)$ is also dense in Φ^{2n+1}. Since Φ^{2n+1} is a complete computable metric space, we can deduce by the computable version of the Baire Category Theorem [Bra01] that the countable intersection

$$\bigcap_{f \in \mathbb{Q}[x_1,\ldots,x_n]\setminus\{0\}} \Omega(f)$$

contains a computable point φ. On the other hand, the points $\varphi = (\varphi_0, \ldots, \varphi_{2n})$ in this intersection still satisfy the equation in Kahane's Theorem 4 (see the proof of Theorem 1.2 in [LMP97]) and thus this directly implies Corollary 1. Following the same line we could even deduce the fully uniform computable version of the Kolmogorov Superposition Theorem 2.

The proof sketched in this section is based on classical proofs of the Kolmogorov Superposition Theorem and just adds the considerations which are required in order to obtain the computable version. Other such proofs are possible as well. In contrast to this, the proof presented in Sect. 13.4 proves both: the classical and the computational version. Another shortcut for a computational version of the theorem which takes the classical version and the existence of a computable φ for granted can be based on the fact that the map $H : C \to C^n : g \mapsto h_{g,\varphi}$ is a computable linear operator on Banach spaces. The classical Kolmogorov Superposition Theorem implies that the operator is surjective and the computable version of the Open Mapping Theorem [Bra03b] states that this operator is effectively open (there is an algorithm, which given a description of an open set $U \subseteq C$ provides a description of $H(U)$). This directly implies that H has a computable multi-valued right inverse $K : C \rightrightarrows C^n$ (and thus we obtain the statement of Theorem 2). However, this alternative proof of Theorem 2 is much less informative: it does not come with any description of an algorithm for K. The existence of the algorithm is guaranteed by a non-constructive proof (since the computable Open Mapping Theorem is only effective in the open subset but not in the operator).

13.7 Applications to feedforward neural networks

In 1987, about 30 years after Kolmogorov's Superposition Theorem has been published, Hecht-Nielsen noticed an interesting application of this theorem to the theory of neural networks [Hec87, Hec90]: each continuous function $f : [0,1]^n \to \mathbb{R}$ can be implemented by a feedforward neural network with continuous activation functions $t : [0,1] \to \mathbb{R}$.

Theorem 5 (Hecht-Nielsen). *The class of functions $f : [0,1]^n \to \mathbb{R}$, implementable by three-layer feedforward neural networks with continuous activation functions $t : [0,1] \to \mathbb{R}$ and weights $\lambda \in \mathbb{R}$, is exactly the class of continuous functions $f : [0,1]^n \to \mathbb{R}$.*

Formally, a *neural network* can be defined as a directed graph where the nodes come associated with some *activation function $t : \mathbb{R} \to \mathbb{R}$* and to the edges there are assigned some *weights $w \in \mathbb{R}$*. Certain nodes are considered as input and output nodes. Such a network is called a *feedforward network*, if its graph is cycle free (otherwise it is called a *feedback network*). Typically, these networks are divided into distinguished *layers* depending on the distance to the input nodes. Such a network implements a real number function $f : [0,1]^n \to \mathbb{R}$ (if there are n input nodes and one unique output node) which is determined as follows. By each input node one number $x_i \in [0,1]$ is introduced $(1 \leq i \leq n)$. These numbers are propagated in the network in the following way. Any node takes the sum of all its inputs multiplied by the weights of the corresponding incoming edges. Then the activation function of the node is applied to the sum and the result is propagated on the outgoing edges. Finally, $f(x_1, ..., x_n)$ is the number arriving at the output node.

A priori it is not clear at all which functions can be implemented by such networks. It is due to the pioneering work of Hecht-Nielsen that the power of such networks has been identified. By Definition 3 we obtain $\varphi_q(x) := c\varphi(x+aq)+bq$ with $c := 1/(4n+2)$, $a := 1/(\gamma(\gamma-1))$ and $b := 1/((2n+1)\lambda)$. Hence, by the Kolmogorov Superposition Theorem 1 we can write f in the normal form

$$f(x_1, ..., x_n) = \sum_{q=0}^{2n} g \left(d_q + \sum_{p=1}^{n} \lambda_p c\varphi(x_p + aq) \right)$$

with $d_q := \lambda b q = \frac{q}{2n+1} = dq$, where $d := \frac{1}{2n+1}$. A similar normal form has first been derived by Sprecher [Spr65] and has been used by Hecht-Nielsen for the proof of his theorem. In case of $n = 2$ we obtain a feedforward network as shown in Fig. 13.6 (the idea of the graphical presentation is essentially taken from [SD02]).

Although by the Theorem of Hecht-Nielsen 5 the power of feedforward neural networks has been characterized in a very useful way, it seems that this observation has not have much impacts on practice in neural computing. Maybe this is at least partially due to the fact that Hecht-Nielsen's Theorem has been considered as non-constructive. In [Hec90] Hecht-Nielsen describes his theorem as follows: *"The proof of the theorem is not constructive, so it does not tell us how to determine these quantities. It is strictly an existence*

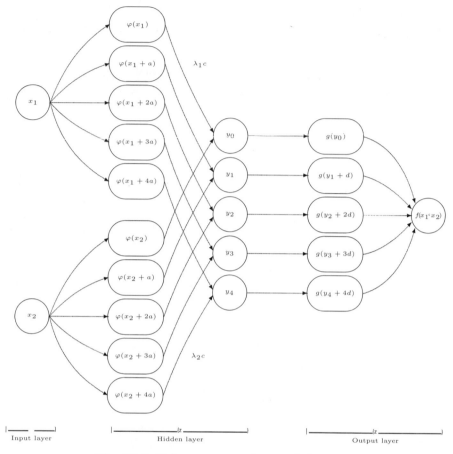

Fig. 13.6. Kolmogorov neural network for $n = 2$

theorem. It tells us that such a three-layer mapping network must exist, but it doesn't tell us how to find it. Unfortunately, there does not appear too much hope that a method for finding the Kolmogorov network will be developed soon." This is one of the reasons why much of the following research has been concentrated on approximative versions of Kolmogorov type results on neural networks (see e.g. [HSW89, LLPS93, Kur91, Kur92]). However, due to the recent work of Sprecher [Spr93, KS94, Spr96, Spr97] we are much closer to a practical usage of the exact representation of continuous functions by Kolmogorov type neural networks today. The network presented in Fig. 13.6 has a fixed hidden layer which only depends on the dimension n but not on the represented function f. Essentially, the processing units of the hidden layer are determined by Sprecher's function φ and they could be encoded in hardware. Only the function g in the output layer depends on f and has to be determined by a learning mechanism. Recently, efficient learning algorithms for such networks have been discussed [NŠD01] and higher degrees of freedom for activation functions which are motivated by Kolmogorov type networks have successfully been implemented in practice [IP03].

While the classical Hecht-Nielsen Theorem characterizes those functions which are implementable by neural networks with arbitrary real weights and continuous activation functions, it is interesting to know what happens if we restrict these components to computable ones. It turns out that we can derive the following computable version of Hecht-Nielsen's Theorem from our computable version of the Kolmogorov Superposition Theorem 2.

Theorem 6 (Computable Hecht-Nielsen Theorem). *The class of functions $f : [0,1]^n \to \mathbb{R}$, implementable by three-layer feedforward neural networks with computable activation functions $t : [0,1] \to \mathbb{R}$ and computable weights $\lambda \in \mathbb{R}$, is exactly the class of computable functions $f : [0,1]^n \to \mathbb{R}$.*

This underlines the practical importance of the Hecht-Nielsen Theorem: even if we restrict our neural networks to such networks with components which can be determined by algorithms, we do not loose the full power of Turing machines.

13.8 Conclusion

We have seen that Hilbert's 13th problem has inspired an impressing line of research which covers interesting areas of mathematics as well as applications. It turned out that behind the problem there are somewhat unexpected but important algorithmic questions (this is similar with some of the other problems in Hilbert's list, see [Kos98]).

Kolmogorov's Superposition Theorem provides a very powerful refutation of Hilbert's conjecture and still deserves a wider popularization. Even among mathematicians it is not widely known and its statement can still evoke surprise. This is within striking contrast to its general intellectual implications. Besides the discussed applications to pure mathematics and neural networks its impact on fuzzy logic, soft computing and even physics has been the subject of mathematical results and speculation [KNS96, NK97, YK99].

While the early research following Kolmogorov's result was mainly devoted to questions of smoothness, we hope that the algorithmic aspects of the result will be much more precisely analyzed. In view of the importance of Kolmogorov's Superposition Theorem with respect to its applications to neural networks, such a better understanding could even have practical implications.

We would like to close our discussion by coming back to the question of universality of addition. While it is known so far that addition is universal for the classes of continuous and computable functions (and not for the classes of continuously differentiable and analytic functions), no one has studied this question with respect to computational complexity (to the best

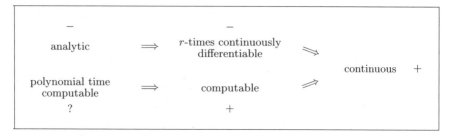

Fig. 13.7. Universality of addition for classes of functions $f : [0,1]^n \to \mathbb{R}$

of our knowledge). We have seen that computability can be considered as a strengthening of continuity which is logically independent to the smoothness conditions mentioned above. A further strengthening such as "polynomial time computability" (see [Ko91, Wei00] for definitions) could be considered and it seems to be an open question whether addition is universal for the class of polynomial time computable functions. Figure 13.7 illustrates the situation (any "+" indicates universality of addition for the corresponding class of functions, any "−" indicates the contrary; the question marks stands for the unknown case).

More generally, one could study the same question for any other complexity class and it is still a challenging open problem for mathematician and computer scientist of our century to identify and characterize those complexity classes for which addition is universal!

13.9 Acknowledgement

The author is grateful to Jiří Wiedermann for the suggestion to analyze the Kolmogorov Superposition Theorem from the point of view of computable analysis. The author has been supported by the National Research Foundation (NRF).

References

[Arn57]   Arnold, V.I.: On functions of three variables. Dokl. Akad. Nauk SSSR, **114**, 679-681 (1957) Translated in: Amer. Math. Soc. Transl. **28**, 51–54 (1963)

[BB85]   Bishop, E., Bridges, D.S.: Constructive Analysis. Grundlehren der Math. Wissen, vol. 279. Springer, Berlin (1985)

[Bra01]   Brattka, V.: Computable versions of Baire's category theorem. In: Sgall, J., Pultr, A., Kolman, P. (eds) Mathematical Foundations of Computer Science. 26th International Symposium, MFCS 2001, Mariánské Lázně, Czech Republic, August 27-31, 2001. Lecture Notes in Computer Science, vol. 2136, 224–235. Springer, Berlin (2001)

[Bra03a]   Brattka, V.: Computability over topological structures. In: Cooper, S.B., Goncharov, S.S. (eds) Computability and Models, 93–136. Kluwer Academic Publishers, New York (2003)

[Bra03b]   Brattka, V.: The inversion problem for computable linear operators. In: Alt, H., Habib, M. (eds) STACS 2003. 20th Annual Symposium on Theoretical Aspects of Computer Science, Berlin, Germany, February 27-March 1, 2003. Lecture Notes in Computer Science, vol. 2607, 391–402. Springer, Berlin (2003)

[Dix93]   Dixmier, J.: Histoire du 13-ième problème de Hilbert. In: Chabert, J.-L. et al. (eds) Analyse diophantienne et géométrie algébrique. Exposés du Séminaire d'Histoire des Mathématiques de l'Institut Henri Poincaré. Paris: Université Pierre et Marie Curie, Lab. de Mathématiques Fondamentales, Cah. Sémin. Hist. Math., **3**, 85–94 (1993)

[d'Oc99]   d'Ocagne, M.: Traité de nomographie. Gauthier-Villars, Paris (1899)

[d'Oc00]   d'Ocagne, M.: Sur la résolution nomographique de l'équation du septième degré. Comptes Rendus Académie des Sciences Paris, **131**, 522–524 (1900)

[Dos63]   Doss, R: On the representation of the continuous functions of two variables by means of addition and continuous functions of one variable. Colloq. Math., **10**, 249–259 (1963)

[Dos76]   Doss, R.: Representations of continuous functions of several variables. Amer. J. Math., **98**(2), 375–383 (1976)

[Dos77]   Doss, R.: A superposition theorem for unbounded continuous functions. Trans. Amer. Math. Soc., **233**, 197–203 (1977)

[Eng78]   Engelking, R.: Dimension Theory. North-Holland Mathematical Library, vol. 19. North-Holland, Amsterdam (1978)

[Fri95]   Fridman, B.L.: A uniqueness result for a generalized Radon transform. SIAM J. Math. Anal., **26**(6), 1467–1472 (1995)

[GP65]   Goffman, C., Pedrick, G.: First Course in Functional Analysis. Prentince-Hall, Englewood Cliffs (1965)

[Gra00]   Grattan-Guinness, I.: A sideways look at Hilbert's twenty-three problems of 1900. Notices of the AMS, **47**(7), 752–757 (2000)

[Hec87]   Hecht-Nielsen, R.: Kolmogorov's mapping neural network existence theorem. In *Proceedings IEEE International Conference On Neural Networks*, volume II, pages 11–13, New York, 1987. IEEE Press.

[Hec90]   Hecht-Nielsen, R.: Neurocomputing. Addison-Wesley, Reading (1990)

[Hed71]   Hedberg, T.: The Kolmogorov superposition theorem. In: Shapiro, H.S. (ed) Topics in Approximation Theory, Lecture Notes in Mathematics, vol. 187. Springer, Berlin (1971)

[Hil00]   Hilbert, D.: Mathematische Probleme. Vortrag, gehalten auf dem internationalen Mathematiker-Kongreß zu Paris 1900. Nachr. König. Gesell. Wissen. Göttingen, 253–297 (1900)

[Hil01]   Hilbert, D.: Mathematische Probleme. Vortrag, gehalten auf dem internationalen Mathematiker-Kongreß zu Paris 1900. Arch. Math. Physik, **1**(3), 44–63, 213–237 (1901)

[Hil27]   Hilbert, D.: Über die Gleichung neunten Grades. Math. Ann., **97**, 243–250 (1927)

[HSW89]   Hornik, K., Stinchcombe, M., White, H.: Multilayer feedforward networks are universal approximators. Neural Networks, **2**, 359–366 (1989)

278   Vasco Brattka

[IP03]   Igelnik, B., Parikh, N.: Kolmogorov's spline network. IEEE Transactions on Neural Networks, **14**(4), 725–733 (2003)
[Jon97]   Jones, S.H.: Applications of the Baire category theorem. Real Anal. Exchange, **23**(2), 363–394 (1997)
[Kah75]   Kahane, J.-P.: Sur le théorème de superposition de Kolmogorov. Journal of Approximation Theory, **13**, 229–234 (1975)
[Kah82]   Kahane, J.-P.: Le 13-ième problème de Hilbert : un carrefour de l'algèbre, de l'analyse et de la géometrie. In: Proceedings of the Seminar on the History of Mathematics 3, Inst. Henri Poincaré, Paris. Cah. Sémin. Hist. Math., **3**, 1–25 (1982)
[KNS96]   Kreinovich, V., Nguyen, H.T., Sprecher, D.A.: Normal forms for fuzzy logic - an application of Kolmogorov's theorem. Internat. J. Uncertain. Fuzziness Knowledge-Based Systems, **4**(4), 331–349 (1996)
[Ko91]   Ko, K.I: Complexity Theory of Real Functions. Progress in Theoretical Computer Science. Birkhäuser, Boston (1991)
[Kol56]   Kolmogorov, A.N.: On the representation of continuous functions of several variables by superposition of continuous functions of a smaller number of variables. Dokl. Akad. Nauk SSSR, **108**, 179–182 (1956) Translated in: Amer. Math. Soc. Transl. **17**, 369–373 (1961)
[Kol57]   Kolmogorov, A.N.: On the representation of continuous functions of many variables by superposition of continuous functions of one variable and addition. Dokl. Akad. Nauk SSSR, **114**, 953–956 (1957) Translated in: Amer. Math. Soc. Transl. **28**, 55–59 (1963)
[Kos98]   Kosheleva, O.M.: Hilbert problems (almost) 100 years later (from the viewpoint of interval computations). Reliable Computing, **4**(4), 399–403 (1998)
[KS94]   Katsuura, H., Sprecher, D.A.: Computational aspects of Kolmogorov's superposition theorem. Neural Networks, **7**(3), 455–461 (1994)
[Kur91]   Kůrková, V.: 13th Hilbert's problem and neural networks. In: Novák, M., Pelikán, E. (eds) Theoretical Aspects of Neurocomputing, 213–216. Symposium on Neural Networks and Neurocomputing, NEURONET, Prague, 1990. World Scientific, Singapore (1991)
[Kur92]   Kůrková, V.: Kolmogorov's theorem and multilayer neural networks. Neural Networks, **5**, 501–506 (1992)
[Kus84]   Kushner, B.A.: Lectures on Constructive Mathematical Analysis. Translations of Mathematical Monographs, **60**. American Mathematical Society, Providence (1984)
[Lev90]   Levin, M.: Dimension and superposition of continuous functions. Israel J. Math., **70**(2), 205–218 (1990)
[LGM96]   Lorentz, G.G., Golitschek, M., Makovoz, Y.: Constructive Approximation. Grund. Math. Wissen., **304**. Springer, Berlin (1996)
[LLPS93]   Leshno, M., Lin, V.Ya., Pinkus, A., Schocken, S.: Multilayer feedforward networks with a nonpolynomial activation function can approximate any function. Neural Networks, **6**, 861–867 (1993)
[LMP97]   Leiderman, A., Morris, S.A., Pestov, V.: The free abelian topological group and the free locally convex space on the unit interval. J. London Math. Soc., **56**(3), 529–538 (1997)
[Lor66]   Lorentz, G.G.: Approximation of Functions. Athena Series, Selected Topics in Mathematics. Holt, Rinehart and Winston, Inc., New York (1966)

[Lor76]   Lorentz, G.G.: The 13-th problem of Hilbert. In: Browder, F.E. (ed)
             Mathematical developments arising from Hilbert problems. Proceed-
             ings of the Symposium in Pure Mathematics of the AMS, **28**, 419–430.
             American Mathematical Society, Providence (1976)
[Nee94]   Nees, M.: Approximative versions of Kolmogorov's superposition the-
             orem. proved constructively. J. Comput. Appl. Math., **54**(2), 239–250
             (1994)
[NK97]   Nguyen, H.T., Kreinovich, V.: Kolmogorov's theorem and its im-
             pact on soft computing. In: Yager, R.R., Kacprzyk, J. (eds) The Or-
             dered Weighted Averaging Operators: Theory and Applications, 3–17.
             Kluwer, Boston (1997)
[NMK93]   Nakamura, M., Mines, R., Kreinovich, V.: Guaranteed intervals for Kol-
             mogorov's theorem (and their possible relation to neural networks).
             Proceedings of the International Conference on Numerical Analysis
             with Automatic Result Verification (Lafayette, LA, 1993). Interval
             Computations, **3**, 183–199 (1993)
[NŠD01]   Neruda, R., Štědrý, A., Drkošová, J.: Implementation of Kolmogorov
             learning algorithm for feedforward neural networks. In: Alexandrov,
             V.N. et al. (eds) Computational Science - ICCS 2001. International
             conference, San Francisco, CA, USA, May 28-30, 2001. Lecture Notes
             in Computer Science, vol. 2074, 986–995. Springer, Berlin (2001)
[Ost65]   Ostrand, P.A.: Dimension of metric spaces and Hilbert's problem 13.
             Bull. Amer. Math. Soc., **71**, 619–622 (1965)
[PR89]   Pour-El, M.B., Richards, J.I.: Computability in Analysis and Physics.
             Perspectives in Mathematical Logic. Springer, Berlin (1989)
[SD02]   Sprecher, D.A., Draghici, S.: Space-filling curves and Kolmogorov
             superposition-based neural networks. Neural Networks, **15**, 57–67
             (2002)
[Spr65]   Sprecher, D.A.: On the structure of continuous functions of several vari-
             ables. Transactions American Mathematical Society, **115**(3), 340–355
             (1965)
[Spr93]   Sprecher, D.A.: A universal mapping for Kolmogorov's superposition
             theorem. Neural Networks, **6**, 1089–1094 (1993)
[Spr96]   Sprecher, D.A.: A numerical implementation of Kolmogorov's super-
             positions. Neural Networks, **9**(5), 765–772 (1996)
[Spr97]   Sprecher, D.A.: A numerical implementation of Kolmogorov's super-
             positions II. Neural Networks, **10**(3), 447–457 (1997)
[Ste79]   Sternfeld, Y.: Superpositions of continuous functions. J. Approx. The-
             ory, **25**(4), 360–368 (1979)
[Ste85]   Sternfeld, Y.: Dimension, superposition of functions and separation
             of points, in compact metric spaces. Israel J. Math., **50**(1/2):, 13–53
             (1985)
[Ste89]   Sternfeld, Y.: Hilbert's 13th problem and dimension. In: Linden-
             strauss, J., Milman, V.D. (eds) Geometric aspects of functional anal-
             ysis. Lecture Notes in Mathematics, vol. 1376, 1–49. Springer, Berlin
             (1989)
[Vit54]   Vitushkin, A.G.: On the Hilbert's thirteenth problem (in Russian).
             Dokl. Akad. Nauk SSSR, **95**, 701–704 (1954)

[Vit69]   Vitushkin, A.G.: On Hilbert's thirteenth problem (in Russian). In: Alexandrov, P.S. (ed) Hilbert's problems, 163–170. Nauka, Moscow (1969) Translated to German as: Die Hilbertschen Probleme. Verlag Harri Deutsch, Frankfurt/Main, 4th edition (1998)

[Vit77]   Vitushkin, A.G.: On representation of functions by means of superpositions and related topics. Enseignement Math. **23**(3-4), 255–320 (1977)

[Wei00]   Weihrauch, K.: Computable Analysis. Springer, Berlin (2000)

[YK99]   Yamakawa, T., Kreinovich, V.: Why fundamental physical equations are of second order? International Journal of Theoretical Physics, **38**(6), 1763–1770 (1999)

Kolmogorov complexity

Bruno Durand[1] and Alexander Zvonkin[2]

[1] Laboratoire d'informatique fondamentale de Marseille, University of Provence
  (Aix-Marseille I), France
  http://www.lif.univ-mrs.fr/∼bdurand
  Bruno.Durand@lif.univ-mrs.fr
[2] Laboratoire Bordelais de Recherche en Informatique (LaBRI), University
  Bordeaux I, France
  http://www.labri.fr/perso/zvonkin
  zvonkin@labri.fr

The term "complexity" has different meanings in different contexts. *Computational complexity* measures how much time or space is needed to perform some computational task. On the other hand, the *complexity of description* (called also *Kolmogorov complexity*) is the minimal number of information bits needed to define (describe) a given object. It may well happen that a short description requires a lot of time and space to follow it and actually construct the described object. However, when speaking about Kolmogorov complexity, we usually ignore this problem and count only the description bits.

As it was common to him, Kolmogorov published, in 1965, a short note [Kol65] that started a new line of research. Aside from the formal definition of complexity, he has also suggested to use this notion in the foundations of probability theory. His idea was quite simple:

> An object is random if it has maximal possible complexity.

The definition of complexity uses the notion of an algorithm; this unexpected marriage of two *a priori* distant domains —in our case, probability theory and theory of algorithms— is also a typical trait of Kolmogorov's work.

14.1 Algorithms

The notion of an algorithm in quite recent. In 1912 (when neither computers nor programming languages existed) Émile Borel (see [US93]) used the phrase "a formal and precise automatic rule" describing an object which we would now call an algorithm[3]. However, a mathematical theory of algorithms was developed only in the 1930s (by Turing, Gödel, Post, Church, Kleene and others). The key observation was the existence of a universal algorithm (see below); it allows to prove easily that some problems (e.g. the so-called "halting problem" that asks whether a given algorithm terminates on a given input) are undecidable (cannot be solved by algorithms). Note that to prove the non-existence of an algorithm that solves a certain problem we need a mathematically precise definition of this notion. When appeared, this notion became a subject of the theory of algorithms, also called *theory of recursive functions* or *theory of computability*.

The remaining part of this section discusses some aspects of the notion of algorithm; the reader not interested in these details may skip it and proceed directly to Sect. 14.2.

It is rather difficult to give a mathematical definition that captures the intuitive idea of an algorithm in its full generality; instead, we may define a specific class of algorithms and claim that this class is representative, i.e. that any algorithm is equivalent to a certain algorithm in this class. (By the way, one of these classes was suggested by Kolmogorov.)

14.1.1 Models of computation

A model of computation formally describes some specific class of algorithms (the class of objects used as input/output data, how they are processed, etc.) Some computational models resemble programming languages while others look more as a hardware description. In any case, we assume that computational resources are unlimited (and forget that in real programming languages integers are usually bounded, processor architecture has a fixed word length, etc.).

(The study of resources −time and space− needed to solve a given problem is a different field called *computational complexity*. Let us note that an important notion in this field, *NP-completeness*, was introduced at the beginning of the 1970s independently by three researchers, one of whom, Leonid Levin, is Kolmogorov's student. The first publications by Levin were about Kolmogorov complexity [ZL70]. His short biography and a brief story how Kolmogorov influenced him may be found in the book [SL95].)

---

[3] The history of the term "algorithm" is interesting in itself. This word is a derivative of the name of a medieval Persian savant Al-Khwārizmī (787 – c. 850) who was the author of a book through which the Europeans learned the positional number system and the rules of arithmetic operations (addition, multiplication, etc.). The name of Al-Khwārizmī (which means "of Khorezm", a town in Uzbekistan today called Khiva) was transliterated in Latin as *Algorithmus*. The term "algorithms" meant at the beginning "the rules of four arithmetic operations". Then by extension it has got the meaning of any systematic method of computation. Leibnitz called "algorithms" the set of rules of computing differentials and integrals. It is only gradually that the word acquired its modern meaning; one hundred years ago this process was not yet finished

Which computational model is "the best one"? This depends on our purposes. If we want to write real programs, it is natural to use a real computer and an appropriate programming language. On the other hand, if we want to prove theorems it would be more convenient to work with an abstract model of computation; a very simple model, with a small number of primitives, would then be better. However, there is no canonical model adapted for proofs since different models are more suitable for different results.

The most popular model is Turing machine. It is rather easy to prove the universality of this model; however, we have to deal with many details concerning tapes, symbols, representation of the transition table, etc. There are many versions of Turing machines; the most common one was, by the way, presented by Post and not by Turing.

Recursive functions "à la Church" give a more mathematical and attractive model though the proofs of certain basic theorems become somewhat discouraging if not frightening.

Markov algorithms are similar to rewriting systems for strings with termination conditions; this is a model difficult to manipulate (but well suited for the proof of the undecidability of word problems).

The RAM (random access machines) model resembles von Neumann-style computers...

Teaching the algorithms theory, one may choose a different approach and not fix any specific model but rely directly on the intuition of algorithms. More formally, it means that we have to accept some properties of algorithms used in the proofs as axioms. Then we do not need to go into cumbersome details of a specific computational model; the price is, however, that the list of axioms is open (e.g. if during the proof we need to establish the computability of some function, we just describe informally its computation and then add a new axiom saying that this function is computable).

14.1.2 All models of computation are equivalent

Why do we believe that this or that computational model correctly reflects the intuitive notion of an algorithm? This statement is usually called "the Church thesis" (for a given computation model): it claims that any computable function (computed by an algorithm in the informal sense) is computable in this model. This assertion is not a mathematical one; it is a belief concerning the notion of intuitive computability. On the other hand, we can prove that these assertions for different computation models are equivalent, since it turns out that the class of computable functions is the same for different existing models (Turing machines, recursive functions, etc.).

The name given to the thesis is rather inappropriate. Church claimed that all intuitively computable *total* functions are computable in his model. A long controversy followed, in which Gödel took sometimes surprising positions [BG03]. The first equivalence theorem for two different models (recursive functions "à la Church" and Turing machines) was established by Turing in his seminal article, and the thesis in its most general form was formulated by Post. Therefore, a more appropriate name would be "Church–Turing–Post thesis".

All this was done in the 1930s, so why Kolmogorov might want to suggest a different computation model in the 1950s? His motivation could be reconstructed as follows. Though all computation models mentioned above are equivalent, the translation between them sometimes replaces one step in one model by a long sequence of steps in another one. For example, an addition may be an elementary operation in some programming language while its implementation by Turing machine requires many steps.

Kolmogorov wanted to find a model whose steps are "elementary" in the sense that they do not allow natural decomposition into a sequence of simpler steps. On the other hand, he tried to find a most general (and natural) model among these models. This means that elementary steps of any other model (if they are indeed elementary according to our intuition) should not require further decomposition when translated into Kolmogorov's model.

14.1.3 Kolmogorov–Uspensky machines

The model suggested by Kolmogorov was later called *Kolmogorov–Uspensky machines*. These machines are not related to Kolmogorov complexity, but they are related to Kolmogorov himself; hence we say a couple of words about them.

The configuration (state of the computation) of a Kolmogorov–Uspensky machine is a graph; some node of this graph is declared to be active. The program for the machine is a list of rules that say how this active part should be transformed and when the processing halts. So the computation step is indeed "local"; it deals with a finite size neighborhood of the active node. On the other hand, the "topological structure" of the computation can become rather complicated. This may be considered as a disadvantage of the model since it allows some actions that are hard to perform in a physical space. (For example, a Kolmogorov–Uspensky machine can create a labeled tree that provides an unreasonably fast access to an exponential amount of information.) So one may want to restrict somehow the class of allowed graphs [US93, Gur88, BG03]. Later a version of this model was considered by Schönhage (who used directed graphs with unlimited in-degrees). It seems pertinent to mention here the development of the GASM (Gurevich Abstract State Machines) which were inspired by Kolmogorov–Uspensky machines but have other goals and do not play a specific role in the classical computability theory. The first complete description of Kolmogorov–Uspensky machines may be found in [KU58]; a more modern presentation is given in [US93].

14.1.4 Universality

Now we are accustomed to the idea that the same processor can be used to perform different tasks if provided with a suitable program. However, this idea of "universal computation" was a nontrivial and very important step in the development of the first real computers.

The same idea can be formally expressed as follows: there exists a *universal* computable function U of two arguments p and x. The universality means that we can obtain any computable function of x by fixing an appropriate first argument p (a *program* for this function).

Why does a universal function exist? Imagine an interpreter of an arbitrary programming language that considers its first argument p as a program and executes this program using x as its input.

14.1.5 Non-computable functions

The existence of a universal computable function immediately brings us to a paradox. Consider the function $F(p) = U(p, p) + 1$. This (unary) function is computable since U is. It should then have a program associated to it (since U is universal); let us denote this program by q. What happens if we apply program q to itself? By definition of U this gives $U(q, q)$. On the other hand, since q is a program for F, the same result must be equal to $F(q) = U(q, q) + 1$. So we get $U(q, q) = F(q) = U(q, q) + 1$, and this seems impossible.

The only way to explain this paradox is to recall that certain computations may never terminate, so a program may compute a non-total function. And the contradiction disappears if $U(q, q)$ is not defined.

A similar argument shows that the halting problem is undecidable: there is no algorithm that gets a program p and input x and tells whether $U(p, x)$ is defined (= whether the program p terminates on input x).

14.1.6 Back to algorithms

Returning to practice, let us note that the notion of a computable function captures only one aspect of algorithmic practice. For example, the behavior of a real-time algorithm (such as an operating system) is a more complicated thing than a mere function. The choice of a correct mathematical model for this class of algorithms (very important for practice) is a well studied but not fully solved problem of theoretical computer science.

14.2 Descriptions and sizes

Any information may be encoded as a bit string (a finite sequence of bits). For this reason, in what follows we assume that our algorithms deal with bit strings. Binary strings are also called *words in the alphabet* $\mathbb{B} = \{0, 1\}$, and the set of all binary strings is denoted as $\mathbb{B}^*$. We identify $\mathbb{B}^*$ with the set $\mathbb{Z}^+ \setminus \{0\} = \{1, 2, 3, \ldots\}$ using the lexicographic order. (The empty word is associated with 1, then $0 \mapsto 2$, $1 \mapsto 3$, $00 \mapsto 4$, $01 \mapsto 5$, etc.: a string u

is associated with a natural number that has binary representation $1u$. For example, the word 00 corresponds to the number 100_2, i.e. 4.)

The length $|u|$ of a binary word u, i.e., the number of letters in it, is then equal to the integral part $\lfloor \log u \rfloor$ of the binary logarithm of the number associated with u. (Note that $|u|$ stands for the length of the word u and not for the absolute value of the corresponding integer.)

Definition 1. *Let $f : \mathbb{B}^* \to \mathbb{B}^*$ be a computable function. We define the complexity of $x \in \mathbb{B}^*$ with respect to f as*

$$K_f(x) = \begin{cases} \min |t| \ \textit{such that } f(t) = x, \\ \infty \qquad \textit{if such } t \textit{ does not exist.} \end{cases}$$

In other terms, we call *descriptions* of x (with respect to f) all strings t such that $f(t) = x$; then the complexity $K_f(x)$ is defined as the length of the shortest description.

The main problem with this definition is that the complexity depends on the choice of f. It is unavoidable, but the theorem stated below (due to Kolmogorov but already present, in an informal way, in the paper of Solomonoff [Sol64]) explains in which way this dependence can be limited. This theorem was later independently proved by Chaitin but does not appear in his first papers on the subject [Cha66a, Cha66b] − the priority claims have provoked a long and futile controversy explained in [LV97].

Theorem 1 (Existence of an optimal function). *There exists a computable function f_0 (called optimal function) such that for any other computable function f there exists a constant C such that*

$$\forall x \quad K_{f_0}(x) \leq K_f(x) + C. \tag{14.1}$$

(Note that the constant C may depend on f but not on x.)

Proof. Let t be a shortest description of x with respect to f, i.e. $f(t) = x$. Then f_0 uses as a description of x the pair (p, t) where p is a program that computes the function f. In this pair p has $|p|$ bits and t has $|t|$ bits, so the total number of bits is $|p| + |t|$, i.e. $|p| + K_f(x)$. So we let $C = |p|$. $\quad\square$

Remark 1. This argument needs some refinement. We cannot use the pair (p, t) directly; we need to encode it by a single string. Not any encoding will work. An appropriate encoding may encode p in a very inefficient way − this only increases the constant C. On the other hand, it is essential to be able to encode t without any loss of space since an encoding of t which demands, say, $\alpha|t|$ bits with $\alpha > 1$ leads to the complexity $\alpha K_f(x) + C$ instead of $K_f(x) + C$.

Corollary 1. *If f_1 and f_2 are two optimal functions then there exists a constant C such that*

$$\forall x \quad |K_{f_1}(x) - K_{f_2}(x)| \leq C. \tag{14.2}$$

Proceeding from this corollary, we choose some optimal function f_0 and fix it. The subscript f_0 in K_{f_0} is then suppressed. However, after doing this we still have in mind that in fact the Kolmogorov complexity is defined only up to a bounded additive term.

Definition 2. *The* Kolmogorov complexity $K(x)$ *is the complexity $K_{f_0}(x)$ with respect to some optimal function f_0. The complexity $K(x)$ is defined up to a bounded additive term.*

Proposition 1.

$$K(x) \leq |x| + C, \quad or,\ equivalently, \quad K(x) \leq \log x + C. \qquad (14.3)$$

Proof. It suffices to let $f(x) = x$ in (14.1), i.e. to use x itself as a description of x.   □

Proposition 2 (Distribution of complexities). *Consider all binary strings of length n. The fraction of strings x of length n such that $K(x) < n - k$ does not exceed 2^{-k}.*

Proof. The number of strings of length n is 2^n while the number of (potential) descriptions of length less than $n - k$ is

$$1 + 2 + \ldots + 2^{n-k-1} < 2^{n-k}.$$

□

There exist strings of length n whose complexity is at least n (they are often called *incompressible* strings). Indeed, there are 2^n strings of length n and at most $1 + 2 + \ldots + 2^{n-1} = 2^n - 1$ potential descriptions of length less than n.

One may ask for an example of an incompressible string. However, it is not possible to find an incompressible string of length n effectively (having n as input). Indeed, if it were possible, a string generated by this algorithm would have complexity $\log n + c$ since we need to specify n (about $\log n$ bits) and the algorithm itself (constant number of bits), and $\log n + c$ is less than n for all sufficiently large n.

Incompressible strings are a useful tool in theoretical computer science (automata theory, formal languages, etc.).

Today everybody uses software for data compression and decompression; this was not the case in the 1960s when Kolmogorov complexity was introduced. However, the Kolmogorov complexity theory may still provide useful hints: for example, if a software advertisement claims that a latest version of the super-compressor compresses every file by a certain factor, you better avoid this product.

Finally, to prepare for the next section (on Gödel's incompleteness theorem), we present a variation on a well known theme of *busy beavers*. Initially the busy beaver numbers were defined as follows. Consider Turing machines that have at most n states and whose tape alphabet consists of two symbols (say, "blank" and "stroke"). We start such a machine on the blank tape. Some machines do not terminate at all. For the machines that terminate we count the number of steps; let $T(n)$ be the maximal number of steps among the terminating machines with at most n states.

Evidently, $T(n)$ is an increasing function of n since we consider all machines that have *at most* n states. It grows very fast; in fact, it grows faster that any computable function (does not have a computable upper bound). Indeed, if a computable upper bound $f(n)$ exists, it may be used to solve the halting problem, since we know that if a machine with n states does not terminate after $f(n)$ steps, it will never terminate. So no computable function, even a fast growing one, like $n!^{n!^{\cdot^{\cdot^{\cdot^{n!}}}}}$ ($n!$ levels), is an upper bound for $T(n)$.

But here we consider a different (but related) fast-growing function. Let us define $\delta(n)$ as the biggest integer that has complexity less than n. It exists since the number of descriptions of size less than n is finite. By definition we have $n \leq K(x)$ for any $x > \delta(n)$, e.g. for $x = \delta(n) + 1$. If the function δ were computable we would have $K(\delta(n) + 1) \leq \log n + C$ since n might serve as a description of $\delta(n) + 1$. The contradiction is evident. Hence, δ is not computable. In a similar way we can prove that δ grows faster than any computable function. (It suffices to replace $\delta(n)$ in the preceding inequalities by any computable upper bound for δ.)

14.3 Gödel's theorem

14.3.1 It is proved that one cannot prove everything

The function $K(x)$ is not computable. How can we use it? For example, to prove theorems. Maybe the most remarkable example is the proof of Gödel's incompleteness theorem. Roughly speaking, this theorem claims that not all the truths are provable. Mathematics has its intrinsic limits: there exist propositions that are true but impossible to prove.

We propose to you a more "concrete" form of a proposition that is true but unprovable; it was suggested by Gregory Chaitin [Cha70].

Theorem 2 (Gödel's incompleteness theorem). *There exists a number m such that for every x the proposition*

$$\boxed{K(x) \geq m}$$

is unprovable.

Note that the set of all x such that $K(x) < m$ is finite. So the proposition $K(x) \geq m$ is *true* for infinitely many values of x. And all these truths have no proof.

Proof of Gödel's theorem.

We use the same argument as in the previous section (when we proved that the busy beaver function $\delta(n)$ is non-computable) with some modifications.

Suppose that the statement is false, i.e. that for any integer m there exists x such that the proposition "$K(x) \geq m$" is provable. Then consider an algorithm that finds this x given m:

- input: an integer m;
- enumerate all the theorems (a theorem is a proposition which has a proof);
- as soon as a theorem "$K(x) \geq m$" is found, return x.

Using this algorithm, we may consider its input m as a description of its output x. Therefore, according to (14.3), $K(x) \leq \log m + C$. But, on the other hand, $K(x) \geq m$ is a theorem (and therefore is true; we assume that all theorems are true, otherwise our notion of proof would be bad). So

$$m \leq \log m + C \,.$$

The constant C is "absolute": it depends neither on m nor on x. So we get a contradiction, since this inequality is false for sufficiently large m.   □

For a neophyte it is difficult to appreciate fully this simple argument. One should know, however, that Gödel's theorem had literally shattered the mathematical community at the beginning of the 1930s. The projects and hopes of great mathematicians, such as David Hilbert, to get a complete formal theory as a framework for mathematics were reduced to nothing. Gödel's theorem became (and remains) one of the basic results and one of the gems of mathematical logic. (Numerous volumes are devoted to this theorem, including philosophical essays and popular expositions; the bestseller by Douglas Hofstadter [Hof79] has 800 pages.) Generations of logicians tried to understand fully *why* and *how* mathematics is incomplete. Due to all their work, we are now able to explain the proof of this theorem in a single paragraph.

A philosopher once remarked that "every profound idea passes through three stages during its development: (1) it is impossible; (2) it is maybe possible but incomprehensible; (3) it is trivial". It seems that Gödel's theorem has already arrived to the third stage.

14.3.2 Formal systems

Of course, our account of the complexity proof of Gödel theorem is quite informal. An informed reader may be worried about this. He had probably heard the words *formal systems*. Indeed, we speak about proofs and theorems but do not say what the axioms or inference rules are (or any other proof machinery). It turns out, however, that in fact we do not need to go into these details. There is only one property of the proof system that is necessary.

Definition 3 (Formal system). *A proof system is an algorithm that generates statements, and all these statements are true.*

All usual proof systems (based on axioms and inference rules) are formal systems according to the above definition. Indeed, theorems can be enumerated in the following way: write all the strings of characters in a certain order; for each of them check whether it is a derivation (starts with axioms, follows inference rules, etc.); if yes, find the statement that has been derived (usually the last statement of the derivation) and output this statement.

This assumes that there is an algorithm that can distinguish derivations from other character strings, but this is true for all reasonable formal systems. Otherwise, how could we check that a proof is correct? (By a vote of members of a jury? By asking an oracle, a prophet or another sort of authority? By a tournament of knights like in Middle Ages? By drawing lots?) This is indeed the basic underlying idea of a formal system; the correctness of proofs should be verified "mechanically", that is, by an algorithm.

14.3.3 Berry paradox

Gregory Chaitin, who suggested this remarkable proof, mentioned that this proof is a formalization of the "Berry paradox" published by Bertrand Russell in 1908. It considers

> the smallest integer N whose description
> needs more than thousand words.

First of all, the integers that need more than one thousand words in order to be described, do exist — just because the number of shorter descriptions is finite. Therefore, the boxed sentence characterizes the integer N without ambiguity; in other words, it is a description of N. But it contains less than thousand words!

Quite often a paradox appears since we refer to a notion not well defined. What is this notion here? The notion of the smallest element (in a set of positive integers) used in the phrase is well defined: the axiom of induction implies that every non-empty subset of $\mathbb{N}$ has the smallest element. On the other hand, the notion of "description" is indeed not well formalized. Kolmogorov complexity provides a formal framework for this notion. Then, replacing words by bits, consider (for every m)

> the smallest integer N such that $K(N) > m$.

Such an integer exists for every m; however, this expression (for a given m) is not a description of N in Kolmogorov's sense, since there is no algorithm that finds this N. But if we change the sentence and say

> the first integer N such that $K(N) > m$ is provable

(where "first" means "first in the sequence of generated proofs"), then it is indeed a description of complexity $\log m + c$, and the only way to avoid the contradiction is to conclude that for some m there are no proofs of statements of the form $K(N) > m$. As you see, we come to the proof of Gödel's theorem explained above.

14.3.4 Gödelian propositions: "concrete" examples

Good students often ask: is it possible to give a concrete example of an unprovable proposition?

This question sets a trap for us. Without doubt, we mean a true but unprovable proposition. But how, then, could we know that some proposition is true if it has no proof? Apparently, we should have other reasons, and very strong reasons, by the way, in order to *believe* that it is true.

Logicians know different ways to address this problem. For example, we can provide two statement "A" and "not A" that are unprovable. Then we know that at least one of them is true but unprovable. (But this probably cannot be considered as a "concrete" example.)

The other possibility is to consider different theories: a weak one (e.g. first-order arithmetic) and a stronger one (second-order arithmetic or set theory). Then we show a statement that is not provable in the weak theory, and this fact as well as the statement itself can be proved in the strong theory.

This is the approach found by Gödel himself. He proved that by using only (first-order) arithmetic it is impossible to prove that this theory (first-order arithmetic) is *consistent*, i. e. that it does not contain a contradiction. But the consistency of the (first-order) arithmetic can be proved in the set theory or second-order arithmetic (and, last but not least, it is confirmed by mathematical practice).

Kolmogorov complexity provides us with another procedure of producing *Gödelian* (that is, true but unprovable) propositions. Let us suppose that the number m in the Gödel theorem is, say, 100. (A careful reasoning can indeed provide some specific value for m. It depends on the formal system we use and the optimal function we choose in the definition of Kolmogorov complexity.)

Then we may toss up a coin, say, 500 times, and then claim that the complexity of the sequence of bits obtained is greater than 100. This statement will be impossible to prove, but we may be practically sure that it is true: the probability of getting a false statement in this way is less than 2^{-400} (see the proposition about the distribution of complexities on p. 287). We thus obtain an arithmetic statement which we believe to be true for probabilistic reasons.

14.4 Definition of randomness

14.4.1 Questions, questions, questions...

The more we think about the notions of probability and randomness, the more difficult is to explain even the most "basic" things. Let us start by an example borrowed from the everyday life. Suppose that you see a car whose number on the licence plate is 7777 ZZ 77. This number seems rather extraordinary, doesn't it? As to the number 7353 NY 42, it seems perfectly "normal". Why?

We would like to say: because the first number has very small probability. Yet, this answer is not valid: the probability of the first number is exactly the same as that of the second. If we suppose that all digits and all letters are equiprobable and independent, then this probability is equal to $1/(10^6 \times 26^2)$. When we toss up a coin 1000 times, the probability to get 1000 heads is 2^{-1000}, but the probability of every other sequence of heads and tails is exactly the same! Why then does the sequence of 1000 identical tosses arouse a suspicion as to its random character?

If we think more about this phenomenon, we finally understand that, in fact, while speaking about car "numbers", we do not mean individual numbers but *sets* of "similar" numbers. The first number is a representative of the set of numbers where "the digits are repeated, and also the letters". This set is simple to describe, and its probability is small. As to the second number, it is "just a number". We are unable to outline its specific simple property which would describe a set of small probability. (And, if you *are*, this was not intended by the authors.) This is related to complexity: a simple property that is true only for few objects makes these objects simple.

Now, let us go further: what is *probability*?

Despite what one might believe, probability theory (whose rigorous mathematical foundation was provided by Kolmogorov himself in 1933) does not answer this question. This theory formulates, in a form of axioms, the properties of probabilities. It also permits to calculate probabilities of certain events when probabilities of other events are known. Thus, it treats probability theory just as any other branch of mathematics, without bothering much about "useless" philosophical questions. People were quite satisfied by this situation − except for Kolmogorov.

How would you explain to an intelligent person with no mathematical background what probability is? You claim that when one tosses a coin, the probability to get a head is $1/2$. Then he starts to question you:

− I don't understand the word "probability" in your sentence.
− I mean that the chances to get a head or a tail are equal.
− Hm... you've replaced the word "probability" by "chance", but what does it mean?
− OK, OK. I would only like to say that in, say, a thousand of coin tosses you get approximately half of heads and half of tails.

- Ah... It seems that I begin to understand something. For the moment, I won't ask you how precise this approximation is. But please tell me: do you really guarantee that the fraction of heads is *always* close to one half?
- Alas, no. It is not always the case, but it is true with a very high probability. However, there remain extremely small chances to get (for example) only the heads.
- "With a very high probability"! Again this word! You started to explain what probability is, but now you use the same notion in a much more complicated context, that of 1000 tosses instead of one. Frankly, all that is not very serious.
- But wait, wait! I can give you axioms which describe the properties of probabilities...
- To know the properties of something is certainly very important. But it would also be good, before speaking about properties, to understand what is the object whose properties we want to study. *The sepulkas are used for sepulation, one puts them in a sepulkary, they can be assembled in beads, and they are able to wistle*[4]: do you understand anything here?
- We've been talking for a long time already, but there still remains an approach by which I could try to convince you. You see, the property of having the proportion of 0's and of 1's close to 1/2 is true not only with a large probability; it is also true for *all* random sequences. The sequences which do not satisfy this property are just not random.
- Is the sequence of alternating zeros and ones random according to you?
- No, it is not. It is obviously too regular to be random.
- Then I don't understand at all what you are speaking about. What does the word *random* mean?
- Mmm...
- Are there at least any axioms which would describe the properties of the objects you call random?
- Mmm...

14.4.2 Random sequences

The approach based on Kolmogorov complexity permits to define the notion of an *individual* random sequence formally without any references to probabilities. For infinite sequences of bits it provides a sharp boundary between random and non-random sequences. For finite sequences (binary strings) we have no hope to achieve this sharp division. (Indeed, changing one bit cannot make a random sequence non-random, but a sequence of changes can.)

For a finite sequence, to be random is a synonym of having a complexity close to the length. In other words, the best (or close to the best) way to describe such a sequence is to present it literally. Then we can prove that in a random sequence the frequency of zeros (and ones) is close to 1/2. For

---

[4] See Stanisław Lem, *Memoirs of a space traveller*, London, 1982

example, consider a sequence of 1000 bits that contains, say, 300 ones and 700 zeros. This fact significantly reduces its complexity, and therefore the sequence is not random. (Indeed, we can say that this is a sequence that has number N in the list of all sequences that have 300 ones and 700 zeros, and one can see that the bit length of N is much smaller than 1000: $N \leq \binom{1000}{300}$ hence $\log N \leq \log \binom{1000}{300} < 877$.) So the random sequence should contain approximately equal number of zeros and ones. However, if we push the same reasoning a little further, we see that if a sequence had exactly the same number of zeros and ones, we would also have some nontrivial information about it, so it could not be perfectly random. For a truly random sequence, zeros and ones must be slightly unbalanced (the difference should be proportional to the square root of the length).

As we have said, according to Kolmogorov's idea a sequence is random if it is "almost" incompressible. However, complexity is defined for finite sequences. Therefore to define randomness for an infinite sequence we need to consider some finite strings related to it. The most natural choice is prefixes.

If an infinite sequence is denoted by x, let $x_{1:n}$ denote a finite string consisting of the first n bits of x. We could try the following definition: x is random if and only if

$$\exists C \ \forall n \ \ K(x_{1:n}) > n - C \,.$$

The constant term C is natural since the complexity K is defined up to an additive constant. Unfortunately, this definition does not work: there is no sequence x that satisfies this requirement. Ten years passed before this difficulty was resolved. The solution is sometimes considered as a "technical trick". However, what is considered as a technical trick by mathematicians corresponds to a reality well known to computer scientists: we should distinguish between a program that reads/writes a bit string (of a specified length) and a program that reads from the (potentially infinite) input stream or writes into the output stream. Storing a file or a string, we should reserve additional place to store its length or reserve some symbol as a terminator. Both solutions require additional space, at least $\log n$ bits for keeping the length of an n-bit string.

There are different technical solutions; one of them is that we require our descriptions to have the prefix property: if a string t is a description of some x, then any continuation of t (i.e. any string that extends t) is also a description of x. So we do not need to say when the description stops, since the trailing bits do not change anything. If we modify the definition of the Kolmogorov complexity in this direction (which requires some precautions but is feasible), the formula suggested in the previous paragraph becomes a reasonable definition of randomness for infinite sequences. Technically speaking, we may switch from the "plain" complexity to "prefix" complexity. This gives some other advantages; for example, for this version of complexity the complexity of a pair of binary strings (under any computable encoding) does not exceed the sum of their complexities. (This is not true for the original "plain" Kolmogorov complexity where an additive logarithmic term is needed.)

Another Kolmogorov's idea was to define a random sequence as *a sequence which escapes from every effectively null set*. In order to define the notion of an "effectively null set" we take a usual definition of a null set (a set of measure zero) and interpret the existential quantifier in an effective way (instead of mere existence we demand that the required object be provided by some algorithm). This gives us the following definition:

A set A is an *effectively null* set if there exists a program p which, for any integer n given as input, produces an infinite series of strings

$$x_0^{(n)}, x_1^{(n)}, \ldots$$

such that for all n

$$\sum_i 2^{-|x_i^{(n)}|} < 1/n$$

and for every $w \in A$ and for every n the sequence w has one of the strings $x_i^{(n)}$ as a prefix.

This idea was developed by Martin-Löf [Mar66], a student of Kolmogorov. The effectively null sets correspond to "non-randomness" tests, and a sequence is random if it resists to all these tests. The existence of a universal algorithm allows us to construct one *universal* test: every sequence which resists to this test resists as well to all other tests and is therefore random.

One of the principal results of the algorithmic information theory is the connection between the incompressibility of prefixes of infinite sequence and its randomness seen as resistance to every algorithmic test. This equivalence is a theorem proved in the 1970s by Levin and Schnorr [Lev73, Sch73] in the contexts of slightly different definitions (they used some version of the so called "monotone" complexity; see [USS90] for the details). Thus a good definition of randomness (for an infinite sequence) was obtained; "good" means here that two different reasonable definitions turn out to be equivalent. Moreover, all basic theorems of probability theory that have the form "for almost all x the property P is true" can be now reformulated as follows: "for every random (in the sense described above) sequence the property P is true". The latter result is not a formal statement; we mean that different authors studied different theorems of this form (e.g. the ergodic theorem) and proved that these theorems remain true for every algo-random sequence. In certain cases (e.g. for ergodic theorem), this is a rather delicate work and about ten years were required to complete it.

The relation between complexity and measure can be used also for finite sequences. For example, we may prove that any incompressible sequence has some property (by showing that sequences which do not have this property can be compressed). Then we conclude that almost all sequences have this property (being incompressible).

14.4.3 Sequences of low complexity

We have seen that the sequences that have prefixes of high complexity are random. It is natural to ask which sequences have prefixes of small complexity. There exists a nice theorem, proved independently by several authors long ago when the theory of Kolmogorov complexity appeared. According to the date of the first publication, this theorem must be attributed to Albert Meyer and it was published in a paper by Loveland [Lov69]. Its proof may be found in [ZL70]. Let us state this result using the notation of the previous section.

Theorem 3. *A sequence x is recursive (i. e. computable by an algorithm) if and only if $\exists C \; \forall n \;\; K(x_{1:n} \mid n) < C$.*

We use here a slightly more general —namely, *conditional*— form of Kolmogorov complexity. In order to simplify our presentation we did not mention it until now, but it is a useful and natural notion. We define $K(x \mid y)$ (complexity of x while knowing y) as the length of the shortest description of x, if descriptions have access to y as input. Formally,

$$K_f(x \mid y) = \begin{cases} \min |t| \text{ such that } f(t, y) = x, \\ \infty \qquad \text{if such } t \text{ does not exist.} \end{cases}$$

The existence of optimal functions is proved in the same way as before. If y is fixed, the complexity $K(x \mid y)$ as a function of x coincides with $K(x)$ up to a constant so we get nothing really new (recall that the complexity is defined up to an additive constant anyway). But this new notion makes sense, e.g. if we let y be the length of x (or the number of zeros in x, the substring formed by bits with even indices, etc.)

The theorem says that the "simplest" infinite sequences are exactly the computable ones. It is important to use $K(x_{1:n}|n)$ and not $K(x_{1:n})$ since even for a computable x the prefix $x_{1:n}$ contains a small amount of information, i.e. the length n. (Why didn't we add a similar term in the characterization of random sequences? In fact this is also possible but not necessary.)

In one direction this theorem is trivial: if a sequence is recursive then complexity is bounded (in fact, bounded by the complexity of a program that produces $x_{1:n}$ given n).

The converse implication is more subtle. It is one of the examples that appear from time to time in theoretical computer science, when it is possible to prove that an algorithm exists but it is impossible to construct it. In this specific case we can prove that the sequence is recursive but there is no computable bound on the size of the program generating x that depends only on C.

We can explain informally why this happens (see [DP02, DSV02] for details) in the following way. Consider a sequence x that starts with a large number N of zeros that are followed by 1, then some string z and then zeros again. Any program that generates x gives us complete information about z (we have only to delete the leading zeros), and its complexity is high if z has high complexity. On the other hand the complexity $K(x_{1:n}|n)$ is low if $n \leq N$ (since only zeros appear in $x_{1:n}$ and can be low for $n > N$ since in this case we know some number n greater than N and this information may be useful for finding z.

14.4.4 Back to the definition of randomness

Our definition of "random sequence" (in this section we say "algo-random") can be criticized from many different viewpoints. First, this definition uses the notion of an algorithm that was never used in probability theory. It leads to a natural question: is the notion of algorithm really necessary to give a reasonable definition of a random sequence?

Second, one could note that some easily definable sequences are algo-random. For example, there exists a sequence defined by G. Chaitin (called ω) that is algo-random. It is defined as follows. Consider an optimal algorithm in the sense described in Sect. 14.2, but in the self-delimiting version, and apply it to random bits obtained by coin tossing. (This algorithm will actually use only finitely many of these bits to produce the output.) The computation may terminate or not, depending on the choice of random bits. Then ω is defined as the probability of termination[5].

This sequence is (as Chaitin noted) an algo-random one, and this raises a question. The proposition "$x \neq \omega$ for almost all x" is true (almost all sequences differ from ω). However, we cannot claim that "$x \neq \omega$ for all algo-random x", since ω is one of them. Even if this example seems to be a little artificial, a true problem is raised.

The first possibility is to change the notion of algo-randomness, allowing a broader class of randomness tests. In this way we obtain a notion of "arithmo-random" sequences. Two formal definitions are possible: one considers the classical theory based on algorithms and then relativizes these algorithms using arithmetical oracles; the other one defines everything directly using arithmetic formulas. These two approaches lead to the same notion, which corresponds to a smaller class of random sequences. The problem is that this definition is not closed: there is no universal test in the class considered. This is due to an important structural difference between the enumerable sets and the arithmetic sets: universal set exists for enumerable sets but not for the arithmetic ones.

Then we may make another, more radical, suggestion [DKUV03]: let us consider all the theorems of the form "for almost all x, $P(x)$", where P is a formula in some language. There are countably many theorems of this form, and their set is recursively enumerable. Each of these theorems corresponds to a set of measure 1 (sequences for which P is true). Consider the intersection of all

---

[5] Similar experiment was performed at the early stages of Unix development. Some standard utilities were taken and sequences of random bits were fed into them. The probability of crash turned out to be embarrassingly large

these sets. The σ-additivity (countable additivity) of the measure guarantees that this intersection also has measure 1. Let us take this intersection as the set of random sequences. Then by definition all the theorems of probability theory (provably true for almost all sequences) are true for the sequences from this set; however, we encounter then other difficulties (related to the basic problems in the foundations of set theory, like the absence of the set of all sets, etc.)

More subtle versions of this approach can be considered, but they are based on rather delicate techniques of the set theory. For example, instead a provably minimal set we may consider a set which would be minimal in a consistent way: this means that it is impossible to prove that it is not minimal. The existence of such a set is not at all evident; the proof makes use of fine techniques of the set theory. To give an informal image of this approach we may compare it with the presumption of innocence. A sequence must always be presumed random; if it is suspected not to be such, it must be taken to court; but in the absence of any proofs whatsoever of its "guilt" the sequence must be exonerated (that is, considered as random) for the benefit of the doubt.

Acknowledgement

The authors are grateful to Alexander Shen for many helpful comments.

References

[BG03]   Blass, A., Gurevich, Y.: Algorithms: a quest for absolute definitions. Bull. EATCS, **81**, 195–225 (Oct. 2003)

[Cha66a]   Chaitin, G.: On the length of programs for computing finite binary sequences by bounded-transfer Turing machines. AMS Notices, **13**, 133 (1966)

[Cha66b]   Chaitin, G.: On the length of programs for computing finite binary sequences by bounded-transfer Turing machines, II. AMS Notices, **13**, 228–229 (1966)

[Cha70]   Chaitin, G.: Computational complexity and Gödel's incompleteness theorem. AMS Notices, **17**, 672 (1970)

[DKUV03]   Durand, B., Kanovei, V., Uspensky, V., Vereshchagin, N.: *Do stronger definitions of randomness exist?* Theoret. Comput. Sci., **290**:3, 1987–1996 (2003)

[DP02]   Durand, B., Porrot, S.: Comparison between the complexity of a function and the complexity of its graph. Theoret. Comput. Sci., **271**, 37–46 (2002)

[DSV02]   Durand, B., Shen, A., Vereshchagin, N.: Descriptive complexity of computable sequences. Theoret. Comput. Sci. **271**, 47–58 (2002)

[Gur88]   Gurevich, Y.: On Kolmogorov machines and related issues. Bull. EATCS, **35**, 71–82 (June 1988)

[Hof79]   Hofstadter, D.: Gödel, Escher, Bach: An Eternal Golden Braid. Basic Books (1979)

[Kol65]   Kolmogorov, A.N.: Three approaches to the definition of the concept of "quantity of information" (in Russian). Problemy Peredachi Informatsii, 1:1, 3–11 (1965)

[KU58]   Kolmogorov, A.N., Uspensky, V.A.: On the definition of an algorithm. Uspekhi Mat. Nauk, 13:4, 3–28 (1958). Translations Amer. Math. Soc., 29, 217–245 (1963).

[Lev73]   Levin, L.: The concept of a random sequence. Soviet Math. Doklady, 212, 1413–1416 (1973)

[LV97]   Li, M., Vitanyi, P. : An Introduction to Kolmogorov Complexity and its Applications (2nd edition). Graduate Texts in Computer Science, Springer-Verlag, New York (1997)

[Lov69]   Loveland, D.W.: A variant of Kolmogorov concept of complexity. Information and Control, 15, 510–526 (1969)

[Mar66]   Martin-Löf, P.: The definition of random sequences. Information and Control, 9, 602–619 (1966)

[Sch73]   Schnorr, C.P.: Process complexity and effective random sets. J. Comput. System Sci., 7, 376–388 (1973)

[SL95]   Shasha, D., Lazere, C.: Out of Their Minds: The Lives and Discoveries of 15 Great Computer Scientists. Copernicus (1995)

[Sol64]   Solomonoff, R.J.: A formal theory of inductive inference, I. Information and Control, 7, 1–22 (1964)

[US93]   Uspensky, V., Semenov, A.: Algorithms: Main Ideas and Applications (transl. from Russian by A. Shen). Kluwer (1993)

[USS90]   Uspensky, V., Semenov, A., Shen, A.: Can an (individual) sequence of zeros and ones be random? Russian Math. Surveys, 45:1, 121–189 (1990)

[ZL70]   Zvonkin, A.K., Levin, L.: The complexity of finite objects and the development of the concepts of information and randomness by means of the theory of algorithms. Russian Math. Surveys, 25:6, 83–124 (1970)

A.N. Kolmogorov with P.S. Alexandrov. Picture taken from the Kolmogorov
Archive with kind permission of A.N. Shiryaev

15

Algorithmic chaos and the incompressibility method

Paul Vitanyi

Centrum voor Wiskunde en Informatica (CWI), Amsterdam, The Netherlands
http://www.cwi.nl/~paulv
Paul.Vitanyi@cwi.nl

Many physical theories like chaos theory are fundamentally concerned with the conceptual tension between determinism and randomness. Kolmogorov complexity can express randomness in determinism and gives an approach to formulate chaotic behavior. As a technical tool to quantify the unpredictability of chaotic systems we use the Incompressibility Method. We introduce the method by examples: the distribution of prime numbers and largest clique size in random graphs.

15.1 Introduction

Ideally, physical theories are abstract representations – mathematical axiomatic theories for the underlying physical reality. This reality cannot be directly experienced, and is therefore unknown and even in principle unknowable. Instead, scientists postulate an informal description which is intuitively acceptable, and subsequently formulate one or more mathematical theories to describe the phenomena.

Deterministic Chaos: Many phenomena in physics (like the weather) satisfy well accepted deterministic equations. From initial data we can extrapolate and compute the next states of the system. Traditionally it was thought that increased precision of the initial data (measurement) and increased computing power would result in increasingly accurate extrapolation (prediction) for futures of lengths that linearly scaled inversely with the precision. But it has turned out that for many (i.e. the chaotic) systems it scales not better than logarithmic inversely with the accuracy. In fact, it turns out that any long range prediction with any confidence better than what we would get by flipping a fair coin is practically impossible: this phenomenon is known as chaos (see [Dev89] for an introduction). There are two, more or less related, causes for this:

— **Instability:** In certain deterministic systems, an arbitrary small error in initial conditions can exponentially increase during the subsequent evolution of the system, until it encompasses the full range of values achievable by the system. This phenomenon of instability of a computation is in fact well known in numerical analysis: computational procedures inverting ill-conditioned matrices (with determinant about zero) will introduce exponentially increasing errors.

— **Unpredictability:** Assume we deal with a system described by deterministic equations which can be finitely represented (see below). Even if fixed-length initial segments of the infinite binary representation of the real parameters describing past states of the system are perfectly known, and the computational procedure used is perfectly error free, for many such systems it will still be impossible to effectively predict (compute) any significantly long extrapolation of system states with any confidence higher than using a random coin flip. This is the core of chaotic phenomena: randomness in determinism.

Remark 1. In the following we use the notion of "effective computation" in the well-known mathematical sense of "computability by Turing machine." Similarly, we use the notion of "(partial) recursive function" and "(partial) computable function" interchangeably. In recursion theory such functions are mappings from a subset of $\mathbb{N}$ into $\mathbb{N}$ (or into $\mathbb{Q}$, after composition by an explicit numbering of rationals). In the current context we may want to consider the extension to real numbers. A function $f : \mathbb{N} \to \mathbb{R}$ is *upper semi-computable* if there is a Turing machine T computing a total function $\phi : \mathbb{N} \times \mathbb{N} \to \mathbb{Q}$ such that $\phi(x, t+1) \leq \phi(x, t)$ and $\lim_{t \to \infty} \phi(x, t) = f(x)$. This means that f can be computably approximated from above. If $-f$ is upper semi-computable, then f is called lower semi-computable. If f is both upper semi-computable and lower semi-computable, then we call f *computable* (or recursive, if the range is integer or rational). (We can similarly consider computable functions with a real domain: $f : \mathbb{R} \to \mathbb{R}$. This requires careful definitions and it turns out that computability implies continuity. But this sophistication is not needed in the current treatment.) The extension of the notion of computable functions to domain and range to vectors is straightforward. For details see any textbook on computability or [LV97].

Remark 2. It is perhaps useful to stress that instability and unpredictability, although close companions, are not always the same. A trivial example of instability without unpredictability is a system that makes a first choice in an instable manner but afterwards sticks to that choice. (Such a system is equivalent, for instance, to a "dictatorial coin" that gives outcome 0 or 1 with equal probability when flipped the first time, but at every next flip will give the same outcome it gave the first time.) An example of unpredictability without instability is a function $f_r : \mathbb{N} \to \{0, 1\}$ defined by $f_r(n) = r_n$, with $r = r_1 r_2 \ldots$ an infinite binary sequence that is random in Martin-Löf's sense (see below) and hence unpredictable. (Here $n \in \mathbb{N}$ plays the rôle of time.)

Probability: Classical probability theory deals with randomness in the sense of *random variables*. The concept of random individual data cannot be expressed. Yet our intuition about the latter is very strong: an adversary claims to have a true random coin and invites us to bet on the outcome. The coin produces a hundred heads in a row. We say that the coin cannot have been fair. The adversary, however, appeals to probability theory which says that each sequence of outcomes of a hundred coin flips is equally likely, $1/2^{100}$, and one sequence had to come up. Probability theory gives us no basis to challenge an outcome *after* it has happened. We could only exclude unfairness in advance by putting a penalty side-bet on an outcome of 100 heads. But what about 1010...? What about an initial segment of the binary expansion of π?

Regular sequence:

$$\Pr(00000000000000000000000000) = \frac{1}{2^{26}}$$

Regular sequence:

$$\Pr(01000110110000010100111001) = \frac{1}{2^{26}}$$

Random sequence:

$$\Pr(10010011011000111011010000) = \frac{1}{2^{26}}$$

The first sequence is regular, but what is the distinction of the second sequence and the third? The third sequence was generated by flipping a quarter. The second sequence is very regular: $0, 1, 00, 01, \ldots$ The third sequence will pass (pseudo) randomness tests.

In fact, classical probability theory cannot express the notion of *randomness of an individual sequence*. It can only express expectation of properties of the total set of sequences under some distribution.

This is analogous to the situation in physics above: *"how can an individual object be random?"* is as much a probability theory paradox as *"how can an individual sequence of states of a deterministic system be random?"* is a paradox of deterministic physical systems.

In probability theory the paradox has found a satisfactory resolution by combining notions of computability and information theory to express the complexity of a finite object. This complexity is the length of the shortest binary program from which the object can be effectively reconstructed. It may be called the *algorithmic information content* of the object. This quantity turns out to be an attribute of the object alone, and recursively invariant. It is the *Kolmogorov complexity* of the object. It turns out that this notion can be brought to bear on the physical riddle too, as we shall see below.

15.2 Kolmogorov complexity

To make this paper self-contained we briefly review notions and properties required. For details and further properties see the textbook [LV97]. We identify the natural numbers $\mathbb{N}$ with the finite binary sequences as

$$(0, \epsilon), (1, 0), (2, 1), (3, 00), (4, 01), \ldots,$$

where ϵ is the empty sequence. The *length* $l(x)$ is the number of bits in the binary sequence x (for instance, $l(\epsilon) = 0$). That defines also the "length" of the corresponding natural integer. If A is a set, then $|A|$ denotes the cardinality of A. Let $\langle \cdot, \cdot \rangle : \mathbb{N} \times \mathbb{N} \to \mathbb{N}$ denote a standard computable bijective "pairing" function. Throughout this paper, we will assume that $\langle x, y \rangle = 1^{l(x)} 0 x y$.

Define $\langle x, y, z \rangle$ by $\langle x, \langle y, z \rangle \rangle$.

We need some notions from the theory of algorithms, see [Rog67]. Let $\phi_1, \phi_2, \ldots$ be a standard enumeration of the partial recursive functions. The (Kolmogorov) *complexity* of $x \in \mathbb{N}$, given y, is defined as

$$C(x|y) = \min\{l(\langle n, z \rangle) : \phi_n(\langle y, z \rangle) = x\}.$$

This means that $C(x|y)$ is the *minimal* number of bits in a description from which x can be effectively reconstructed, given y. The unconditional complexity is defined as $C(x) = C(x|\epsilon)$. These notions were originally introduced in [Kol65].

An alternative definition is as follows. Let

$$C_\psi(x|y) = \min\{l(z) : \psi(\langle y, z \rangle) = x\} \tag{15.1}$$

be the conditional complexity of x given y with reference to a decoding function ψ. Then $C(x|y) = C_\psi(x|y)$ for a universal partial recursive function ψ that satisfies $\psi(\langle y, n, z \rangle) = \phi_n(\langle y, z \rangle)$.

We need the following properties. For each $x, y \in \mathbb{N}$ we have[1]

$$C(x|y) \leq l(x) + O(1). \tag{15.2}$$

For each $y \in \mathbb{N}$ there is an $x \in \mathbb{N}$ of length n such that $C(x|y) \geq n$. In particular, we can set $y = \epsilon$. Such x's may be called *random*, since they are without regularities that can be used to compress the description: intuitively, the shortest effective description of such an integer x is x itself. In general, for each n and y, there are at least $2^n - 2^{n-c} + 1$ distinct x's of length n with

$$C(x|y) \geq n - c. \tag{15.3}$$

In some cases we want to encode x in *self-delimiting* (s.d.) form x', in order to be able to decompose $x'y$ into x and y. Then we will make use of the

---

[1] Throughout $O(1)$ (resp. $o(1)$) will denote a bounded quantity (resp. a quantity that converges to 0), whatever its sign, and $O(f(n))$ will mean $f(n) \times O(1)$ (resp. $o(f(n))$ will mean $f(n) \times o(1)$)

prefix complexity $K(x)$, introduced in [Lev74], which denotes the length of the shortest *self-delimiting* description. To this end, we consider so called *prefix* Turing machines, which have only 0's and 1's on their input tape, and thus cannot detect the end of the input. Instead we define an input as that part of the input tape which the machine has read when it halts. When $x \neq y$ are two such input, we clearly have that x cannot be a prefix of y (that is, y cannot have the form xz), and hence the set of inputs forms what is called a *prefix code* or *prefix-free code*. We define $K(x|y), K(x)$ similarly to $C(x|y), C(x)$ above, but with reference to a universal prefix machine that first reads $1^n 0$ from the input tape and then simulates prefix machine n on the rest of the input.

Good upper bounds on the prefix complexity of x are obtained by iterating the simple rule that a self-delimiting description of the length of x followed by x itself is a s.d. description of x. For example, $x' = 1^{l(x)}0x$ and $x'' = 1^{l(l(x))}0l(x)x$ are both s.d. descriptions for x, and this shows that $K(x) \leq 2l(x) + O(1)$ and $K(x) \leq l(x) + 2l(l(x)) + O(1)$.

Similarly, we can encode x in a self-delimiting form of its shortest program $p(x)$ (of length $l(p(x)) = C(x)$) in $2C(x) + 1$ bits. Iterating this scheme, we can encode x as a self-delimiting program of $C(x) + 2 \log C(x) + 1$ bits[2], which shows that $K(x) \leq C(x) + 2 \log C(x) + 1$, and so on.

15.2.1 The incompressibility method

The secret of the successful use of Kolmogorov complexity arguments as a proof technique lies in a simple fact: the overwhelming majority of strings have almost no computable regularities. We have called such a string "random." There is no shorter description of such a string than the literal description: it is incompressible.

Traditional proofs often involve all instances of a problem in order to conclude that some property holds for at least one instance. The proof would be more simple, if only that one instance could have been used in the first place. Unfortunately, that instance is hard or impossible to find, and the proof has to involve all the instances. In contrast, in a proof by the incompressibility method, we first choose a random (that is, incompressible) individual object that is known to exist (even though we cannot construct it). Then we show that if the assumed property did not hold, then this object could be compressed, and hence it would not be random. Let us give some simple examples.

Distribution of prime numbers. A prime number is a natural number that is not divisible by natural numbers other than itself and 1. In the nineteenth century, Chebychev showed that the number of primes less than n grows

---

[2] Throughout log denotes the binary logarithm

asymptotically like $n/\log n$.[3] Using the incompressibility method we cannot (yet) prove this statement precisely, but we can come remarkably close with a minimal amount of effort. We first prove that for infinitely many n, the number of primes less than or equal to n is at least $\log n/\log \log n$. The proof method is as follows. For each n, we construct a description from which n can be effectively retrieved. This description will involve the primes less than n. For some n this description must be long, which will give the desired result.

Assume that $p_1, p_2, \ldots, p_m$ is the list of all the primes less than n. Then,

$$n = p_1^{e_1} p_2^{e_2} \cdots p_m^{e_m}$$

can be reconstructed from the vector of the exponents. Each exponent is at most $\log n$ and can be represented by $\log \log n$ bits. The description of n (given the maximal order of magnitude $\log n$ of the exponents) can be given in (approximately) $m \log \log n$ bits. But it can be shown that each n that is random (given $\log n$) cannot be described in fewer than $\log n$ bits, whence the result follows. Can we do better? This is slightly more complicated. Let $l(x)$ denote the length of the binary representation of x. We shall show that p_m (the m-th prime number) is $\leq m \log^2 m$. (One can show that this is equivalent to state that the number of primes less or equal to n is greater than $n/\log^2 n$.)

Firstly, any given integer n is completely determined by giving the index m of its greatest prime factor p_m, together with the (integral) quotient n/p_m. Thus we can describe n by $E(m)n/p_m$, where $E(m)$ is a prefix-free encoding of m. (The description of m needs to be self-delimiting or else we wouldn't know where the description of m ends, and where the description of n/p_m starts.) For random n, the length of this description, $l(E(m)) + \log n - \log p_m$, must exceed $\log n$. Therefore, $\log p_m < l(E(m))$. It is known (and straightforward from the earlier discussion) that we can set $l(E(m)) \leq \log m + 2 \log \log m$. Hence, $p_m < m \log^2 m$: we have proven our claim.

Random graphs. The interpretation of strings as more complex combinatorial objects leads to a new set of properties and problems that have no direct counterpart in the "flatter" string world. Here we derive topological, combinatorial, and statistical properties of graphs with high Kolmogorov complexity. Every such graph possesses simultaneously all properties that hold with high probability for randomly generated graphs. They constitute "almost all graphs" and the derived properties a fortiori hold with probability that goes to 1 as the number of nodes grows unboundedly.

---

[3] More precisely, Chebychev showed (in 1850) that the quotient of $\pi(n)$ (the number of primes $\leq n$) by $n/\log n$ is bounded by explicit positive constants. In fact, this ratio tends to $\log e$ when n tends to ∞: this is the "prime number theorem" proved (in 1896) by Hadamard and La Vallée Poussin. One can show that this theorem is equivalent to this statement: *if p_n denotes the n-th prime number, then $\frac{p_n \log e}{n \log n} \to 1$ when $n \to \infty$.* [Note added by the translator in the French edition (Michel Balazard).]

Each labeled graph $G = (V, E)$ on n nodes $V = \{1, 2, \ldots, n\}$ (at most one non-oriented edge for each pair of different nodes) can be represented (up to an automorphism) by a binary string $E(G)$ of length $\binom{n}{2}$: we simply assume a fixed ordering of the $\binom{n}{2}$ possible edges in an n-node graph, e.g. lexicographically, and let the ith bit in the string indicate presence (1) or absence (0) of the i'th edge. Conversely, each binary string of length $\binom{n}{2}$ encodes an n-node graph. Hence we can identify each such graph with its binary string representation.

We are going to prove that G does not contain a clique (complete graph) on more than $2 \log n + 1 + o(1)$ nodes.

Let m be the number of nodes of the largest clique $\mathcal{K}$ in G. We try to compress $E(G)$, to an encoding $E'(G)$, as follows:

1. Prefix the list of nodes in $\mathcal{K}$ to $E(G)$, each node using $\lceil \log n \rceil$ bits[4], adding $m \lceil \log n \rceil$ bits.
2. Delete all redundant bits from the $E(G)$ part, representing the edges between nodes in $\mathcal{K}$, saving $m(m-1)/2$ bits.

Then

$$l(E'(G)) = l(E(G)) + m \lceil \log n \rceil - \binom{m}{2}. \tag{15.4}$$

Let p be a program which, from n and $E'(G)$, reconstructs $E(G)$. Then,

$$C(E(G)|n, p) \leq l(E'(G)). \tag{15.5}$$

Since there are $2^{\binom{n}{2}}$ labeled graphs on n nodes and at most $2^{\binom{n}{2}} - 1$ binary descriptions of length less than $\binom{n}{2}$, we can choose a labeled graph G on n nodes that satisfies

$$C(E(G)|n, p) \geq \binom{n}{2} + o(\log n). \tag{15.6}$$

Equations (15.6), (15.4), and (15.5) are true only when $m \leq 2 \log n + 1 + o(1)$.

In fact, the discerning reader will by now understand that while the information in the new prefix of $E'(G)$ is used by the program p to insert "1"s in the appropriate slots in the old suffix of $E'(G)$ to reconstruct the edges of the clique, using another program p' we show that the largest set of nodes with no pairwise edges is bounded by the same upper bound. Indeed, every easily (in $O(\log n)$ bits, given the labeled nodes) describable subgraph of G cannot have more than $2 \log n + 1$ nodes. This includes virtually all properties we can conceivably be interested in. Moreover, the set of graphs G that satisfy (15.6) is very large: an easy counting argument shows that of the $2^{\binom{n}{2}}$ labeled graphs on n nodes, at least $(1 - 1/n)2^{\binom{n}{2}}$ graphs do so. That is, flipping a fair coin to determine presence or absence of the $\binom{n}{2}$ edges of a labeled graph on n nodes,

---

[4] $\lceil x \rceil$ denotes the smallest integer that is greater than x

with an overwhelming probability of $1 - 1/n$ we will flip a graph satisfying (15.6). Hence our conclusion about the maximal size of easily describable subgraphs holds with probability almost 1 (that is, probability $\geq 1 - 1/n$) for randomly generated graphs.

15.2.2 Random sequences

We would like to call an infinite sequence $\omega \in \{0, 1\}^{\infty}$ random if $C(\omega_{1:n}) \geq n + O(1)$ for all n (where $\omega_{1:n}$ denotes the sequence composed by the n first bits of ω). It turns out that such sequences do not exist. This remark led P. Martin-Löf [Mar66] to create its celebrated theory of randomness. That ω is random in Martin-Löf's sense means, roughly, that it will pass *all* effective tests for randomness: both the tests which are known now and the ones which are as yet unknown [Mar66].

Later it turned out, [Cha75], that we can yet precisely define the Martin-Löf random sequences, but using prefix Kolmogorov complexity:

Theorem 1. *An infinite binary sequence ω is random in the sense of Martin-Löf iff there is an n_0 such that $K(\omega_{1:n}) \geq n$ for all $n > n_0$,*

Similar properties hold for high-complexity finite strings, although in a less absolute sense. For every finite set $S \subseteq \{0, 1\}^*$ containing x we have $K(x|S) \leq \log |S| + O(1)$. Indeed, consider the self-delimiting code of x consisting of its $\lceil \log |S| \rceil$ bit long index of x in the lexicographical ordering of S. This code is called *data-to-model code*. The lack of typicality of x with respect to S is the amount by which $K(x|S)$ falls short of the length of the data-to-model code. The *randomness deficiency* of x in S is defined by

$$\delta(x|S) = \log |S| - K(x|S), \qquad (15.7)$$

for $x \in S$, and ∞ otherwise. If $\delta(x|S)$ is small, then x may be considered as a *typical* member of S. There are no simple special properties that single it out from the majority of elements in S. This is not just terminology: if $\delta(x|S)$ is small, then x satisfies *all* properties of low Kolmogorov complexity that hold with high probability for the elements of S. For example: Consider strings x of length n and let $S = \{0, 1\}^n$ be the set of such strings. Then $\delta(x|S) = n - K(x|n) + O(1)$. Let $\delta(n)$ be an appropriate function like $\log n$ or $\sqrt{n}$. Then, the following properties are a finitary analog of Martin-Löf randomness of infinite sequences, [LV97]:

(i) If P is a property satisfied by all x with $\delta(x|S) \leq \delta(n)$, then P holds with probability at least $1 - 1/2^{\delta(n)}$ for the elements of S.

(ii) Let P be any property that holds with probability at least $1 - 1/2^{\delta(n)}$ for the elements of S. Then, every such P holds simultaneously for every $x \in S$ with $\delta(x|S) \leq \delta(n) - K(P|n) + O(1)$.

Let us go one step further. The notion of randomness of infinite sequences and finite strings can only exist in the context of a probabilistic ensemble with respect to which they are a random element. In the above case, for the infinite sequences this ensemble is the set $\{0,1\}^\infty$ supplied with the uniform measure λ, also called the "coin-flip" measure, since $\lambda(\omega_1 \ldots \omega_n) = 1/2^n$ is the probability of producing the n-bit string $\omega_1 \ldots \omega_n$ with n flips of a fair coin. One can generalize the randomness approach to an arbitrary computable measure μ. This is a measure such that there is a Turing machine T such that for every n and $\varepsilon > 0$, on every input $\omega_1 \ldots \omega_n, \varepsilon$, the machine T halts with output r such that $|\mu(\omega_1 \ldots \omega_n) - r| \leq \varepsilon$. (For "Turing machine" we can also read "computer program" in a universal computer language like LISP or Java.) We can now talk about μ-*random sequences*, that is, sequences that satisfy every property (in the appropriate effective Martin-Löf sense) that holds with μ-probability 1 for the sequences in $\{0,1\}^\infty$. The following theorem is taken from [LV97]:

Theorem 2. *Let μ be a computable measure. An infinite binary sequence ω is μ-random in the sense of Martin-Löf iff there is an n_0 such that $K(\omega_{1:n}) \geq -\log \mu(\omega_{1:n})$ for all $n > n_0$.*

Note that for $\mu = \lambda$, the uniform distribution, we have $-\log \lambda(\omega_{1:n}) = n$, retrieving Theorem 1 again. We can extend the notion of μ-randomness to finite strings in the appropriate manner.

15.3 Algorithmic chaos theory

When physicists deal with a chaotic system, they believe that the whole thing is based on an underlying deterministic system but that the *fixed trajectory* is *unpredictable* on the basis of the observable states. Unfortunately, in the classical framework this cannot be expressed and therefore one has to use the kludge of an ensemble of states and trajectories, and to go through the rigmarole of probabilistic reasoning which is essentially besides the point. But using Kolmogorov complexity we can express directly the chaoticity of the system and the unpredictability properties of single trajectories, which is the intuition one wants to express. This is this idea that we shall advocate now.

For convenience assume that time is discrete: $\mathbb{N}$. In a deterministic system X the *state* of the system at time t is X_t. The *orbit* of the system is the sequence of subsequent states $X_0, X_1, X_2, \ldots$. For convenience we assume the states are elements of $\{0,1\}$. The definitions below are easily generalized. For each system, be it deterministic or random, we associate a measure μ with the space $\{0,1\}^\infty$ of orbits. That is, $\mu(x)$ is the probability that an orbit starts with $x \in \{0,1\}^*$.

Given an initial segment $X_{0:t}$ of the orbit we want to compute X_{t+1}. Even if it would not be possible to compute X_{t+1}, we would like to compute a prediction of it which does better than a random coin flip.

Definition 1. Let the set of orbits of a system X be $S = \{0,1\}^\infty$ with measure μ. Let ϕ be a partial recursive function and let $\omega \in S$. Define

$$\zeta_i = \begin{cases} 1 \text{ if } \phi(\omega_{1:i-1}) = \omega_i \\ 0 \text{ otherwise} \end{cases}$$

A system is *chaotic* if, for every computable function ϕ, we have

$$\lim_{t \to \infty} \frac{1}{t} \sum_{i=0}^{t-1} \zeta_i = \frac{1}{2},$$

with μ-probability 1.

Remark 3. For a chaotic system, no computable function ϕ predicts the outcomes of the system better than a fair coin flip. In this definition of chaoticity the essential requirement has been formulated as algorithmic unpredictability of the orbits. The instability properties of the system are expressed by the measure μ (as in Definition 1) the system induces. For example, let μ be the uniform measure (usually denoted as λ). An orbit like $\omega = \omega_1 \ldots \omega_n 11 \ldots$ is perfectly predictable after the first n bits. In fact, predictability by an appropriate computable function holds for all ω that are computable sequences, such as the binary expansions of the rationals but also transcendental numbers such as $\pi = 3.14\ldots$. However, for every $\omega \in S$, and every $\varepsilon > 0$, the ω' such that $|\omega - \omega'| \le \varepsilon$, that is, the ω' in the ε-ball around ω, are unpredictable with uniform probability 1. This is because the set of Martin-Löf random sequences in the ε-ball has the same uniform measure as the set of all sequences in the ε-ball. So the slightest random perturbation of an orbit will result in an unpredictable orbit, with probability 1.

It is not the case, however, that unpredictability implies instability. For example, if μ concentrates all its probability like $\mu(\omega) = 1$ for an $\omega = \omega_1 \omega_2 \ldots$ such that $K(\omega_{1:n}) \ge n$ for all n, that is, ω is Martin-Löf random, then the subsequent elements of ω are completely unpredictable given the preceding elements. Yet the orbit is completely stable, in fact, it is deterministic. The crux is of course that the orbit is a fixed sequence albeit a quite noncomputable one. We leave it to the reader to construct similar examples where the orbit is not completely fixed, not instable, but yet completely unpredictable.

Remark 4. In chaos theory one typically considers deterministic systems X with states x from some domain R evolving in discrete steps according to $x_{n+1} = f(x_n)$, where the x_n's are real numbers, or vectors of real numbers, x_0 is given as initial value and f is a function mapping the current system state to the next system state. Physically it makes no sense to consider arbitrary real numbers since they require infinite precision – and that is not accessible to physical measurement. Moreover, no physical law or constant is known to hold with a precision of more than, say ten, decimals, and therefore the same will hold for the precision of the system evolution operator f.

Hence, in analysis of the system behavior one replaces the actual values x_n by a finitary approximation represented by an equivalence class containing x_n. These equivalence classes represent the different states we can actually "distinguish", "observe" or "measure." For example, if the x_n are taken from the domain $[0, 1]$ then we can choose to divide the domain $[0, 1]$ into two equal parts, $R_0 = [0, \frac{1}{2})$ and $R_1 = [\frac{1}{2}, 1]$. Subsequently, we consider a system (X_n) defined by $X_n = i$ if $x_n \in R_i$ ($n = 0, 1, \ldots$ and $i \in \{0, 1\}$). Note that this defines both the initial value X_0 and the subsequent system states $X_1, X_2, \ldots$, from the original system X with initial value x_0. "Chaos" is defined for the derived system (X_n) that represents the evolution of "distinguishable" states. Now it becomes clear that for different initial states x_0 and x_0', even if they are taken from the same equivalence class, say R_0, so $X_0 = 0$, the orbits $X_0 = 0, X_1, \ldots$ may be quite different from X_1 onwards. If this happens so that the orbit $X_0 = 0, X_1, \ldots$ is in the appropriate sense unpredictable, even though x_0 and the evolution operator f are known, then we call the system "chaotic."

Our Definition 1 is based on the following: Let X be a system defined by $x_{n+1} = f(x_n)$ Suppose we randomly select the initial state x_0 from its domain R according to a measure ρ. That is, if $x_0 = x_{0,1}x_{0,2}, \ldots$ then the probability of selecting $x_{0,1} \ldots x_{0,r}$ is $\rho(x_{0,1} \ldots x_{0,r})$. (This still allows us to select a particular x_0 with probability 1 by concentrating all probability of ρ on x_0.) Considering the derived system of distinguishable states, the probability of the initial state $X_0 = i$ is $\rho(R_i)$ ($i \in \{0, 1\}$), but although the probability of the next state X_1 is determined completely by ρ and f, it is sensitive to change of either, and similarly for the states $X_2, X_3, \ldots$ after that. Nonetheless, f, ρ completely determine the probability of every initial segment of every orbit of distinguishable states. That is, for an initial segment $\omega_0 \ldots \omega_n$ we denote this probability as $\mu(\omega_0 \ldots \omega_n)$, and this defines the measure μ in Definition 1. Note that if f, ρ are computable in an appropriate manner, then so is μ. For example, if $f(\omega_0\omega_1 \ldots) = \omega_0'\omega_1' \ldots$ is such that there exists a computable monotonic increasing function g and a computable function h such that $h(\omega_0 \ldots \omega_n) = \omega_0' \ldots \omega_{g(n)}'$, and moreover ρ is computable, then

$$\mu(X_0 \ldots X_m) = \rho\{\omega_0 \ldots \omega_n \ : \ g(n) = m \text{ and } h(\omega_0 \ldots \omega_n) = \omega_0' \ldots \omega_m'$$
$$\text{and } \omega_j' \in R_i \text{ if and only if } X_j = i \ (j = 0, \ldots, m, i \in \{0, 1\})\}.$$

The system is *uniformly instable* if for every ω and every $\varepsilon > 0$, the ε-ball $\{\omega' \ : \ |\omega - \omega'| \leq \varepsilon\}$ has a corresponding set of orbits $\{X_0'X_1' \ldots\}$ that is in an appropriate sense "dense" in the set of all possible orbits of the system. For example, by that set being equal to the set of possible orbits of the system.

The system is *uniformly unpredictable* if for every ω and every $\varepsilon > 0$, the ε-ball $\{\omega' \ : \ |\omega - \omega'| \leq \varepsilon\}$ produces a set of orbits in which the subset of Martin-Löf random sequences has full measure. (Here we mean Martin-Löf randomness with respect to the uniform distribution, and the "full measure" with respect to the induced measure μ on the set of orbits $X_0, X_1, \ldots$ of distinguishable states of the system.)

Clearly, systems can be both uniformly unpredictable and uniformly instable, but they can also be either one without being the other. *Chaoticity* as in Definition 1 can be the result of any of these three possibilities.

15.3.1 Doubling map

A well-known example of such a chaotic system is the *doubling map*, see [For83]. Consider the *deterministic* system D with initial state $x_0 = 0.w$ a real number in the interval $[0, 1]$ where $w \in S$ is the binary representation.

$$x_{n+1} = 2x_n \quad (\text{mod } 1) \tag{15.8}$$

where (mod 1) means droping the integer part. Thus, all iterates of x_0 under the transformation 15.8 lie in the unit interval $[0, 1]$. This interval corresponds to what is called the "phase space" in physics. We can partition this phase space into two cells, a left cell $R_0 = [0, \frac{1}{2})$ and a right cell $R_1 = [\frac{1}{2}, 1]$. Thus x_n lies in the left cell R_0 if and only if the nth digit of w is 0.

One way to derive the doubling map is as follows: In chaos theory, [Dev89], people have for years being studying the discrete-time logistic system L_α

$$y_{n+1} = \alpha y_n (1 - y_n)$$

which maps the unit interval upon itself when $0 \leq \alpha \leq 4$. When $\alpha = 4$, setting $y_n = \sin^2 \pi x_n$, we obtain:

$$x_{n+1} = 2x_n \quad (\text{mod } 1).$$

Theorem 3. *There are chaotic systems (like the doubling map D and the logistic map L_α for certain values of α, like alpha $= 4$), with μ in Definition 1 being the uniform distribution (the coin-flip measure λ, with $\lambda(x) = 2^{-l(x)}$: the probability of flipping the finite binary string x with a fair coin).*

Proof. We prove that D is a chaotic system. Since L_4 reduces to D by specialization, this shows that L_4 is chaotic as well. Assume w is random. Then by Theorem 1,

$$C(w_{1:n}) > n - 2\log n + O(1). \tag{15.9}$$

Let ϕ be any partial recursive function. Construct ζ from ϕ and w as in Definition 1.

Assume by way of contradiction that there is an $\varepsilon > 0$ such that

$$\left| \lim_{n \to \infty} \frac{1}{n} \sum_{i=1}^{n} \zeta_i - \frac{1}{2} \right| \geq \varepsilon.$$

Then, there is a $\delta > 0$ such that

$$\lim_{n \to \infty} \frac{C(\zeta_{1:n})}{n} \leq 1 - \delta. \tag{15.10}$$

We prove this as follows. The number of binary sequences of length n where the numbers of 0's and 1's differ by at least an εn is

$$N = 2 \cdot 2^n \sum_{m=(\frac{1}{2}+\varepsilon)n}^{n} b(n, m, \frac{1}{2}) \tag{15.11}$$

where $b(n, m, p)$ is the probability of m successes out of n trials in a $(p, 1-p)$ Bernoulli process: the binomial distribution. A general estimate of the tail probability of the binomial distribution, with m the number of successful outcomes in n experiments with probability of success $0 < p < 1$ and $q = 1-p$, is given by Chernoff's bounds, [ES74, CLR90],

$$\Pr(|m - np| \geq \varepsilon n) \leq 2e^{-(\varepsilon n)^2/3n}. \tag{15.12}$$

Therefore, we can describe any element $\zeta_{1:n}$ concerned by giving n and εn in $2 \log n + 4 \log \log n$ bits self-delimiting descriptions, and pointing out the string concerned in a constrained ensemble of at most N elements in $\log N$ bits, where

$$N = 2^n \Pr(|m - np| \geq \varepsilon n) \leq 2^{n+1} e^{-(\varepsilon n)^2/3n}.$$

Therefore,

$$C(\zeta_{1:n}) \leq n - \varepsilon^2 n \log e + 2 \log n + 4 \log \log n + O(1).$$

That is, we can choose

$$\delta = \varepsilon^2 \log e - \frac{2 \log n + 4 \log \log n + O(1)}{n}.$$

Next, given ζ and ϕ we can reconstruct ω as follows:

```
for i := 1, 2, ... do:
if φ(ω_{1:i−1}) = a and ζ_i = 0 then ω_i := ¬a
else ω_i := a.
```

Therefore,

$$C(\omega_{1:n}) \leq C(\zeta_{1:n}) + K(\phi) + O(1). \tag{15.13}$$

Now (15.9), (15.10), (15.13) give the desired contradiction. By Theorem 1, one has

Claim. The set of ω's satisfying (15.9) has uniform measure one.

In the definition of the doubling map we have already noted that: Starting from an initial state $x_0 = 0.\omega_1\omega_2 \ldots$ the doubling map yields the trajectory $X_0, X_1, \ldots$ with $X_0 = i$ iff $x_0 \in R_i$ and $X_j = \omega_j$ for $j = 1, 2, \ldots$ and $i \in \{0, 1\}$. Therefore,

Claim. If we select the initial state x_0 with uniform probability from $[0, 1]$, then the induced measure on the resulting set of trajectories of distinguishable states $X_0, X_1, \ldots$ of the doubling map is the uniform measure.

Together, Claims 15.3.1, 15.3.1, prove the theorem.

In [For83] the argument is as follows. Assuming that the initial state is randomly drawn from $[0, 1)$ according to the uniform measure λ, we can use complexity arguments to show that the doubling map's observable orbit cannot be predicted better than a coin toss. Namely, with λ-probability 1 the drawn initial state will be a Martin-Löf random infinite sequence. Such sequences by definition cannot be effectively predicted better than a random coin toss, see [Mar66].

But in this case we do not need to go to such trouble. The observed orbit essentially consists of the consecutive bits of the initial state. Selecting the initial state randomly from the uniform measure is isomorphic to flipping a fair coin to generate it. The approach we have taken above, however, allows us to treat chaoticity under nonuniform measures of selecting the initial condition. Moreover, we can think of initial states that are computable but pseudorandom versus prediction algorithms that are polynomially time bounded. Such extensions will be part of a future work.

From a practical viewpoint it may be argued that we really are not interested in infinite sequences: in practice the input will always be finite precision. Now an infinite sequence which is random may still have an arbitrary long finite initial segment which is completely regular. Therefore, we analyse the theory for finite precision inputs in the following section.

15.3.2 Chaos with finite precision input

In the case of finite precision real inputs, the distinction between chaotic and non-chaotic systems can be precisely drawn, but it is necessarily a matter of degree. This occasions the following definition.

Definition 2. Let S, μ, ϕ, ω and ζ be as in Definition 1. A deterministic system with *input precision* n is (ε, δ)-*chaotic* if, for every computable function ϕ, we have

$$\left| \frac{1}{n} \sum_{i=1}^{n} \zeta_i - \frac{1}{2} \right| \leq \varepsilon,$$

with μ-probability at least $1 - \delta$.

So systems are chaotic in the sense of Definition 1, like the doubling map above, iff they are $(0,0)$-chaotic with precision ∞. The system is *probably approximately unpredictable*: a *pai*-chaotic system.

Theorem 4. *Systems D and L_α (for certain values of α, like $\alpha = 4$) above are $(\sqrt{(\delta(n) + O(1))}\ln 2/n, 1/2^{\delta(n)})$-chaotic for every function δ such that $0 < \delta(n) < n$, with μ in Definition 2 being the uniform measure λ.*

Proof. We prove that D is (ε, δ)-chaotic. Since L_4 reduces to D, this implies that L_4 is (ε, δ)-chaotic as well. Assume that x is a binary string of length n with

$$C(x) \geq n - \delta(n). \tag{15.14}$$

Let ϕ be a polynomial time computable function, and define z by:

$$z_i = \begin{cases} 1 \text{ if } \phi(x_{1:i-1}) = x_i \\ 0 \text{ otherwise} \end{cases}$$

Then, x can be reconstructed from z and ϕ as before, and therefore:

$$C(x) \leq C(z) + K(\phi) + O(1).$$

By (15.14) this means

$$C(z) \geq n - \delta(n) - K(\phi) + O(1). \tag{15.15}$$

We analyse the number of zeros and ones in z (we shall denote by $\#\mathrm{ones}(z)$ the number of ones in z). Using Chernoff's bounds, Equation 15.12, with $p = q = \frac{1}{2}$, the number N of z's which have an excess of ones over zeros such that

$$\left|\#\mathrm{ones}(z) - \frac{n}{2}\right| \geq \varepsilon n,$$

is such that:

$$N \leq 2^{n+1} e^{-(\varepsilon n)^2/n}.$$

Then, we can give an effective description of z by giving a description of ϕ, δ and z's index in the set of size N in this many bits

$$n - \varepsilon^2 n \log e + K(\phi) + K(\delta) + O(1). \tag{15.16}$$

From (15.15), (15.16) we find

$$\varepsilon \leq \sqrt{\frac{\delta(n) + 2K(\phi) + K(\delta) + O(1)}{n \log e}}. \tag{15.17}$$

Making the simplifying assumption that $K(\phi), K(\delta) = O(1)$ this yields

$$\left|\#\mathrm{ones}(z) - \frac{n}{2}\right| \leq \sqrt{(\delta(n) + O(1))n \ln 2}. \tag{15.18}$$

The number of binary strings x of length n with $C(x) < n - \delta(n)$ is at most $2^{n-\delta(n)} - 1$ (there are not more programs of length less than $n - \delta(n)$). Therefore, the uniform probability of a real number starting with an n-length initial segment x such that $C(x) \geq n - \delta(n)$ is given by:

$$\lambda\{\omega : C(\omega_{1:n}) \geq n - \delta(n)\} > 1 - \frac{1}{2^{\delta(n)}}. \tag{15.19}$$

Therefore, since we use the same doubling map D as in Theorem 3, the initial uniform distribution on the inputs induces a uniform distribution $\mu = \lambda$ on the corresponding set of trajectories of distinguishable states. Moreover, each such trajectory is the same bit sequence as the decimal expansion of the initial state. Then, by the almost equivalent two claims as in the proof of Theorem 3, the system D is (ε, δ) chaotic with $\varepsilon = \sqrt{(\delta(n) + O(1)) \ln 2 / n}$ and $\delta = 1/2^{\delta(n)}$.

Acknowledgement

This paper is based on a talk by the author at the University of Waterloo, Canada, in 1991. Partially supported by EU through the NeuroColt II Working Group, the PASCAL Network of Excellence and the QAIP, RESQ Projects.
The idea of connecting primality and prefix code-word length, in the Incompressibility method proof of (a weaker version of) the Prime Number Theorem, is due to P. Berman, and the presented proof is due to J. Tromp.

References

[Cha75]   Chaitin, G.J.: A theory of program size formally identical to information theory. J. Assoc. Comp. Mach., **22**, 329-340 (1975)

[CLR90]   Cormen, T.H., Leiserson, C.E., Rivest, R.L.: Introduction to algorithms. MIT Press, Cambridge (1990)

[Dev89]   Devaney, R.L.: An Introduction to Chaos Dynamical Systems. Addison-Wesley, 2nd Edition (1989)

[ES74]   Erdös, P., Spencer, J.: Probabilistic Methods in Combinatorics. Academic Press, New York (1974)

[For83]   Ford, J.: How random is a random coin toss? Physics Today, **36**, 40-47 (April 1983)

[Kol65]   Kolmogorov, A.N.: Three approaches to the definition of the concept of 'quantity of information'. Problems in Information Transmission, **1**, 1-7 (1965)

[Lev74]   Levin, L.A.: Laws of information conservation (non-growth) and aspects of the foundation of probability theory. Problems in Information Transmission, **10**, 206-210 (1974)

[LV97]   Li, M., Vitanyi, P.M.B.: An Introduction to Kolmogorov Complexity and its Applications, 2nd Edition. Springer-Verlag, New York (1997)

[Mar66]   Martin-Löf, P.: On the definition of random sequences. Information and Control, **9**, 602-619 (1966)

[Rog67]   Rogers, H.J. Jr.: Theory of Recursive Functions and Effective Computability. McGraw-Hill (1967)

Printing: Krips bv, Meppel
Binding: Stürtz, Würzburg